Lecture Notes in Civil Engineering

Volume 16

Lecture Notes in Civil Engineering (LNCE) publishes the latest developments in Civil Engineering - quickly, informally and in top quality. Though original research reported in proceedings and post-proceedings represents the core of LNCE, edited volumes of exceptionally high quality and interest may also be considered for publication. Volumes published in LNCE embrace all aspects and subfields of, as well as new challenges in, Civil Engineering. Topics in the series include:

- Construction and Structural Mechanics
- Building Materials
- Concrete, Steel and Timber Structures
- Geotechnical Engineering
- Earthquake Engineering
- Coastal Engineering
- Hydraulics, Hydrology and Water Resources Engineering
- Environmental Engineering and Sustainability
- Structural Health and Monitoring
- Surveying and Geographical Information Systems
- Heating, Ventilation and Air Conditioning (HVAC)
- Transportation and Traffic
- Risk Analysis
- Safety and Security

To submit a proposal or request further information, please contact the appropriate Springer Editor:

- Mr. Pierpaolo Riva at pierpaolo.riva@springer.com (Europe and Americas);
- Ms. Swati Meherishi at swati.meherishi@springer.com (India);
- Ms. Li Shen at li.shen@springer.com (China);
- Dr. Loyola D'Silva at loyola.dsilva@springer.com (Southeast Asia and Australia/NZ).

More information about this series at http://www.springer.com/series/15087

V. K. Stalin · M. Muttharam
Editors

Geotechnical Characterisation and Geoenvironmental Engineering

IGC 2016 Volume 1

Springer

Editors
V. K. Stalin
Department of Civil Engineering,
College of Engineering Guindy
Anna University
Chennai
India

M. Muttharam
Department of Civil Engineering,
College of Engineering Guindy
Anna University
Chennai
India

ISSN 2366-2557 ISSN 2366-2565 (electronic)
Lecture Notes in Civil Engineering
ISBN 978-981-13-0898-7 ISBN 978-981-13-0899-4 (eBook)
https://doi.org/10.1007/978-981-13-0899-4

Library of Congress Control Number: 2018942934

Printed on acid-free paper

This Springer imprint is published by the registered company Springer Nature Singapore Pte Ltd.
The registered company address is: 152 Beach Road, #21-01/04 Gateway East, Singapore 189721, Singapore

Foreword

The Indian Geotechnical Society (IGS) was started as the Indian National Society of Soil Mechanics and Foundation Engineering in the year 1948, soon after the Second International Conference on Soil Mechanics and Foundation Engineering held at Rotterdam. The Society was affiliated to the International Society in the same year, and since then, it has strived to fulfil and promote the objectives of the International Society. In December 1970, the name 'Indian Geotechnical Society (IGS)' was adopted.

The Society conducted several symposia and workshops in different parts of India since its inception in 1948. In the year of 1983, the Indian Geotechnical Society organized its first annual conference IGC 1983 in Indian Institute of Technology Madras.

Several local chapters of the Society were established over the years, and gradually, the annual conferences were held in different cities under the leadership of the respective local chapters. The conferences were well utilized as the venue for showcasing the research works and the case studies on geotechnical engineering and geoenvironmental engineering, and the papers presented during the deliberations were being published as conference proceedings volume.

The annual conferences IGC 1996 and IGC 2006 were organized by IGS Chennai Chapter, and the papers were published as printed volumes. The practice of printing the volumes was later discontinued in view of large expenditure involved and also because of the easy access to the electronic storage media. The papers then became not accessible to the rest of the geotechnical community.

The responsibility of organizing the annual conference of 2016 was taken up by IGS Chennai Chapter, and the conference was held during 15–17 December 2016. It was felt necessary to publish selected papers for the benefit of the entire geotechnical engineering community as well as of the contributing authors. The Chapter approached Springer with a request to take up this responsibility, and with the great help of Springer, about 150 selected papers are being published in four volumes under the series 'Lecture Notes in Civil Engineering'. The selected papers are distributed among these volumes depending upon the topic of discussion.

Volume 1—Geotechnical Characterisation and Geoenvironmental Engineering
Volume 2—Ground Improvement Techniques and Geosynthetics
Volume 3—Soil Dynamics and Earthquake Geotechnical Engineering
Volume 4—Geotechnical Applications

I am thankful to all the general editors and the reviewers of these papers who provided valuable review comments for the betterment of the papers published in these volumes. The support provided by the former President of the Indian Geotechnical Society, the late Prof. Sreerama Rao, and the former Honorary Secretary of IGS, Shri Jai Bhagwan, is gratefully acknowledged here.

Bringing out these volumes under the banner of prestigious publishers, Springer will immensely improve the visibility of these conference proceedings volumes. The Chennai Chapter of the Indian Geotechnical Society places on record its acknowledgement of the efforts put forth by Springer for bringing out these volumes for the benefit of the geotechnical engineering community.

Chennai, India

Prof. A. Boominathan
Chairman, Indian Geotechnical Conference 2016
Department of Civil Engineering IIT Madras

Preface

Stability of any structures resting on soil is primarily governed by the response of foundation soil with reference to shear strength and volume change characteristics which include compressibility, swelling and shrinkage. Unlike the man-made synthesized material, the disintegrated weathered product of rock referred as soil is a multi-complex system often altered by compositional and environmental factors from time to time, which depends on the initial state of soil.

Even with all the advent of technologies and innovations, it is still a challenging task for a geotechnical engineer to understand and predict the accurate engineering properties of soils owing to their influential changes by various parameters such as grain size, mineralogy, void ratio, pH, water content, pore fluid concentration and valency, dielectric constant. The complexity of understanding soil is further hindered by the time-bound variation of these parameters. In a way, geotechnical characterization is a day-to-day affair to upgrade and upkeep the knowledge of soil to behave as per the design requirements of a geotechnical engineer.

Because of the industrialization in developing countries like India, the quantum of liquid and solid waste generation has been increasing phenomenally at an alarming rate and leads to severe ground modification. The soil–pollutant interactions are largely governed by chemical constituents of waste, nature of soil and duration of contamination, which further creates complexity over the other compositional and environmental factors. Geoenvironmental engineering addresses all the geotechnical issues of soil–waste interaction with reference to modification of index and engineering behaviour of soil.

Thus, the evaluation of geotechnical characterization of multi-complex soil system is an essential and integral part of a geotechnical engineer. The technical papers presented under the theme 'Geotechnical Characterisation and Geoenvironmental Engineering' are highlighting the modified geotechnical properties of soil admixed industrial waste and also the characteristics of soil with different pore fluids under varying test conditions.

We would like to place on their thanks to the reviewers of the papers for their timely review and to the technical committee of IGC 2016 for entrusting us the responsibility of editing the volume.

Chennai, India

Dr. V. K. Stalin
Dr. M. Muttharam
General Editors

Contents

About the Editors

Dr. V. K. Stalin is a professor at the Department of Civil Engineering, CEG, Anna University. He obtained his bachelor's in civil engineering and his master's in soil mechanics and foundation engineering from Annamalai University in 1987 and 1989, respectively. He subsequently obtained his Ph.D. in the field of soil mechanics and foundation engineering from the Indian Institute of Science, Bangalore, in 1995. He has teaching and research experience of 22 years at Anna University. He has published over 110 papers in national and international journals and at conferences and seminars.

He also serves as a consultant on various challenging field problems such as pavement failures, railway embankments, monitoring land reclamation, modification of soil and recommendation of foundations as well as soil testing for various private and government organizations. He was an executive committee member of the Chennai Chapter of the Indian Geotechnical Society and is currently the chairman of the same chapter.

Dr. M. Muttharam is a professor at the Department of Civil Engineering, CEG, Anna University. She obtained her bachelor's in civil engineering from Madras University in 1992 and her master's from Anna University in the field of hydrology and water resources engineering in 1995. She subsequently obtained her doctoral degree in the field of soil mechanics and foundation engineering from the Indian Institute of Science, Bangalore, India, in 2001.

She pursued postdoctoral research at Institut National Polytechnique de Grenoble, France, from 2002 to 2003, which was funded by French Higher Education Ministry and has worked as a CSIR research associate for about 3 years. She started her academic career as a lecturer at CEG, Anna University, in 2003 and was promoted to associate professor in 2006 and then professor in 2014.

She has published more than 40 papers in national and international journals and at conferences. She also served as the secretary of the Chennai Chapter of the Indian Geotechnical Society for 5 years and is currently an executive committee member of the same chapter.

Part I
Characterization of Soils, Rocks and Synthesized Materials

Static Compaction Test and Determination of Equivalent Static Pressure

Binu Sharma and Animesh Deka

Abstract Soil compaction is a vital part of the construction process. The compaction curve between moisture content and dry unit weight should be prepared for all types of soil in the project area, before earthwork commences as it forms the basis for specification and field compaction control. The relationship between moisture content and dry unit weight of the soil is a function of the compactive effort. The modified Proctor test, the reduced modified Proctor test, the standard Proctor test and the reduced standard Proctor test are dynamic methods which use different compactive efforts. These tests require considerable time and effort, and also these have some imperfections. In the present study, a laboratory procedure is devised to determine the relationships between moisture content and dry unit weight by using static compaction method for different static pressures. The static compaction pressure test is devised in the Proctor mould itself to statically compact the soil at different water contents. This method is less laborious, and the time involved is less. For this, seven fine-grained soil samples of various plasticity characteristics were tested. The analysis shows that the relation between water content and dry unit weight in static compaction for different static pressure is parabolic in nature. In this study, the equivalent static pressures to the modified Proctor test, the reduced modified Proctor test, the standard Proctor test and the reduced standard Proctor test are determined, to obtain the maximum dry unit weight and the optimum water content corresponding to the four different compactive efforts.

Keywords Static compaction test · Standard Proctor test · Modified Proctor test
Equivalent static pressure

B. Sharma (✉) · A. Deka (✉)
Department of Civil Engineering, Assam Engineering College, Guwahati, India
e-mail: binusharma78@gmail.com

A. Deka
e-mail: animeshdeka412@gmail.com

V. K. Stalin and M. Muttharam (eds.), *Geotechnical Characterisation and Geoenvironmental Engineering*, Lecture Notes in Civil Engineering 16,
https://doi.org/10.1007/978-981-13-0899-4_1

1 Introduction

The modified Proctor test (MP), the reduced modified Proctor test (RMP), the standard Proctor test (SP) and the reduced standard Proctor test (RSP) are dynamic methods of soil compaction which uses different compactive efforts. These dynamic Proctor tests require considerable time and effort. An attempt has been made to obtain compaction characteristic using static compaction method. It is also attempted to determine the equivalent static pressures by applying which the static compaction curves equivalent to the four Proctor's curve of fine-grained soils can be obtained.

Different research workers intended to predict the compaction properties of soil by static compaction. However, the literature in static compaction is very scarce. Kenneth and Steven (1968), Reddy and Jagadish (1993), Oliver and Mesbah (1999), Kenai et al. (2006), Hafez et al. (2010), Yuce and Kayabali (2010) are a few research workers in this direction. Talukdar et al. (2014) in their studies reproduced the compaction characteristics through static compaction test and derived an equivalent static pressure by applying which a static compaction curve equivalent to the standard Proctor's curve of fine-grained soils could be obtained. In this study, the equivalent static pressures to the MP, RMP, SP and the RSP are determined, to obtain the maximum dry unit (MDU) weight and the optimum water content equivalent to the four different Proctor's compactive efforts.

2 Test Programme and Test Results

Altogether, seven fine-grained inorganic soil samples were tested to determine static compaction characteristics and their relevant physical properties. The MP, RMP, SP and the RSP tests were also performed to determine the dynamic compaction characteristics. The SP test has been carried out according to IS 2720 (part 7) 1980, and MP test has been carried out according to IS 2720 (part 8) 1983. The procedure and equipment for RSP and RMP tests are essentially same as that used for SP and MP test, respectively. However each layer was given 15 number of blows of a rammer instead of 25. The liquid limits of the soils varied from 30 to 79%, and the plastic limits varied from 16 to 23%. There is a wide range of variation in the plastic properties of the soils. Table 1 shows the physical properties of the soil samples. The static compaction test consisted of placing a known weight of soil with known water content into the standard Proctor mould of 1000 mL capacity.

The maximum height to which soil could be filled in the compaction mould was 100 mm. Experiments were carried out varying the layer thickness; however, it was observed that compacting the soil in three different thicknesses did not result in any significant variation in the dry density. Two metal plates of diameter 98 mm and thicknesses 5 and 16 mm, respectively, were placed one above the other, on top of the soil sample in the mould. The entire assembly was placed under a cylindrical

Table 1 Physical properties of the soil samples

Soil No.	Sp. Gr.	LL (%)	PL (%)	Sand (%)	Silt + clay (%)	IS classification
1	2.63	46	19	27	73	CI
2	2.80	30	23	23	77	ML
3	2.65	57	16	21	79	CH
4	2.75	72	19	23	77	CH
5	2.77	79	20	24	76	CH
6	2.73	40	18	24	76	CI
7	2.72	49	18	20	80	CI

plunger of diameter 50 mm of the loading frame. Load was then statically applied to the soil through a proving ring having a proving ring constant of 0.99 kg/div. The height of penetration of the metal plate from the top surface of the mould was measured corresponding to different load levels. Since the soil was filled at a known water content, the corresponding dry unit weight of the soil gets determined. The static pressure was gradually increased till no further increase in dry unit weight occurred with further increase in static pressure. The static pressure was calculated by dividing the applied load by the area to which it was applied. The procedure was repeated for more than five different water contents.

The relationship between static pressure and dry unit weight, corresponding to different water contents, has been plotted in the form of curves. The relationship is found to be nonlinear. A typical plot of static pressure versus dry unit weight at a particular water content is shown in Fig. 1. Similar results have been obtained for other water contents. It is observed from the curves that the dry unit weight increases to its maximum value and then remains constant with further increase in static pressure.

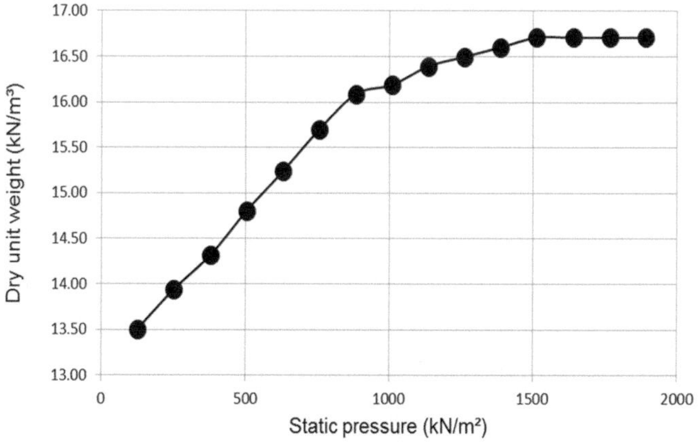

Fig. 1 Static pressure versus dry unit weight curve of sample 2 at 22.45% water content

Next, for a particular static pressure and water content, the dry unit weight was obtained. The relationship between water content and dry unit weight for a particular soil at a particular static pressure is found to be parabolic in nature. Similarly, at all other static pressures, this relationship is found to be parabolic. This parabolic relationship is shown in Fig. 2 for the soil sample 6. One common characteristic in all the curves in all the soil samples was that, beyond a static pressure of around 1513 kN/m^2, further increase in dry unit weight was not possible. In order to understand this behaviour, plot of static pressure versus degree of saturation value at each water content for all the seven number of soil samples was made. These plots are found to be nonlinear in nature. It has been observed that

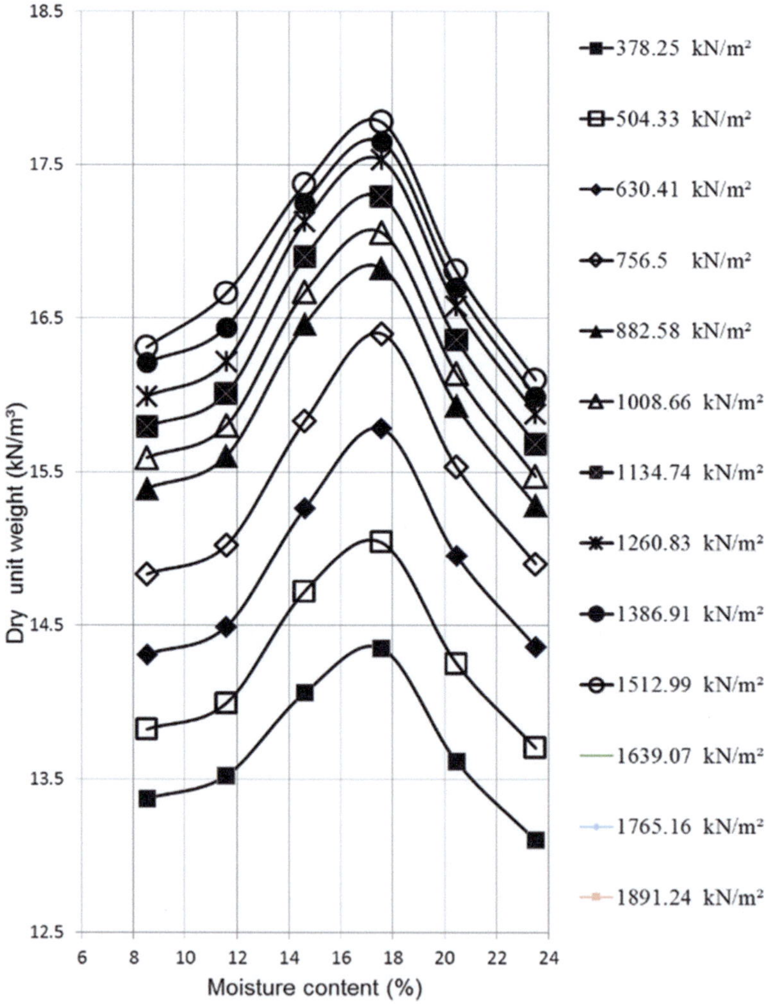

Fig. 2 Parabolic relationship between moisture content and dry unit weight curves for sample 6

degree of saturation can reach a maximum value of 93–98% depending on the soil sample at around the optimum water contents. This behaviour is shown in Fig. 3. This point is further highlighted by the plot of degree of saturation and water content at the static pressure of 1513 kN/m^2 (i.e. the static pressure after which dry unit weight does not increase further) shown in Fig. 4. It is observed that the degree of saturation reaches a maximum value of 93–98% depending on the soil sample at around the optimum water contents. This explains why beyond a static pressure of around 1513 kN/m^2, further increase in dry unit weight was not possible. Similar observation was made by Talukdar et al. (2014) for eight soil samples having classification of CH, CI, CL, MI and ML. Next, the static compaction curves of a particular soil sample corresponding to different static pressures were superimposed with the dynamic compaction curves of the standard Proctor test, modified Proctor

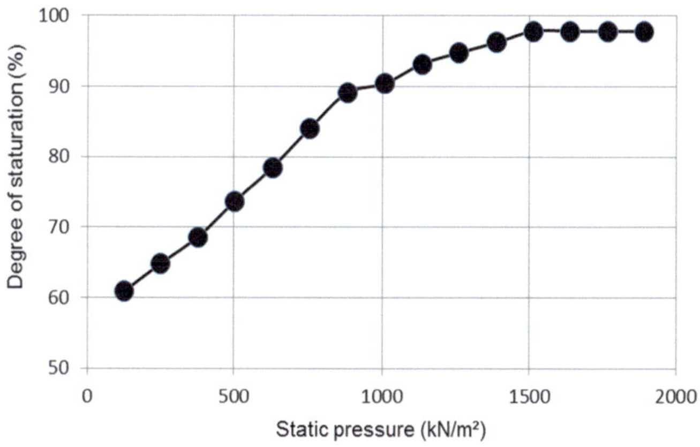

Fig. 3 Static pressure versus degree of saturation curve of sample 2 at 22.45% water content

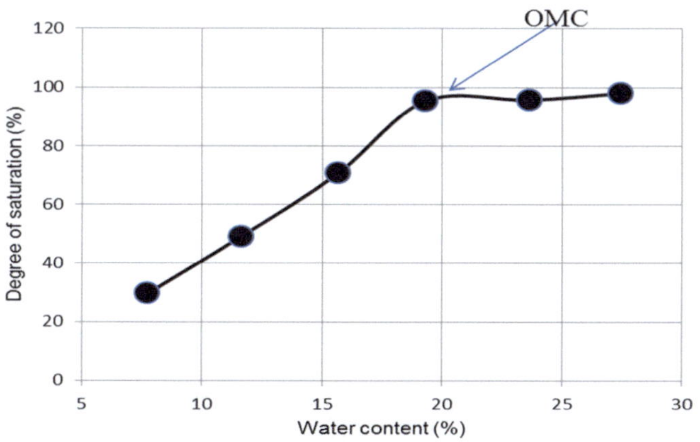

Fig. 4 Water content versus degree of saturation curve

test, reduced standard Proctor test and reduced modified Proctor test. It was attempted to ascertain the static pressure value which gives the nearest value of maximum dry unit weights at optimum water contents from standard Proctor test, reduced standard Proctor test, modified Proctor test and reduced modified Proctor test. A typical plot of moisture content versus dry unit weight is shown in Fig. 5.

From the curves, it is observed that a static pressure in the range of 750–875 kN/m^2 is required to obtain the maximum dry unit weight value at OMC for standard Proctor test and reduced standard Proctor test, and a static pressure in the range of 1375–1500 kN/m^2 is required to obtain the maximum dry unit weight value for reduced modified Proctor test curves in all the seven soil samples. Further, it is observed that the parabolic curve corresponding to the maximum static pressure of 1513 kN/m^2 lies below the modified Proctor test curve in all the seven soil samples.

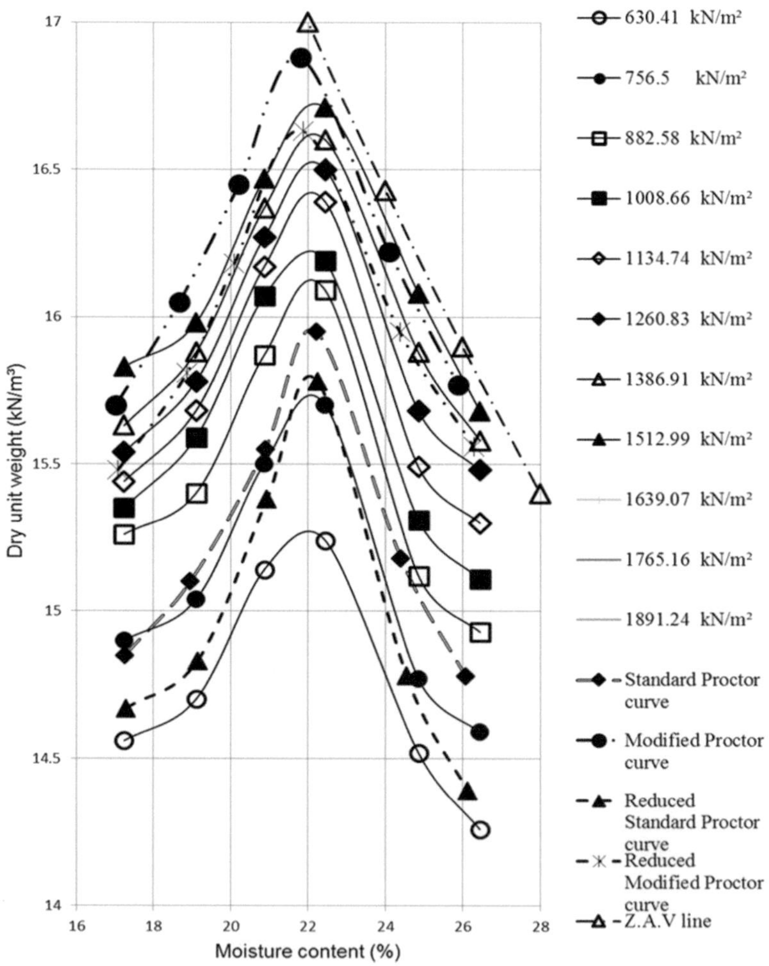

Fig. 5 Moisture content versus dry unit weight curve of sample 2

2.1 Determination of Equivalent Static Pressure

Next, it is attempted to determine the equivalent static pressure at which standard Proctor maximum dry unit weight value can be obtained. For this, the method described by Talukdar et al. (2014) has been used. Two maximum dry unit weights were chosen in such a manner that one is above the maximum dry unit weight as obtained from standard Proctor's test and the other is below it. Assuming linear variation of dry unit weights with the two pressures, the pressure corresponding to the standard Proctor's dry unit weight was determined. Similar method was applied to find the equivalent static pressure for reduced standard Proctor test and reduced modified Proctor test, respectively.

The values of equivalent static pressures as obtained are shown in Table 2. The average equivalent static pressure comes out to be around 845, 788 and 1426 kN/m^2 for SP test, RSP test and RMP test, respectively. Hence, it can be concluded that in the static compaction test, when a static pressure of around 845 kN/m^2 is applied, a static compaction curve equivalent to the SP curve can be obtained. Similar trend is envisaged for the RSP test and RMP test respectively. Talukdar et al. (2014) found an equivalent static pressure of around 820 kN/m^2 for the SP test which is in agreement with the present work. Since the MP test curves for all the seven samples lie above the maximum static pressure curves corresponding to static pressure of 1513 kN/m^2, it is not possible to determine the equivalent static pressure for the MP curve. Hence, it is seen that static compaction curves equivalent to the SP test, RSP test and RMP test can be obtained when static pressures of around 845, 788 and 1426 kN/m^2 are applied, respectively.

Table 2 Equivalent static pressure (ESP) for seven different soil samples

Soil No.	ESP for standard Proctor test (kN/m^2)	ESP for reduced standard Proctor test (kN/m^2)	ESP for reduced modified Proctor test (kN/m^2)
1	855	780	1420
2	836	778	1423
3	842	800	1415
4	855	805	1410
5	858	804	1430
6	840	777	1445
7	830	770	1440
Average ESP of all the samples	845.14	787.71	1426.14

3 Conclusion

A static compaction method is described to determine equivalent static pressure equivalent to the standard, reduced standard and the reduced modified Proctor test. The relationship between water content and dry unit weight corresponding to different static pressure is found to be parabolic in nature. The averaged equivalent static pressure comes out to be around 845, 788 and 1426 kN/m^2 for standard Proctor test, reduced standard Proctor test and reduced modified Proctor test, respectively, of the fined grained soils having IS soil classification of CH, CI, CL, MI and ML. In the static compaction test, it was found that beyond a static pressure of around 1513 kN/m^2, further increase in maximum dry unit weight was not obtained and the dry unit weight remained almost constant. It is not possible to determine an equivalent static pressure required to obtain the maximum dry unit weight value as obtained from the modified Proctor's test

References

Bureau of Indian Standards. (1980). *Methods of test for soils: Determination of water content-dry density relation using light compaction*. New Delhi, India: IS 2720-7.

Bureau of Indian Standards. (1983). *Methods of test for soils: Determination of water content-dry density relation using heavy compaction*. New Delhi, India: IS 2720-8.

Hafez, M. A., Doris Asmani, M., & Nurbaya, S. (2010). Comparison between static and dynamic compaction methods, *Electronic Journal Geotechnical Engineering (EJGE), 15*, 1641–1650.

Kenai, S., Bahar, R., & Benazzoug, M. (2006). Experimental analysis of the effect of some compaction methods on mechanical properties and durability of cement stabilized soil. *Journal Materials Science, 2006*(41), 6956–6964.

Kenneth and Steven. (1968). Change in water content and density of compacted kaolinite due to consolidation, *Journal of Soil Mechanics and Foundation Div, 95*, 1564–1567.

Oliver, M., & Mesbah, A. (1999). Clayey soil behavior under static compaction test. *Materials and Structures, 32,* 687–694.

Proctor, R. R. (1933). Fundamental principles of soil compaction, *Engineering News Record, 111*.

Reddy, B. V. V., & Jagadish, K. S. (1993). The static compaction of soils. *Geotechnical Journal, The Institution of Civil Engineers London, 43*(2), 337–341.

Talukdar, P., Sharma, B., & Sridharan, A. (2014). Determination of compaction characteristics of soil by static compaction method. In *Proceedings of I.G.C, 2014*, December 18–20, 2014, Kakinada, India, pp. 70–76.

Yuce, E., & Kayabali, K. (2010). The use of static compaction method. *Geological Congress of Turkey. Book of Abstracts*, 174–175.

Effect of Confining Pressure on Post-Peak Behaviour of Igneous Rock

Rakesh Kumar, K. G. Sharma and A. Varadarajan

Abstract Many hydroelectric, metro rail and nuclear repository projects are under construction in the country. The structures for these projects, which include dams, tunnels, powerhouse caverns, are mostly constructed on/in rocks. The characterization of the behaviour of rock forms essential part of the analysis and design of these structures. Many of the rocks around tunnels and underground caverns exhibit strain softening behaviour under loading. Strain softening is defined as the progressive loss of strength when material is compressed beyond peak. The present paper deals with the testing of an igneous rock, i.e. basalt from hydroelectric project located in Madhya Pradesh in India. The testing has been conducted using closed-loop servo-controlled testing machine under strain-controlled loading. The specimens were prepared as per ISRM standards. The diameter of the specimen was 54 mm, and the length of the specimen was 108 mm. The index properties of the rock were determined as per ISRM standards. Triaxial tests were conducted under various confining pressures at a constant axial strain rate of 9.259×10^{-6} s. The strains were measured with the help of strain gauges and specially designed extensometers. The stress–strain–volume change relationships of the tests are presented and discussed in this paper. The behaviour of basalt is also predicted using an elastoplastic constitutive model, i.e. Mohr-Coulomb strain softening model using FLAC. The material parameters for the model are determined from the experimental results. The stress–strain–volume change response of the basalt is then predicted

R. Kumar (✉)
AF Consult India Pvt. Ltd., Noida 201301, UP, India
e-mail: rakesh_iitd@rediffmail.com

K. G. Sharma
Department of Civil Engineering, Indian Institute of Technology Delhi,
New Delhi 110016, Delhi, India
e-mail: kgsharma@civil.iitd.ernet.in

A. Varadarajan
Indian Institute of Technology Delhi, New Delhi 110016, Delhi, India
e-mail: avr2466@yahoo.co.in

A. Varadarajan
Dr. MGR Educational and Research University, Chennai, India

V. K. Stalin and M. Muttharam (eds.), *Geotechnical Characterisation and Geoenvironmental Engineering*, Lecture Notes in Civil Engineering 16,
https://doi.org/10.1007/978-981-13-0899-4_2

using the material parameters and is compared with the observed results. The predictions using the constitutive modes are found satisfactory and comparable with the experimentally observed results.

Keywords Strain softening · Closed-loop servo-controlled · Strain-controlled Constitutive modelling

1 Introduction

The understanding of the behaviour of rocks plays an important role in the analysis and design of structures constructed in/on rocks. Many of the rocks around tunnels and underground caverns exhibit strain softening behaviour under loading. Strain softening is defined as the progressive loss of strength when material is compressed beyond peak (Karstunen et al. 1997).

The use of strain softening behaviour is of paramount importance in arriving at rational design of the structures constructed on the rocks. The rock behaviour can be obtained only after testing the rock specimens in the laboratory under various loading conditions. However, it is not possible to capture strain softening behaviour with conventional equipments which are useful to characterize behaviour up to peak only. The strain softening can be captured only by using stiff testing machine or closed-loop servo-controlled testing machine (Hudson et al. 1972). The constitutive models used to characterize the strain softening behaviour are elastic, elastoplastic and elasto-viscoplastic (Shang and Yao 1999). The paper deals with characterization of material behaviour of basalt in the laboratory under confining pressures of 0, 10, 20 and 30 MPa, using closed-loop servo-controlled testing machine.

2 Index Properties of Rock

The rock used for testing is basalt. The rock is from a hydroelectric project located in Madhya Pradesh in India. The rock is fine-grained and greenish black in colour. The X-ray diffraction and microscopic studies have been carried out to determine the mineral composition of the rock. The mineral composition is quartz (52%), iron oxide (1%), amphiboles (26%), pyroxenes (16%), olivine and serpentine (5%). The specific gravity of the rock is 2.89. The index properties of the rock as per ISRM standards are determined in the laboratory and are presented in Table 1.

Table 1 Index properties of basalt

Property	Value
Dry density (kN/m^3)	28.100
Saturated density (kN/m^3)	28.100
Void ratio (*e*)	0.0041
Porosity (*n*) %	0.4100
Tensile strength (MPa)	13.700
UCS (MPa)	69.900

3 Experimental Programme

The strain-controlled compression tests under confining pressures of 0, 10, 20 and 30 MPa were conducted on basalt using closed-loop servo-controlled testing machine. The machine works on the principle of closed-loop control which is described elsewhere (Hudson et al. 1972). The machine used for testing has a loading capacity of 1000 kN and has loading rate capability in the range of 0.001–10.000 mm/s. The stiffness of the machine is 1700 kN/mm. High-pressure triaxial cell which has the capability of applying 0–140 MPa confining pressure and has sufficient space to accommodate extensometers inside, is used for the testing. The confining pressure was applied through hydraulic oil with the help of high-pressure confining unit. The specimens were prepared with 54 mm diameter and 108 mm height as per ISRM. The strains were measured with the help of HMM strain gauges and specially designed extensometers. The axial and lateral extensometers having a gauge length of 82.0 and 25.4 mm, respectively, were placed on the specimen and held in position with the help of springs. The tests were conducted under a constant axial strain rate of 9.259×10^{-6} s. The stress–strain–volume change response of the rock is presented in Figs. 1 and 2.

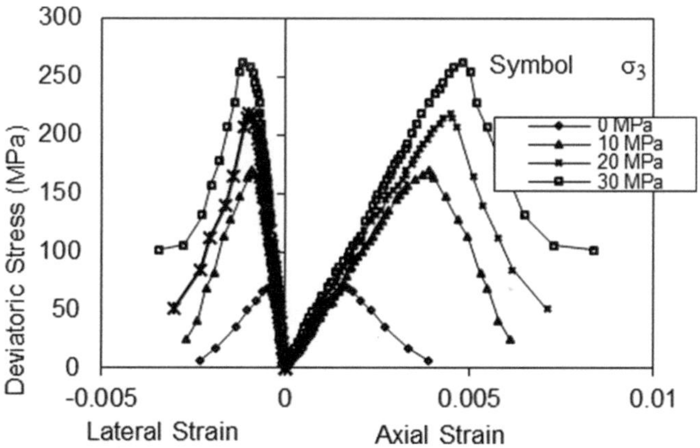

Fig. 1 Stress–strain behaviour for basalt

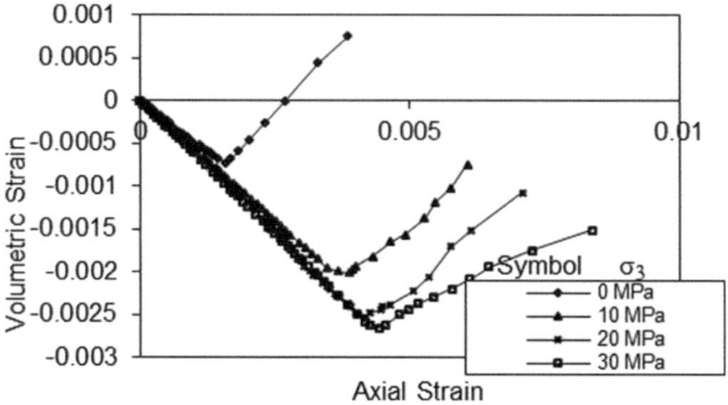

Fig. 2 Volume change response for basalt

As seen from Figs. 1 and 2, it is observed that there is gradual loss of strength from peak to residual strength with increase in strains. The effect of confining pressure is evidently noticed on the stress–strain–volume change behaviour of rock. The effect of softening and dilatancy decreases with increase in confining pressure.

4 Constitutive Modelling

The material behaviour can be characterized by various constitutive models like local models, continuum damage mechanics models (CDM), smeared crack models, Cosserat continuum models, viscoplastic models and models based on the disturbed state concept (Desai 1995, 2001; Varadarajan et al. 2001). In the present paper, Mohr-Coulomb strain softening model has been used to characterize the material behaviour of basalt. The complete details of the Mohr-Coulomb model are given in FLAC Version 4.0 manual (2002) and Kumar (2007). The salient features of the model are discussed herein.

4.1 Mohr-Coulomb Strain Softening Model

In this model, the non-associated shear and associated tension flow rules are used. The difference, however, lies in the possibility that the cohesion, friction, dilation and tensile strength may harden or soften after the onset of plastic yield. In the conventional Mohr-Coulomb model, the properties are assumed to remain constant. Here, the user can define the cohesion, friction and dilation as piecewise-linear functions of a softening parameter measuring the plastic shear strain. The yield and potential functions, plastic flow rules and stress corrections are identical to those of the Mohr-Coulomb model.

The failure envelope for shear is defined by the Mohr-Coulomb yield function

$$f^s = \sigma_1 - \sigma_3 N_\phi + 2c\sqrt{N_\phi} \tag{1}$$

and tension yield function of the form

$$f^t = \sigma_t - \sigma_3 \tag{2}$$

where ϕ is the friction angle; c the cohesion; σ_t the tensile strength and

$$N_\phi = \frac{1 + \sin\phi}{1 - \sin\phi} \tag{3}$$

Plastic shear strain is measured by the shear softening parameter e^{ps}, whose incremental form is defined as

$$\Delta e^{ps} = \left\{\frac{1}{2}\left(\Delta e_1^{ps} - \Delta e_m^{ps}\right)^2 + \frac{1}{2}\left(\Delta e_m^{ps}\right)^2 + \frac{1}{2}\left(\Delta e_3^{ps} - \Delta e_m^{ps}\right)^2\right\}^{\frac{1}{2}} \tag{4}$$

where,

$$\Delta e_m^{ps} = 1/3(\Delta e_1^{ps} + \Delta e_3^{ps}) \tag{5}$$

and $\Delta e_j^{ps}, j = 1, 3$ are the major and minor principal plastic shear strain increments, respectively.

The other details of the model are given in FLAC Version 4.0 manual (2002).

5 Material Parameters

The material parameters of basalt were determined using one set of experimental results, viz. confining pressures of 0, 10 and 20 MPa, and predictions are made for these confining pressures (Group A) and remaining test result, i.e. confining pressure of 30 MPa (Group B). The details of the determination of the parameters are described elsewhere (FLAC Version 4.0 manual 2002; Kumar 2007). The material parameters for basalt based on Mohr-Coulomb model are presented in Figs. 3 and 4. The other details are given in Kumar (2007).

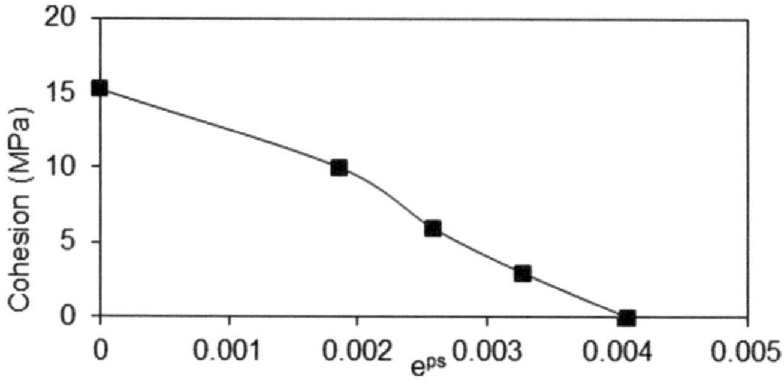

Fig. 3 Variation of cohesion (c) with e^{ps} for basalt

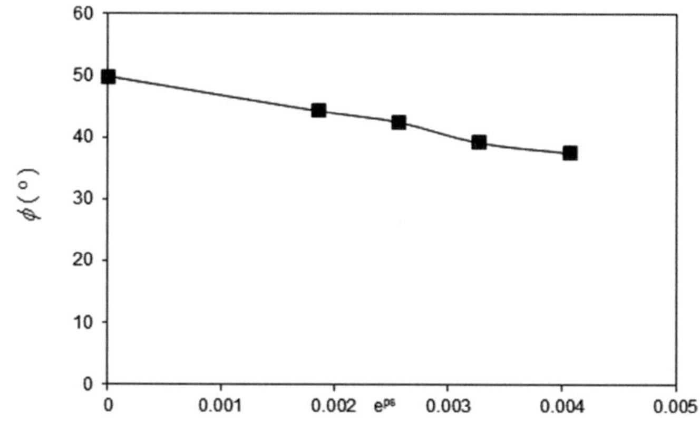

Fig. 4 Variation of friction angle (ϕ) with e^{ps} for basalt

6 Predictions

The stress–strain–volume change response of the basalt was carried out on the simulated test sample using Mohr-Coulomb strain softening model using FLAC for various confining pressures, i.e. 0, 10, 20 and 30 MPa. The results of the analyses are presented in terms of deviatoric stress, axial, lateral and volumetric strains.

7 Results and Discussion

The predicted results of basalt for $\sigma_3 = 0$, 10, 20 and 30 MPa in terms of stress–strain–volume change response using FLAC along with experimental results are presented in Figs. 5 and 6. The predictions are found to be generally satisfactory and comparable with experimental results.

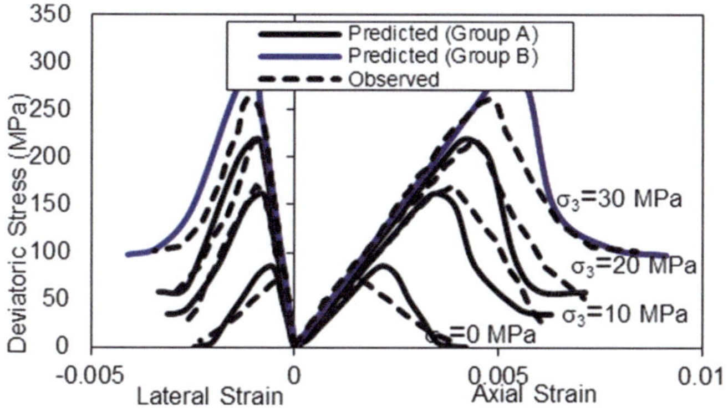

Fig. 5 Predicted stress–strain behaviour for basalt

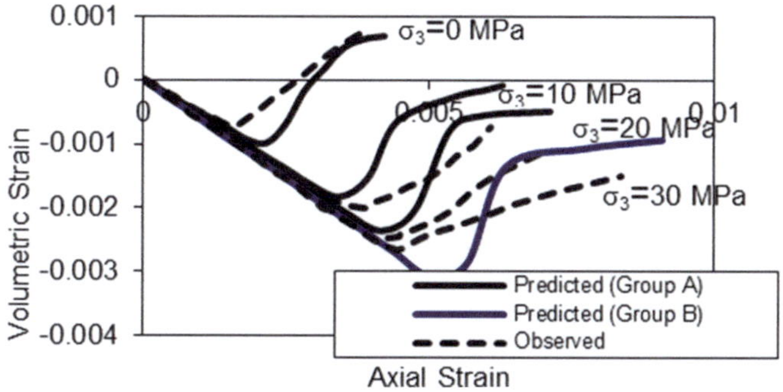

Fig. 6 Predicted volume change response for basalt

8 Conclusions

The strain-controlled compression tests have been conducted on basalt under unconfined state and confining pressures of 10, 20 and 30 MPa in the laboratory using closed-loop servo-controlled testing machine. It is observed that there is gradual loss of strength from peak to residual strength with increase in strains. The effect of confining pressure is evidently noticed on the stress–strain–volume change behaviour of rock. The effect of softening and dilatancy decreases with increase in confining pressure.

The Mohr-Coulomb strain softening model in FLAC has been used to predict the rock behavior. The predicted behaviour and observed behaviour for basalt were found to be in satisfactory agreement for both Group A and Group B predictions.

References

Desai, C. S. (1995). Constitutive modelling using the disturbed state as microstructure self-adjustment concept (Chap. 8). In H. B. Muhlhaus (Ed.), *Continuum models for materials with microstructure*. UK: Wiley.

Desai, C. S. (2001). *Mechanics of materials and interfaces: The disturbed state concept*. Boca Raton, FL, USA: CRC Press.

FLAC (version 4.0). (2002). *Theory and background manual*.

Hudson, J. A., Crouch, S. L., & Fairhurst, C. (1972). Review-soft, stiff and servo-controlled testing machines: A review with reference to rock failure. *Engineering Geology, 6,* 155–189.

Karstunen, M., Pande, G. N., & Desures, J. (1997). Strain localisation and rotation of principal stress axis in biaxial test. In: J. X. Yuan (Ed.), *Proceedings of 9th International Conference Computer Methods and Advanced Geomechanics*, Wuhan, China.

Kumar, R. (2007). Testing and constitutive modelling of the strain-softening behaviour of some rocks, Ph.D. thesis, submitted to Indian Institute of Technology, Delhi.

Shang, D. G., & Yao, W. X. (1999). A non-linear cumulative model for uniaxial fatigue. *International Journal of Fatigue, 21,* 187–194.

Varadarajan, A., Sharma, K. G., Desai, C. S., & Hashemi, M. (2001). Constitutive modeling of a schsitose rock in the Himalaya. *International Journal of Geomechanics, 1*(1), 83–107.

Utility of Lime and Red Mud in Clay Soil Stabilization

M. Aswathy, U. Salini and V. G. Gayathri

Abstract Large amount of industrial wastes are generated from various factories in India and are simply being disposed without any major applications. Alternative materials are to be adopted as a construction material so as to minimize the use of natural resources. Red mud (RM) is generated as a waste material during the production of alumina from bauxite, and it comes to around 40% of the bauxite used for the production. The paper presents the possibility of using red mud as a stabilizing material. The study examines the effect of red mud on behavior of clays by carrying out compaction test, CBR tests, and UCC tests for different percentages of red mud. The red mud soil mix is further treated with lime to understand whether more soil can be replaced by red mud on lime treatment. It was seen that the highest maximum dry density (MDD), UCS, and CBR values were obtained for 15% red mud in soil. On further treatment with 2% of lime, 20% red mud in soil showed more strength compared to 15% red mud in soil ensuring that more soil can be replaced by red mud on treating with lime.

Keywords Compressive strength · Red mud · Stabilization · Lime

1 Introduction

The major drawback of industrialization and urbanization is the production of large quantities of wastes and the problems related to their safe management and disposal. Innovative methods for the utilization of these waste products have great

M. Aswathy (✉) · V. G. Gayathri
Sarabhai Institute of Science and Technology, Thiruvananthapuram, India
e-mail: achu3070@gmail.com

V. G. Gayathri
e-mail: Gayathrivg1990@gmail.com

U. Salini
National Transportation Planning and Research Centre, Thiruvananthapuram, India
e-mail: saliniu@gmail.com

V. K. Stalin and M. Muttharam (eds.), *Geotechnical Characterisation and Geoenvironmental Engineering*, Lecture Notes in Civil Engineering 16,
https://doi.org/10.1007/978-981-13-0899-4_3

demand all over the world due to the scarcity of land to dispose them. Red mud (RM) is the waste generated during the process of extracting alumina from bauxite. It is composed of alkalis, aluminum hydroxides, iron oxides and hydroxides, calcium carbonates, titanium, etc. (Pera et al. 1997; Yalcin and Sevinc 2000). Red mud finds its application in making bricks and cements and even in decontaminating mining sites (Amritphale and Patel 1987; Deng et al. 1980; Kohno et al. 1993; Kara and Emrullahoglu 1994; Singh et al. 1996). The geotechnical properties of red mud were studied by many researchers as to use it as an alternative material to soil (Li 1998; Newson et al. 2006; Sundaram and Gupta 2010; Rout et al. 2012; Sutar et al. 2014). Kehagia (2008) studied the effect of fly ash on soil stabilized with red mud and found that more percentage of soil can be replaced with red mud when it is stabilized fly ash. Chemical stabilization of the soil using lime, cement, fly ash, and their combinations is very common nowadays. Among them lime is the most widely used admixture as it forms cementing products which bond the clay particles, thereby reducing the plasticity, shrinkage, swelling and improving the strength characteristics (Bell 1988; Sherwood 1993; Greaves 1996). Shallow stabilization and deep stabilization using lime (Puppala et al. 2007; Rao and Shivananda 2005; Rao and Thyagaraj 2003) and in situ lime precipitation techniques (Thyagaraj et al. 2012; Thyagaraj and Zodinsanga 2014) are currently being practiced in stabilizing weak clayey soils. Bhuvaneshwari et al. (2014) brought out the role played by lime modification optimum in soil–lime reactions which affected the physical properties as well as the structure of the lime-stabilized soil.

The present study involves experimental investigation to characterize red mud waste from HINDALCO and to study its effect on compaction behavior, California Bearing ratio (CBR), and unconfined compressive strength (UCS) of clayey soil. The soil–red mud mix was further treated with lime to understand if more soil can be replaced by red mud on lime treatment. All the tests were performed in accordance with the procedures specified by Indian standards.

2 Properties of Soil and Red Mud

Clay soil used in this study was collected from Aakulam, Kerala. The soil was air-dried, pulverized, and passed through 425-μ sieve before carrying out the tests. The index properties and engineering properties of soil used are given in Table 1. The red mud sample was obtained from HINDALCO, Belgam, Karnataka. It is the solid waste residue obtained during the production of alumina (Al_2O_3) from bauxite ores. The properties of red mud are given in Table 2.

Table 1 Properties of soil

Property	Value
Specific gravity	2.5
Liquid limit (%)	36
Plastic limit (%)	21
Plasticity index (%)	15
Optimum moisture content (%)	20
Maximum dry density (kN/m^3)	16.4
IS classification	CI
Sand (%)	46
Silt and clay (%)	54

Table 2 Properties of red mud

Property	Value
Specific gravity	3.2
Liquid limit (%)	37
Plastic limit (%)	31
Plasticity index (%)	6
Optimum moisture content (%)	32
Maximum dry density (kN/m^3)	15.1
IS classification	MI
Sand (%)	15
Clay and silt (%)	85

3 Red Mud and Lime Stabilization of Soil

3.1 Soil Stabilization Using Red Mud (RM)

The specimens were prepared by adding 5, 10, 20, 25, and 30% of RM to soil separately. The variation of MDD, OMC, CBR, and UCC was determined for different percentage of RM in soil–RM mixes, and the best mix was obtained when 15% of the soil was replaced with red mud.

3.2 Lime Stabilization of Soil–Red Mud Mix

The soil–RM mix was treated with lime in order to find whether there was any marked improvement in the properties of the mix. Based on Eades and Grim pH test (ASTM-D 6276–99a), the initial consumption of lime (ICL) of the soil–red mud mix was determined. The ICL is the amount of lime in the soil at a pH value of 12.45. It can be observed from Fig. 1 that the pH of the lime soil mixture reached 12.45 at 2% lime. In the present study, the effect of lime treatment at ICL, i.e., at 2% lime, only is considered.

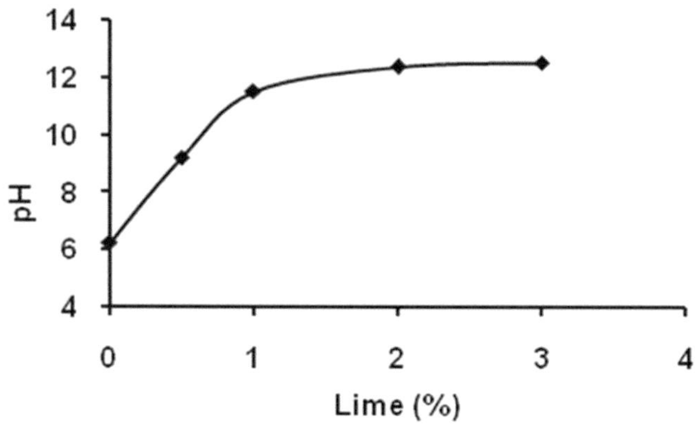

Fig. 1 Variation of pH with lime in soil

For the 2% lime-treated soil–RM mixes also various parameters like MDD, UCS, and CBR values were studied for different percentages of red mud, and the best mix was obtained when 20% of the soil was replaced with red mud.

4 Results and Discussion

4.1 Compaction Behavior

Standard Proctor Compaction tests were conducted to study the effect of different percentages of red mud on compaction behavior of the clay soil. From Fig. 2 it can be seen that with increase in the percentage of red mud, maximum dry density (MDD) initially increased from 16.38 to 16.84 kN/m^3 and then decreased with further increase in percentage of red mud. The increase in MDD was due to the filling of voids in the soil–RM mix by the red mud. The MDD increased till the addition of 20% of RM beyond which the addition of red mud resulted in the decrease of MDD due to the presence of more fines in the mix.

On addition of lime the MDD increased for all the mixes as the soil–RM mix became less plastic. A maximum value of 17.4 kN/m^3 was obtained for 25% of RM in the lime-treated soil–RM mix beyond which the MDD gradually decreased.

4.2 Unconfined Compressive Strength (UCS)

Unconfined compression test was conducted to analyze the effect of addition of red mud to soil. Figure 3 shows the comparison of the UCS of untreated and lime-treated soil–RM mixes. For the untreated mix, it can be seen that there is an

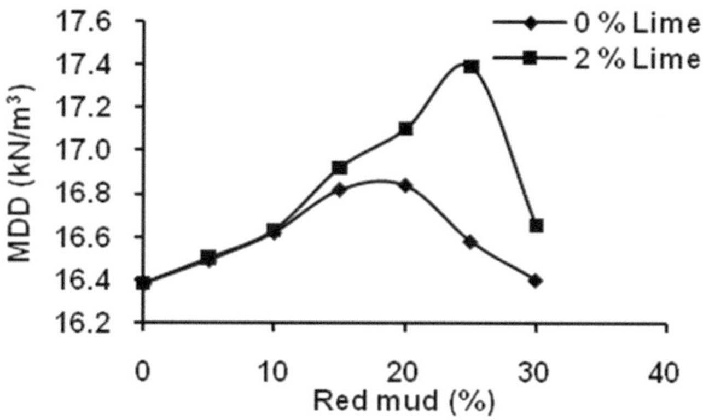

Fig. 2 Variation of MDD with increase in RM in the mix for different percentages of lime

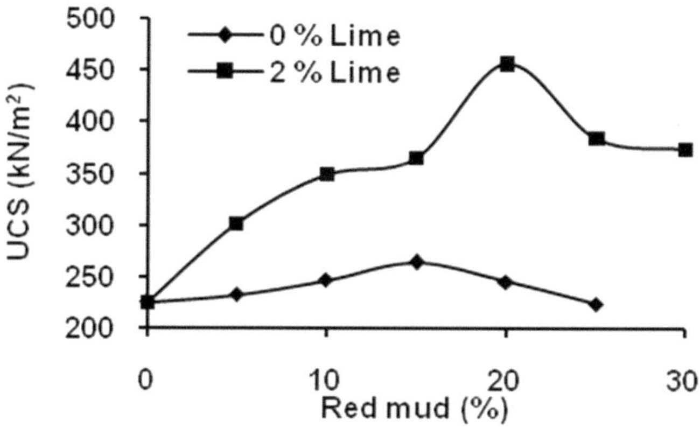

Fig. 3 Variation of UCS with increase in RM in the mix for different percentages of lime

increase in UCS initially till 15% of soil was replaced by red mud and thereafter the UCS decreased with increase in the percentage of red mud. On addition of lime, the UCS increased up to 20% RM in the mix. The tremendous increase in UCS of the soil–RM mix is due to the reactions occurring between soil and lime, thereby forming cementitious bonds which hold the mix together leading to an increase in strength.

4.3 CBR Value

The variation in unsoaked and soaked CBR values with different percentages of red mud is shown in Fig. 4. It can be seen that the CBR value increased till 15% of red

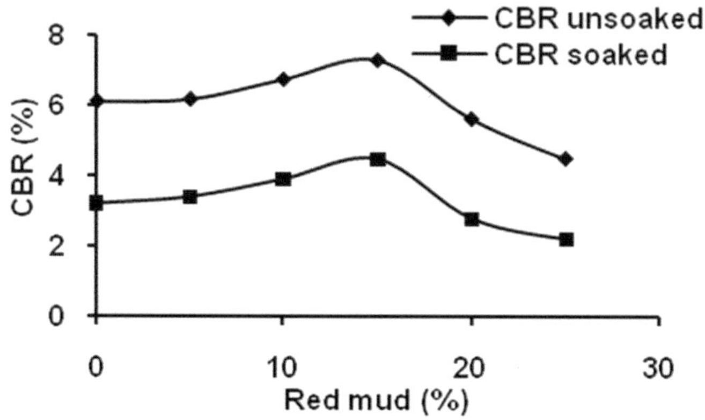

Fig. 4 Variation of CBR for untreated soil–RM mixes

mud and beyond which it decreased. Above this limit the soil became more plastic and finely grained.

The soil–RM mix when treated with lime showed a marked increase in unsoaked as well as soaked CBR value as can be seen in Fig. 5. Hydrated calcium aluminates and calcium silicates are formed when soil is treated with lime which bonds the clay particles making it more resistant to penetration and therefore improves its CBR value. The maximum value of CBR was obtained for 20% of RM in the soil–RM mix.

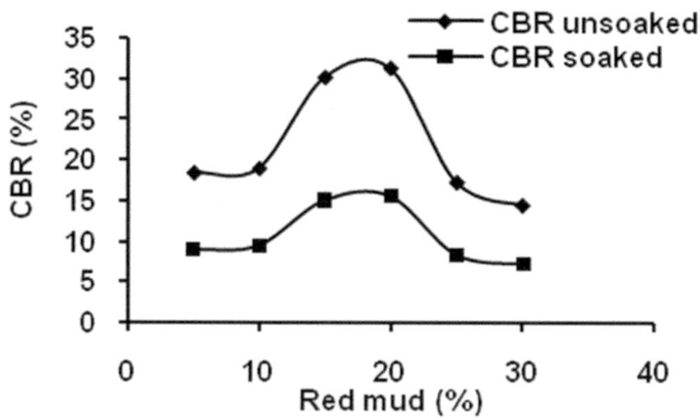

Fig. 5 Variation of CBR for soil–RM mixes treated with 2% lime

5 Conclusion

- The maximum dry density of soil–RM mixes improved with the replacement of soil by red mud up to RM percentage of 20%. For the soil–RM mix treated with 2% lime, maximum MDD was obtained for mix with 25% red mud.
- The maximum UCS was obtained for 15% RM and 20% RM in untreated and treated soil–RM mixes, respectively.
- The unsoaked and soaked CBR value of soil–red mud mix showed improvement until 15% replacement of soil by red mud in case of untreated soil–RM mix and 20% replacement of soil by red mud in case of treated soil.
- The clayey soil when replaced by 15% red mud showed the best improvement in its properties (MDD, UCS, and CBR), and therefore it can be concluded that the weak clayey soil can be improved by addition of red mud which is an industrial waste.
- The treatment of soil–RM mixes with 2% lime could further improve the properties of the mix, and the maximum improvement was seen when 20% of soil was replaced by red mud. Hence, the addition of lime brought about further utilization of red mud in soil by 5%.
- Hence, it can be concluded that red mud can be used in stabilization of weak clayey soil instead of simply being disposed on the land.

References

Amritphale, S. S., & Patel, M. (1987). Utilization of red mud, fly ash for manufacturing bricks with pyrophyllite. *Silicates Industriels, 3–4,* 31.

Bell, F. G. (1988). Stabilization and treatment of clay soils with lime. *Ground Engineering, 21*(1), 10–15.

Bhuvaneshwari, S., Robinson, R. G., & Gandhi, S. R. (2014). Behaviour of lime treated cured expansive soil composites. *Indian Geotechnical Journal, 44*(3), 278–293.

Deng, J., Ge, W., Su, M., Li, X. (1980). Sulfoaluminate cement series. In *Proceedings of 7th International Congress on the Chemistry of Cement* (Vol. 3, pp. 381–386), Paris.

Greaves, H. M. (1996). An introduction to lime stabilization. In *Proceedings of Seminar on Lime Stabilization* (pp. 5–12), Loughborough University, Thomas Telford, London.

Kara, M., & Emrullahoglu, F. (1994). The utilization of Seydisehir red mudas the construction material. In *Proceedings of 2nd International Ceramic Congress* (Vol. 1, pp. 181–189), Istanbul, Turkey.

Kehagia, F. (2008). An innovative geotechnical application of bauxite residue. *Journal of Geotechnical Engineering, 13*(G), 1–9.

Kohno, K., Amo, K., Ogawa, Y., & Ikezoe, Y. (1993). Effect of admixture consisting of finely ground silica and red mud on properties of high strength concrete for products. In *Proceedings of the 4th EASEC* (Vol. 1437), Seoul I.

Li, L. Y. (1998). Properties of red mud tailings produced under varying process conditions. *Journal of Environmental Engineering, 124*(3), 254–264.

Newson, T., Dyer, T., Adam, C., & Sharp, S. (2006). Effect of structure on the geotechnical properties of bauxite residue. *Journal of Geotechnical and Geoenvironmental Engineering, 132*(2), 143–151.

Pera, J., Boumaza, R., & Ambroise, J. (1997). Development of pozzolanic pigment from red mud. *Cement and Concrete Research, 27*(10), 1513–1522.

Puppala, A. J., Madhyannapu, R. S., Nazarian, S., Yuan, D., & Hoyos, L. (2007). *Deep soil mixing technology for mitigation of pavement roughness*. Texas Department of Transportation, Austin, TX, USA, Report No. FHWA/TX-08/0–5179–1.

Rao, S. M., & Shivananda, P. (2005). Role of curing temperature in progress of lime–soil reactions. *Geotechnical and Geological Engineering, 23*(1), 79–85.

Rao, S. M., & Thyagaraj, T. (2003). Lime slurry stabilization of an expansive soil. In *Proceedings of the Institution of Civil Engineers—Geotechnical Engineering* (Vol. 156, No. 3, pp. 139–146).

Rout, S., Sahoo, T., & Das, S. K. (2012). Utility of red mud as an embankment material. *International Journal of Earth Sciences and Engineering, 06*, 1645–1651.

Sherwood, P. T. (1993). *Soil stabilisation with cement and lime*. London: HMSO.

Singh, M., Upadhayay, S. N., & Prasad, P. M. (1996). Preparation of special cements from red mud. *Waste Management, 6*(8), 665–670.

Sundaram, R., & Gupta, S. (2010). Constructing foundations on red mud. In *6th International Congress on Environmental Geotechnics* (pp. 1172–1175). New Delhi, India.

Sutar, H., Mishra, S. C., & Sahoo, S. K. (2014). Progress of red mud utilization: An overview. *American Chemical Science Journal, 4*(3), 255–279.

Thyagaraj, T., Rao, S. M., Suresh, P. S., & Salini, U. (2012). Laboratory studies on stabilization of an expansive soil by lime precipitation technique. *Journal of Materials in Civil Engineering ASCE, 24*(8), 1067–1075.

Thyagaraj, T., & Zodinsanga, S. (2014). Swell–shrink behaviour of lime precipitation treated soil. *Ground Improvement, 167*, 260–273.

Yalcin, N., & Sevinc, V. (2000). Utilization of bauxite waste in ceramic glazes. *Ceramics International, 26,* 485–493.

Geotechnical Characterization of Construction and Demolished (C&D) Waste

J. Y. V. Shiva Bhushan, Partha Sarathi Parhi and Balunaini Umashankar

Abstract Natural well-graded granular materials are conventionally used as a fill material for various geotechnical applications. However, due to high demand for granular material, illegal dredging of sands from the river beds is on the rise. On the other hand, a study carried out by Centre for Science and Technology, New Delhi, has reported that huge amounts to the tune of 530 million tonnes of Construction and Demolished (C&D) Waste were generated during the year 2013 in India. This C&D waste includes waste from demolished structures due to renovations of existing structures and repair of roads, flyovers, bridges, etc. In addition, enormous debris is generated following disasters such as that during the Uttarakhand floods in 2013. Typically, C&D waste is dumped in municipal landfills. This study presents the efficacy of crushed concrete and concrete rubble for use as an alternate backfill material in mechanically stabilized earth (MSE) walls. Crushed concrete (CC) is collected locally from the demolished structures of Ordinance Factory Campus, Yeddumailaram, Telangana, and processed before testing. Various index and engineering properties of CC such as particle-size distribution, specific gravity, water absorption, permeability, compaction characteristics, and crushability are ascertained and compared with TxDOT and MORTH specifications for fill materials. Corrosion resistivity of CC is determined by pH test, and the chemical composition of CC is obtained from X-ray fluorescence (XRF) studies. CC is classified as well-graded gravel (GW) as per IS: 2720 (Part 4)-1985. The study indicates that C&D waste, traditionally considered as waste material, can be reused viably as backfill for MSE walls.

Keywords Construction and demolished waste · Fill material · Mechanically stabilized earth (MSE)

J. Y. V. Shiva Bhushan · P. S. Parhi (✉) · B. Umashankar
Department of Civil Engineering, Indian Institute of Technology Hyderabad, Kandi 502285, India
e-mail: ce14resch11004@iith.ac.in

J. Y. V. Shiva Bhushan
e-mail: ce14mtech11016@iith.ac.in

B. Umashankar
e-mail: buma@iith.ac.in

V. K. Stalin and M. Muttharam (eds.), *Geotechnical Characterisation and Geoenvironmental Engineering*, Lecture Notes in Civil Engineering 16,
https://doi.org/10.1007/978-981-13-0899-4_4

1 Introduction

Infrastructure development in India is on a steep rise. In Union Budget for 2016–17, Government of India had allocated a huge amount of Rs. 2.21 lakh crore for infrastructure sector and about 97,000 crores toward roads sector alone. High-speed railways, new road projects, widening of existing road networks, and affordable housing are major sectors where the Ministry of Roads Transport and Highways (MORTH) is currently focusing on. With such growth envisioned in the coming years, large amount of raw materials such as virgin aggregates and sands are required for the construction activity. To meet these requirements, illegal dredging of sands from the river beds has increased resulting in catastrophic hazards, like slope failure and lowering of groundwater table. On the other hand, studies have indicated that Construction and Demolished (C&D) Waste generated from renovation and repair of existing structures can alternatively be used as a fill material. Hence, there is an urgency for recycling these construction wastes globally, which can be considered a possible solution to the above-stated problems (Landris 2007; Aatheesan et al. 2010; Disfani et al. 2011; Hoyos et al. 2011).

Construction and Demolished (C&D) Waste is a non-hazardous industrial solid waste that includes recycled aggregate concrete (RAC/RCA), recycled asphalt pavement (RAP), crushed brick (CB), plastic, steel, etc. It is widely accepted that recycling and subsequent reuse of C&D materials will reduce the demand for scarce virgin natural resources and simultaneously reduce the quantity of this waste material destined for landfills (Arulrajah et al. 2011; Disfani et al. 2011; Hoyos et al. 2011).

During Commonwealth Games held in 2010 at New Delhi, India, the companies have collected and transported approximately 0.2 million tons of C&D waste from the streets of New Delhi.

Currently as per the Union Ministry of Forests and Environment (MoEF), there is no systematic database on generation and utilization of Construction and Demolished (C&D) Waste. A study carried out by Centre for Science and Technology, New Delhi, had reported that 530 million tons of C&D waste was generated in India, during the year 2013. This C&D waste includes waste from demolished structures, renovations in the real estate sector and construction and repair of roads, flyovers, bridges, etc. In addition, the debris that follows disasters such as that during the Uttarakhand floods in 2013, Chennai floods in 2015 is enormous.

The codal provisions of construction agencies like CPWD indicate the use of only natural granular materials as fill material. Unlike in India, other agencies like Texas Department of Transportation (TxDOT), Texas, USA, had successfully used CC as backfill material in mechanically stabilized earth (MSE) walls (Rathje et al. 2006). Santos et al. (2013) studied the behavior of a geogrid-reinforced wall built with recycled Construction and Demolished (C&D) Waste backfill on a collapsible foundation. Poon and Chan (2006), Tam and Tam (2007), Hoyos et al. (2011), Puppala et al. (2011), Arulrajah et al. (2012) have conducted studies on the use of C&D materials for pavement applications.

In the present study, an experimental investigation is carried out to study the efficacy of crushed concrete as an alternative material for backfill in mechanically stabilized earth wall (MSEW). Various index and engineering properties of CC such as particle-size distribution, specific gravity, water absorption, permeability, compaction characteristics, and crushability are ascertained. Corrosion resistivity of CC is determined by pH test, and the chemical composition of CC is obtained from X-ray fluorescence (XRF) studies.

2 Methodology

2.1 Material Characterization

The C&D waste used in this experimental research was collected locally from the demolished residential structures of Ordinance Factory Estate, Yeddumailaram, Medak, Telangana. It was dumped in the form of blocks on the roadside, and the average length, width, and height of block were equal to 29.5, 25, and 11 cm, respectively. Prior to conducting studies on this C&D waste, they were first processed in a stone-crushing plant, where they were reduced into smaller sizes and segregated into different sizes. The stone crusher is a mechanical/gravity feeding type with single jaw and was fitted with vibratory screens of 40, 20, 12, and 6 mm sizes. The crushed concrete was transported to the vibratory screens with the help of conveyer belts. A small amount of sample from each sizes were collected and sealed in a plastic bag to determine the natural moisture content. The guidelines proposed by ASTM practice for sampling were followed during sampling, and adequate precautions were taken to ensure that each sample contains representation of each particle sizes. Geotechnical characterization was done for the collected CC samples. The investigation includes particle-size distribution, modified Proctor density, hydraulic conductivity, pH, specific gravity, water absorption, and crushability tests. XRF studies were done to ascertain the chemical composition of CC.

Particle-size distribution was performed in accordance with IS: 2720 (Part 4), and CC was classified accordingly. Specific gravity of fine-grained particles as well as coarse-grained particles was ascertained based on the pycnometer method and wire basket method, in accordance to the guidelines laid by IS: 2720 (Part 3) and IS: 2386 (Part 3), respectively. pH values of the CC were ascertained using pH meter following the guidelines presented in IS: 2720 (Part 26). Similarly, modified Proctor compaction tests were performed on the samples in accordance with IS: 2720 (Part 8) to determine the maximum dry density (MDD) and the optimum moisture content (OMC). Samples for hydraulic conductivity tests were prepared at optimum moisture content (OMC) and with modified Proctor compaction effort to achieve at least 98% of maximum dry density (MDD). Constant head permeability test in accordance to IS: 2720 (Part 36) was conducted to the CC samples. Sieve analysis of the samples was done before and after the compaction to get the crushability values. Furthermore, the chemical composition of the CC was ascertained by conducting X-ray fluorescence (XRF) studies.

3 Results and Discussions

3.1 Index Properties of Crushed Concrete

Figure 1 shows the grain-size distribution of crushed concrete used in this study, and the gradation was compared with that reported in the literature. Table 1 gives the grain-size distribution parameters, viz. the effective diameter of particles (D_{10}), the average particle size (D_{50}), the coefficient of uniformity (C_u), and the coefficient of curvature (C_c). CC was classified as well-graded gravel (GW). Liquid limit and plastic limit tests were performed on fine fractions (particles <0.425 mm), and it was observed that CC is non-plastic in nature.

Water absorption and specific gravity tests were performed on both coarse (retained on 4.75 mm sieve) and fine (passing 4.75 mm sieve) fractions of C&D materials. It can be inferred from Table 1 that the specific gravity of coarse aggregates is slightly higher than those of the fine aggregates. Similarly, the water absorptions of coarse aggregates are higher than those of the fine aggregates.

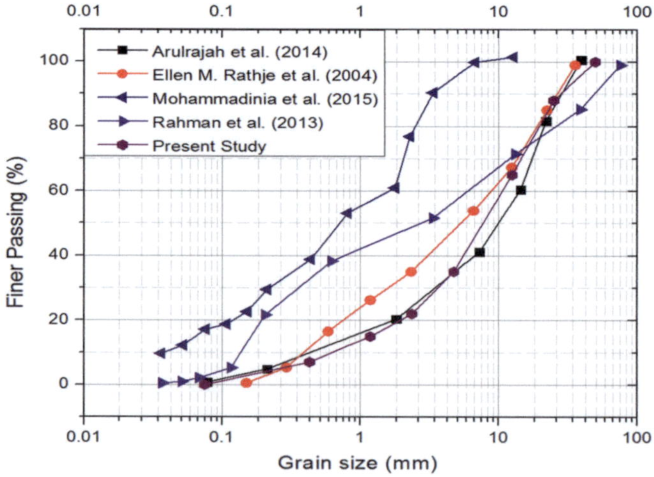

Fig. 1 Grain size distribution curve of CC used in present study along with comparison of distribution from other studies

Table 1 Basic geotechnical properties of CC

Geotechnical parameters	Typical values obtained
Particle-size analysis	
D_{10} (mm)	0.62
D_{30} (mm)	3.8
D_{50} (mm)	8.0
D_{60} (mm)	10.0
C_u	16.1
C_c	2.3
IS classification	GW
Specific gravity testing	
Specific gravity—fine fraction	2.60
Specific gravity—coarse fraction	2.67
Water absorption—fine fraction (%)	0.91
Water absorption—coarse fraction (%)	3.67
Compaction characteristics (modified Proctor testing)	
MDD (kN/m^3)	19.35
OMC (%)	10.8
pH testing	
pH	9.52
Permeability testing	
Hydraulic conductivity (cm/s)	2.1×10^{-5}

3.2 X-ray Fluorescence Test

X-ray fluorescence test was used to determine the chemical composition and their quantity in percentage. The elements predominantly present in C&D wastes were found to be zinc, silica, calcium, iron, and magnesium. These elements were found to exist in the permissible limits and did not possess any detrimental properties.

3.3 Modified Proctor Compaction Test

Figure 2 shows the compaction curve of CC using modified Proctor compaction testing. A comparison of compaction curve from the present study with the results from studies available in the literature is also shown. The compaction curve obtained on CC is similar to that observed for natural aggregates. The maximum dry density and the optimum moisture content of the CC were equal to 19.35 kN/m^3 and 10.8%, respectively. The variation in dry density with water content was not significant. For granular soils, the dry side of OMC is typically flat for well-graded materials, and it usually shows another peak at low moisture content.

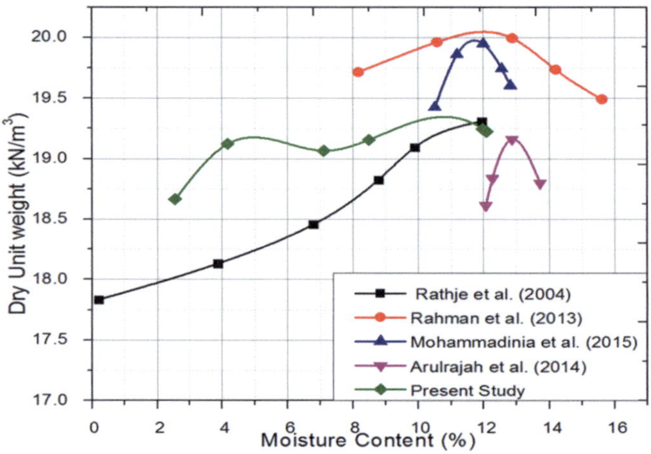

Fig. 2 Modified compaction curve of present study along with results from studies available in the literatures

To assess the crushability of CC during compaction, grain-size analysis was performed on the CC before and after compaction and their gradation is compared (Fig. 3). CC did show a significant change in the gradation indicating minimal crushing of particles during compaction. This can be due to process production that particles were already subjected to breakage during demolition, transportation, storage, crushing, and sieving. This might have led to particles

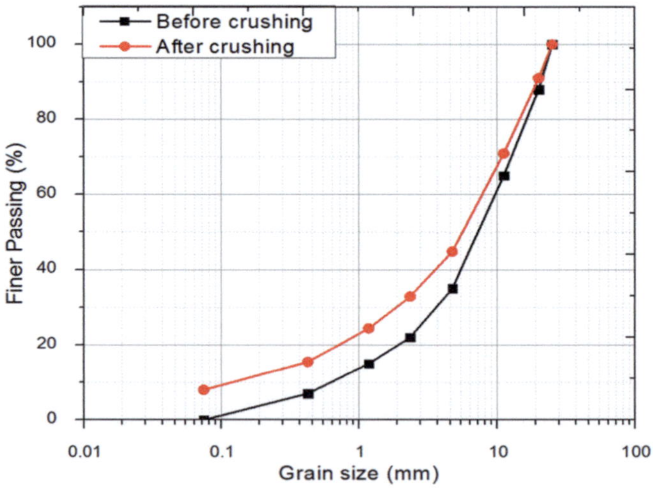

Fig. 3 Gradation of samples before and after compaction

offering enough resistance to breakage when compaction energy imparted during compaction testing.

4 Conclusions

A detailed laboratory investigation was taken up to calculate the index as well as engineering properties of crushed concrete with the aim of calculating the efficacy of CC as a potential backfill material for MSEW. The modified compaction test results indicate that maximum dry density is as par with the virgin coarse, and it is of the order of 19.35 kN/m^3. The pH of CC is more than 7, which indicates that CC is alkaline in nature. The hydraulic conductivity of CC is found out of the order of 2.1×10^{-5}, which indicates that CC can be classified as marginally poor draining material. However, further study is required to evaluate and compare the shear strength properties of CC with that of available literature.

References

Aatheesan, T., Arulrajah, A., Bo, M. W., Vuong, B., & Wilson, J. (2010). Crushed brick blends with crushed rock for pavement systems. *Waste Resource Management, 163*(1), 29–35.

Arulrajah, A., Piratheepan, J., Aatheesan, T., & Bo, M. W. (2011). Geotechnical properties of recycled crushed brick in pavement applications. *Journal of Materials in Civil Engineering, 23*(10), 1444–1452.

Arulrajah, A., Piratheepan, J., Bo, M. W., & Sivakugan, N. (2012). Geotechnical characteristics of recycled crushed brick blends for pavement sub-base applications. *Canadian Geotechnical Journal, 49*(7), 796–811.

Disfani, M. M., Arulrajah, A., Bo, M. W., & Hankour, R. (2011). Recycled crushed glass in road work applications. *Waste Management, 31*(11), 2341–2351.

Hoyos, L. R., Puppala, A. J., & Ordonez, C. A. (2011). Characterization of cement-fiber-treated reclaimed asphalt pavement aggregates: Preliminary investigation. *Journal of Materials in Civil Engineering, 23*(7), 977–989.

Landris, T. L. (2007). *Recycled glass and dredged materials*. Rep. No. ERDC TNDOER-T8, US Army Corps of Engineers, Engineer Research and Development Center, Vicksburg, MS.

Poon, C. S., & Chan, D. (2006). Feasible use of recycled concrete aggregates and crushed clay brick as unbound road subbase. *Construction and Building Materials, 20*(8), 578–585.

Puppala, A. J., Hoyos, L. R., & Potturi, A. K. (2011). Resilient moduli response of moderately cement-treated reclaimed asphalt pavement aggregates. *Journal of Materials in Civil Engineering, 23*(7), 990–998.

Rathje, E. M., Alan F., Rauch David Trejo Kevin, J., Michael, F. C. V., Jain, A., & Ogalla, M. (2006). *Evaluation of crushed concrete and recycled asphalt pavement as backfill for mechanically stabilized earth walls*. Technical report by TxDOT.

Santos, E. C. G., Palmeira, E. M., & Bathurst, R. J. (2013). Behavior of a geogrid reinforced wall built with recycled construction and demolition waste backfill on a collapsible foundation. *Geotextiles and Geomembranes, 39,* 9–19.

Tam, V. W. Y., & Tam, C. M. (2007). Crushed aggregate production from centralized combined and individual waste sources in Hong Kong. *Construction and Building Materials, 21*(4), 879–886.

Lightweight Deflectometer for Compaction Quality Control

Chennarapu Hariprasad, Balunaini Umashankar and Thejesh Kumar Garala

Abstract Quality assessment and control (QA/QC) of compacted pavement layers involve regular monitoring of density and moisture content during compaction. In situ density is traditionally determined using sand-cone test method. However, many recent studies have indicated that stiffness- or strength-based quality measurements are easy to determine and more reliable than the density-based quality measurements. In this study, lightweight deflectometer (LWD) is used as a quality control device to assess the quality of compacted pavement layers. As a part of this study, an extensive LWD field testing program is undertaken on the expressway along the Outer Ring Road (ORR) located in Hyderabad, India, to determine the modulus of deformation (E_{LWD}) of base and surface pavement layers. E_{LWD} of compacted base and surface layers was found to commonly range from 37.6 to 58.6 and 89.3 to 125.7 MPa, respectively. In addition, a case study on a low-volume road is presented to demonstrate the relationship between the E_{LWD} and in situ density obtained from the sand-cone test. LWD is found to be simple to operate and provides quick test results on any pavement layer. Hence, the frequency of quality control tests can be increased leading to an improvement in the overall quality of compacted pavement layers.

Keywords Pavements · Quality control · Stiffness · LWD · Modulus of deformation

---

C. Hariprasad (✉)
Department of Civil Engineering, Vardhaman College of Engineering (Autonomous), Hyderabad 501218, India
e-mail: hariprasadiith@gmail.com

B. Umashankar
Department of Civil Engineering, Indian Institute of Technology Hyderabad, Hyderabad 502205, India
e-mail: buma@iith.ac.in

T. K. Garala
Schofield Centre, Department of Engineering, University of Cambridge, Cambridge CB3 0EL, UK
e-mail: thejeshgarala@gmail.com

V. K. Stalin and M. Muttharam (eds.), *Geotechnical Characterisation and Geoenvironmental Engineering*, Lecture Notes in Civil Engineering 16,
https://doi.org/10.1007/978-981-13-0899-4_5

1 Introduction

Conventionally, in situ density and moisture content tests are performed to assess the quality of compacted pavement layers. Maximum dry density (MDD) and optimum moisture content (OMC) of the pavement materials are determined from laboratory tests such as standard or modified Proctor compaction tests depending on the intensity of loading on to the pavement layers. In the field, the pavement layers are compacted to meet in situ density of greater than or equal to 98% of MDD with the moisture content around ±1% of OMC. However, McCook and Shanklin (2000) concluded that the sand-cone test results are least consistent compared to rubber balloon, drive-cylinder, and nuclear gauge measurements by performing numerous tests on several embankment construction sites in USA. Even though several researchers (Rahman et al. 2008; Berney and Kyzar 2012; Meehan et al. 2012) appreciated the role of nuclear gauge as a pavement quality control device for its ease to operate, frequent readings, and the reliability of the test results, this device failed to gain wide acceptance owing to safety concerns due to radioactive emissions during testing. Quality control or assessment (QC/QA) procedures of compacted pavement layers have moved from density-based to stiffness-/strength-based criteria because of the shift from empirical to mechanistic-empirical methods in pavement design over the last decade (Uz et al. 2015). Several innovative techniques, such as Clegg hammer test, soil stiffness gauge, dynamic cone penetrometer, and lightweight falling deflectometer, have been developed and used to measure the strength/stiffness of compacted layers. Among various methods available, Fleming et al. (2007), Rahman et al. (2008), Vennapusa and White (2009), and Gosk (2016) investigated LWD as a quality control device by performing numerous laboratory and field experiments. In this study, field LWD testing has been performed on four test pads covering 64 test locations to assess the quality of compacted pavement layers. LWD device supplied by Zorn Instruments of Germany has been used in this study. Correlation between in situ density and modulus of deformation (E_{LWD}) of pavement layer of a low-volume road is also proposed.

2 LWD Test Device

Lightweight deflectometer (LWD) is a portable device used to measure the in situ modulus of deformation of the pavement layers (Fig. 1). It consists of a hammer of 10 kg falling mass with a height of fall equal to 72 cm, and it is dropped on 30-cm-diameter bearing plate consisting of inbuilt accelerometer sensor (Zorn 2011).

Equation (1) based on theory of elasticity is used to calculate the deformation modulus of the compacted material.

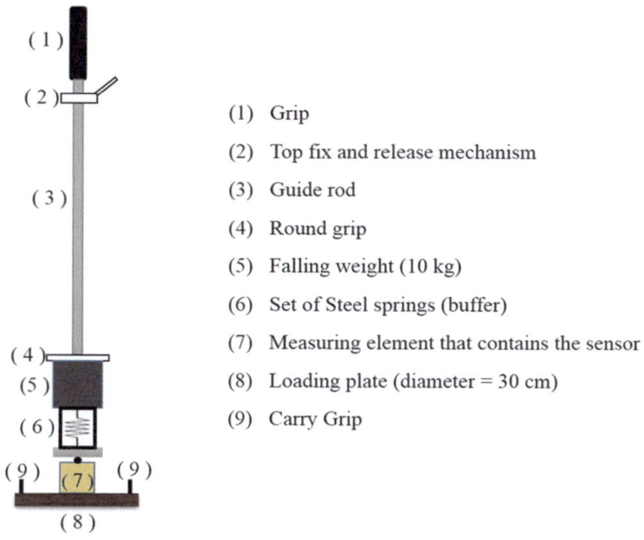

Fig. 1 Schematic of lightweight deflectometer

$$E_{\mathrm{LWD}} = \frac{qr(1-\upsilon^2)}{w} f_{\mathrm{R}} \tag{1}$$

where E_{LWD} is the modulus of deformation of pavement layer; f_{R} is the plate rigidity factor (taken as $\pi/2$ for a rigid plate); q is the maximum contact pressure; r is the radius of the bearing plate; υ is the Poisson's ratio of the soil; and w is the settlement of the bearing plate measured at the center. However, in this study, the inbuilt data processor directly shows the average central deformation and the average modulus of deformation of pavement layer considering three blows of impulse load.

3 Materials

The main objective of the study is to evaluate the quality of compacted base and surface layers using LWD. Base layer consists of graded aggregates and granular material mixed with water, known as wet mix macadam (WMM), and it is laid on a sub-base layer. The MDD and OMC of WMM material were found to be equal to 2.23 g/cc and 4.5%. Further, WMM layer was compacted in two lifts of thickness equal to 125 mm each following MORTH (2001) regulations. The final layer of the pavement is the asphalt surface layer which is constructed using dense-graded bituminous macadam (DBM), and it is laid on a prepared base layer and compacted in 50–100-mm-thick lifts (MORTH 2001).

4 Site Description and Test Pad Construction

LWD tests were performed on four test pads constructed on Outer Ring Road (ORR), Hyderabad, which is an 8-lane ring road expressway of 158 km length. The test pads site is located at Ghatkesar, Hyderabad, along National Highway 202. In order to place and spread the base layer (WMM), a bulldozer was used to spread the material in uniform thickness along the roadway, and the layer is compacted using a smooth drum-vibratory roller (Make: HAMM, model: BW 212-2) to the maximum dry density at optimum moisture content. Dense bituminous layer was laid on a base layer as a surface layer by using a spreader and allowed to dry after compaction. LWD tests were conducted on compacted base and surface layers at selected points. Test points were chosen in a rectangular grid pattern along length and width of test pads with the spacing between the points in the range of 10–30 m, and the variation in modulus of deformation values was assessed from LWD test measurements at these points. Test locations on test pads 1 and 2 correspond to LWD testing on base layer, whereas test locations on test pads 3 and 4 correspond to LWD testing on surface layer. Figure 2 shows the typical layout of test locations on test pad 3, and test locations on other test pads also resemble the similar pattern.

5 Results and Discussion

5.1 Base Layer

The modulus of deformation of base layer was found to commonly range from about 37.6 to 58.6 MPa (Table 1). The modulus of deformation values was found to vary along the length and width, possibly due to variation in the compaction energy imparted due to compactor passes.

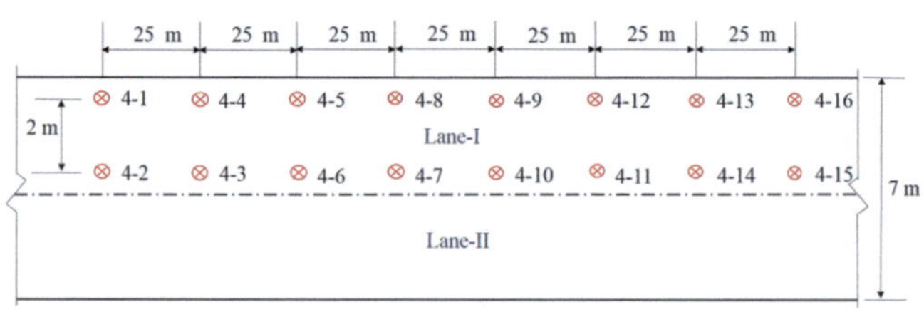

Fig. 2 Schematic view of test locations on test pad 3

Table 1 Summary of test results on base layer

Test point	E_{LWD} (MPa)	Test point	E_{LWD} (MPa)
1-1	45.60	2-4	53.44
1-2	48.60	2-5	43.10
1-3	52.60	2-6	53.44
1-4	52.20	2-7	40.69
1-5	49.00	2-8	44.21
1-6	49.70	2-9	43.69
1-7	37.60	2-10	44.38
1-8	44.20	2-11	42.53
1-9	48.30	2-12	40.83
1-10	39.70	2-13	39.82
1-11	58.60	2-14	46.68
1-12	50.80	2-15	48.39
2-1	47.17	2-16	45.09
2-2	50.11	2-17	44.47
2-3	47.17	2-18	45.09

5.2 *Surface Layer*

The LWD deformation modulus of the surface layer (DBM) was found to range from 89.3 to 125.7 MPa (Table 2).

Table 2 Summary of test results on surface layer

Test point	E_{LWD} (MPa)	Test point	E_{LWD} (MPa)
3-1	104.17	4-2	110.84
3-2	109.76	4-3	111.39
3-3	99.12	4-4	100.90
3-4	97.40	4-5	125.70
3-5	89.29	4-6	103.69
3-6	97.40	4-7	104.17
3-7	94.94	4-8	109.22
3-8	117.19	4-9	117.19
3-9	108.70	4-10	116.58
3-10	107.14	4-11	111.39
3-11	104.17	4-12	109.22
3-12	119.05	4-13	122.95
3-13	119.05	4-14	125.70
3-14	104.65	4-15	125.00
3-15	110.84	4-16	109.22
3-16	103.69	4-17	105.60
4-1	105.63	4-18	117.80

The modulus of deformation values for base and surface layers determined in this study are in good agreement with the published literature. It is obvious that the difference in weight of hammer or height of fall will result in different E_{LWD} values for the same test location. The influence of bottom pavement layers on measured E_{LWD} values was also investigated using finite element analysis and published elsewhere (Umashankar et al. 2015).

6 Low-Volume Road: Case Study

In order to illustrate the significance and usage of LWD as a quality control device, a low-volume road is selected. Native materials having sufficient strength to support loads from vehicles are the key materials in the construction of low-volume roads. As per MORTH statistics, low-volume roads occupy more than 80% of the total road network in India.

Low-volume road site was located in Shankarpalli, Ranga Reddy, Telangana, a rural village near to Ordnance Factory Estate, Telangana. LWD device was used to perform the base layer compaction quality control tests. Base layer was constructed by using the WMM material. Prior to QC testing, a test pad of base layer was constructed, and both sand-cone and LWD tests were performed at two locations separated by a distance of 10 m. Tests were performed corresponding to various compaction energy levels as shown in Table 3. It is evident from Table 3 that E_{LWD} increases with the increase in the compacted density of the pavement layer. Figure 3 is the calibration chart showing the variation of deformation modulus with the relative compaction. In this case, E_{LWD} corresponding to relative compaction of 98% was found to be 32.5 MPa, and this value was used to perform QC check of the compacted base layer.

Thus, similar correlations between in situ density from traditional sand-cone test and E_{LWD} from LWD can be developed for any compacted embankments or pavement layers in order to have more frequent quality control tests based on LWD testing. The target value of modulus of deformation (E_{LWD}) can be obtained corresponding to target relative compaction (e.g., relative compaction of 98%), and this deformation modulus value can then be used to perform QC check of pavement

Table 3 LWD testing on low-volume road

Description	Field density (g/cc)	Relative compaction (%)	E_{LWD} (MPa)
Without passes	2.10	93.3	28.0
One plain pass	2.15	95.6	30.5
Two plain passes	2.20	97.8	32.0
Two low vibratory passes	2.25	100.0	34.5
Two high vibratory passes	2.28	101.3	35.0

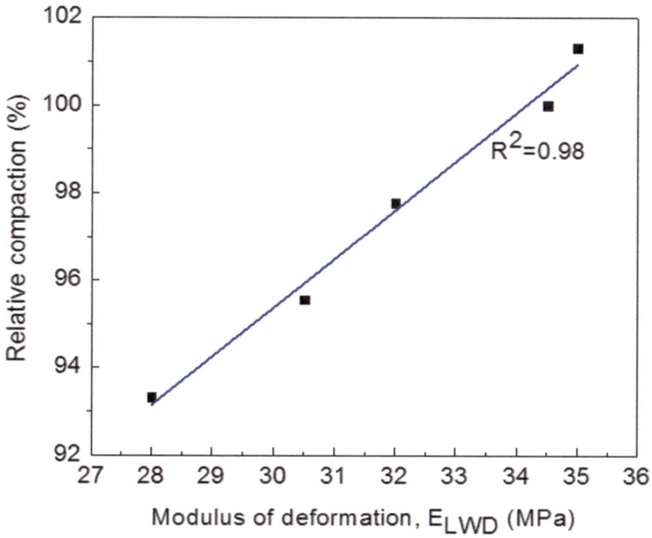

Fig. 3 Relative compaction versus LWD modulus of base layer of a low-volume road

layers using the lightweight deflectometer. The site engineers or field personnel must give proper importance to the parameters like type of LWD device, size of loading plate, height of fall of hammer, type and location of deflection transducer, contact stress beneath plate, plate rigidity and importantly maintain the moisture contents of the test pads close to the targeted levels during construction as the moisture content has significant effect on the measured modulus from LWD testing (Fleming et al. 2007; Vennapusa and White 2009).

7 Conclusions

LWD testing is adopted to assess the quality control of pavement layers. Testing is performed on two surface layers and two base layers. From the study, LWD device is found to be as a simple device to operate and provide quick test results, and hence, the frequency of QC tests can be increased leading to an improvement in the overall quality and performance of compacted pavement layers.

The modulus of deformation of compacted base and surface layers found to be in the range of 37.6–58.6 and 89.3–125.7 MPa, respectively, corresponding to a relative compaction of 98%. A case study on low-volume road showed a good correlation between the modulus of deformation and the compacted density obtained from sand-cone tests with a regression coefficient of 0.98. Similar calibration charts can be developed to perform compaction quality control of pavement layers and embankments based on LWD and sand-cone measurements on test pads. LWD is found to be a reliable device to perform compaction quality control of fill or pavement materials.

References

Berney, E. S., & Kyzar, J. D. (2012). Evaluation of non-nuclear soil moisture and density devices for field quality control. *Transportation Research Record 2310* (pp. 18–26). Washington, D.C.: Transportation Research Board.

Fleming, P. R., Frost, M. W., & Lambert, J. P. (2007). Review of lightweight deflectometer for routine in situ assessment of pavement material stiffness. *Transportation Research Record 2004* (pp. 80–87). Washington, D.C.: Transportation Research Board.

Gosk, W. (2016). Stiffness estimation of the soil built-in road embankment on the basis of light falling weight deflectometer test. In *The 3rd International Conference on Transportation Geotechnics (ICTG 2016)* (Vol. 143, pp. 395–402). Procedia Engineering, Elsevier.

McCook, D. K., & Shanklin, D. W. (2000). NRCS experience with field density test methods including the sand-cone, nuclear gauge, rubber balloon, drive-cylinder, and clod test. In D. W. Shanklin, K. R. Rademacher, & J. R. Talbot (Eds.), *Constructing and controlling compaction of earth fills* (pp. 72–92). West Conshohocken, PA: ASTM.

Meehan, C. L., Tehrani, F. S., & Vahedifard, F. (2012). A comparison of density-based and modulus based in situ test measurements for compaction control. *Geotechnical Testing Journal, 35*(3), 387–399.

MORTH (Ministry of Road, Transport and Highways). (2001). *Specifications for road and bridge works* (4th ed.). India: Indian Road Congress.

Rahman, F., Hossain, M., Hunt, M. M., & Romanoschi, S. A. (2008). Soil stiffness evaluation for compaction control of cohesionless embankments. *Geotechnical Testing Journal, 31*(5), 1–13.

Umashankar, B., Hariprasad, C., & Kumar, G. (2015). Compaction quality control of pavement layers using LWD. *Journal of Materials in Civil Engineering ASCE, 28*(2), 1–9.

Uz, V. E., Saltan, M., & Gökalp, I. (2015). Comparison of DCP, CBR, and RLT test results for granular pavement materials and subgrade with structural perspective. In *International Symposium on Non-Destructive Testing in Civil Engineering* (NDT-CE), Berlin, Germany.

Vennapusa, P. K. R., & White, D. J. (2009). Comparison of light weight deflectometer measurements for pavement foundation materials. *Geotechnical Testing Journal, 32*(3), 239–251.

Zorn Instruments. (2011). User manual for the light weight deflectometer (LWD). *ZFG 3.0*, Hansestadt Stendal, Germany.

A Study on Compressibility, Swelling and Permeability Behaviour of Bentonite–Sand Mixture

Binu Sharma and Priyanka Deka

Abstract In geotechnical engineering field, bentonite–sand mixtures have been proposed and used as engineered barriers for containing the waste. This paper presents the laboratory evaluation of compressibility by performing one-dimensional consolidation tests on six different mixtures of bentonite with sand. The bentonite–sand mixtures were formed by varying sand content in bentonite in increments of 5% from 5 to 25% by dry weight. Dry bentonite–sand mixtures were placed initially in the consolidation cell at their loosest dry state and then allowed to saturate. Swelling characteristics and swelling pressures of the bentonite–sand mixtures were also evaluated. This paper also presents the laboratory evaluation of permeability of the bentonite–sand mixtures by performing falling head test after every load increment during the consolidation test. This study arrived at the conclusion that amount of swelling (expressed in percentage) and swelling pressure decreased with addition of sand. Moreover, changes were observed for the consolidation parameters upon addition of sand to bentonite. The void ratio versus log of permeability plots was found to be linear, and permeability was found to increase with increase in sand content.

Keywords Bentonite–sand mixture · Compressibility · Swelling pressure
Permeability · Consolidation test

1 Introduction

Bentonite–sand mixtures are used as engineered barriers for containing the leachate produced from waste. Bentonite has high swelling capacity, and this makes it suitable for producing low permeability barriers. If clay content is too high,

B. Sharma · P. Deka (✉)
Department of Civil Engineering, Assam Engineering College, Guwahati, India
e-mail: priyankadekaanee@gmail.com

B. Sharma
e-mail: binusharma78@gmail.com

V. K. Stalin and M. Muttharam (eds.), *Geotechnical Characterisation and Geoenvironmental Engineering*, Lecture Notes in Civil Engineering 16,
https://doi.org/10.1007/978-981-13-0899-4_6

shrinkage occurs in bentonite. To avoid shrinkage in bentonite on drying, sand is added to it, thus providing mechanical stability to the mixture (Mollins et al. 1996). Bentonite-enhanced sand (BES) mixtures are widely used as barriers to control the movement of liquid from waste disposal facilities because BES can combine relatively high strength and low compressibility with very low hydraulic conductivity (Stewart et al. 2003). Based on studies conducted on swelling behaviour of bentonite and coarser fractions of different shapes and sizes, Sivapullaiah et al. (1996) reported that the swelling occurs only after the voids of the non-swelling particles are filled up with swollen clay particles. Permeability of sand–bentonite mixture was found to reduce after addition of bentonite to sand (Otoko and Otoko 2014), and they also observed that the permeability values from falling head and one-dimensional consolidation tests were in good agreement.

Cui et al. (2012) have reported that even for the same ratio of sand, in sand–bentonite mixture, considerable differences exist in the nature of time and swelling pressure relationships if initial dry densities are different. At low-density range, rate of swelling is low but increases gradually with increase in initial dry density. Most of the works on swelling, compressibility and permeability behaviour in the literature are at optimum moisture content and at the maximum density state of the soils. In this research work, the swelling, compressibility and permeability characteristics of bentonite–sand mixtures were studied using one-dimensional consolidation test at very low density.

2 Materials and Testing Methods

Commercially available bentonite and dry sand were used for the study. The sand is uniformly graded, fine silica sand and is classified as SP according to the Unified Soil Classification System (USCS). The X-ray diffraction spectra for the bentonite (Fig. 1) show that it is predominantly a montmorillonite with some amount of sand, i.e. the quartz mineral. The liquid limits and the plastic limits of the bentonite sample were determined according to the code IS 2720 (Part 5) 1985. The physical properties of the bentonite–sand mixtures are shown in Table 1. The bentonite–sand mixture was mixed in the following proportion by percentage of dry weight as B:S = 100:0, 95:5, 90:10, 85:15, 80:20 and 75:25 (B: bentonite, S: sand).

2.1 Consolidation Tests

The consolidation test was done in standard fixed-ring consolidometers using stainless steel rings of 60 mm diameter and 20 mm height (Part 15, method 2720, Indian Standard Institution 1986). In this work, the dry mixture was placed in the cutter of the consolidometer ring at the loosest state up to 2/3rd height of the cutter, i.e. 13.33 mm, and seating load of 5 kN/m^2 was placed. The sample was then

Fig. 1 X-ray diffraction pattern for the bentonite

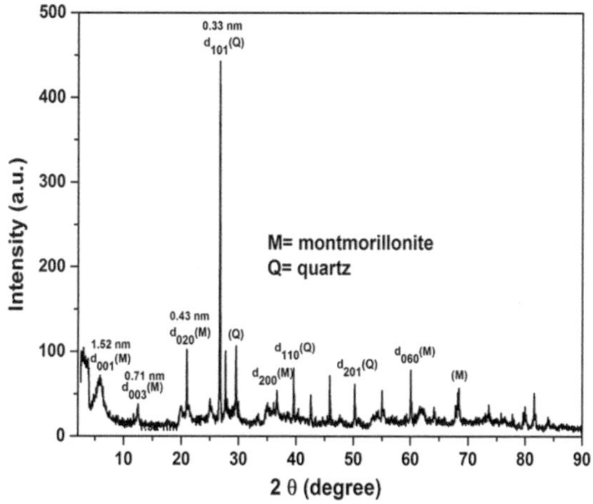

Table 1 Properties of bentonite–sand mixture

Bentonite: sand ratio	Liquid limit (%)	Plastic limit (%)	Plasticity index	Specific gravity	Initial dry density (g/cm^3)
100:0	195	72	123	2.75	0.60
95:5	160	65	95	2.75	0.65
90:10	140	60	80	2.72	0.72
85:15	137	60	77	2.7	0.8
80:20	133	57	76	2.7	0.95
75:25	127	52	75	2.6	1.1

allowed to saturate. As a result of saturation, the sample started swelling. The dial gauge reading was taken after every 24 h till the dial gauge reading showed a constant value. The swollen sample was then subjected to small pressure increments until the swelling pressure was obtained which is indicated by returning back of dial gauge to its initial reading before saturation. After obtaining the swelling pressure, double incremental loading was applied up to 640 kN/m^2 as per IS code of practice.

2.2 *Permeability Test*

Permeability characteristics of the mixtures were determined by performing falling head test in the conventional fixed-ring consolidation cell in the laboratory. At each pressure, after equilibrium was achieved, falling head permeability tests were performed to determine the coefficient of permeability of the mixture. To prevent evaporation from the burette, a thin layer of kerosene over the water was placed. The flow was assumed to be vertical, and permeability coefficient was determined from the experimental data.

3 Results and Discussions

The swelling percentages of the various mixtures were calculated using Eq. (1)

$$\text{Swelling}\ (\%) = \Delta H / H_0 \times 100\%. \tag{1}$$

where $\Delta H = H_f - H_0$; H_f = final height after swelling after every 24 h.

$H_f = H_0 + 0.01\times$ (dial gauge reading after every 24 h). H_0 = initial height before swelling = 13.33 mm.

The swelling percentage (%) versus log of time plot is shown in Fig. 2. Bentonite shows very high swelling because of its prominent cation exchange capacity. Moreover, it had been observed that for all the mixtures, increase in swelling with time is slow initially, increases steeply and then reaches an asymptotic value. The time required to reach an asymptotic value varies considerably, depending upon the percentage of sand. Table 2 shows the swelling pressure, total swelling percentage (*G*) and other swelling characteristics of the various mixtures. As reported by Cui et al. (2012) that even for the same ratio of sand in bentonite, considerable differences exist in the nature of time and swelling pressure relationships if initial dry densities are different. The values shown in this work are for very low-density ranges.

The coefficient of consolidation, C_v, was obtained by the square root of time fitting method. The C_v values (to the power of 10^{-4} cm/s) plotted against the percentage of bentonite are shown in Fig. 3.

There is a slight increase in C_v value with addition of sand for low-pressure ranges, but at high pressure of 320 kN/m^2, the C_v values remain the same. Compression index has been determined from the linear portion of the void ratio

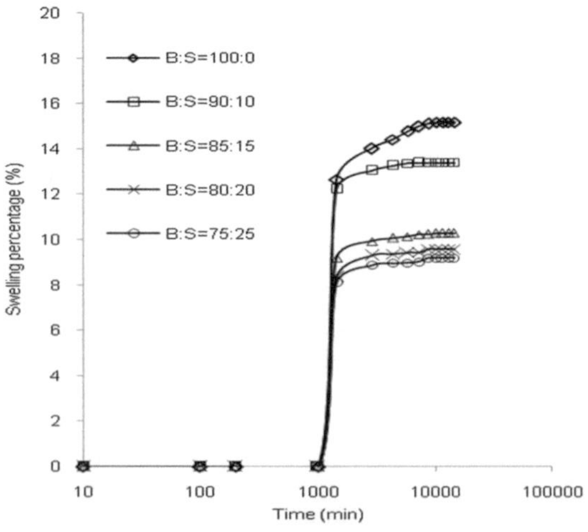

Fig. 2 Swelling percentage (%) versus log of time plot

Table 2 Swelling characteristics of various mixtures

Bentonite: sand	Swelling pressure (kN/m^2)	Total days taken to obtain full swelling	Total swelling percentage (G) (%)
100:0	77.5	7	15.2
95:5	Not done	Not done	Not done
90:10	57.5	6	13.4
85:15	55	6	10.3
80:20	47.5	5	9.6
75:25	Not done	5	9.2

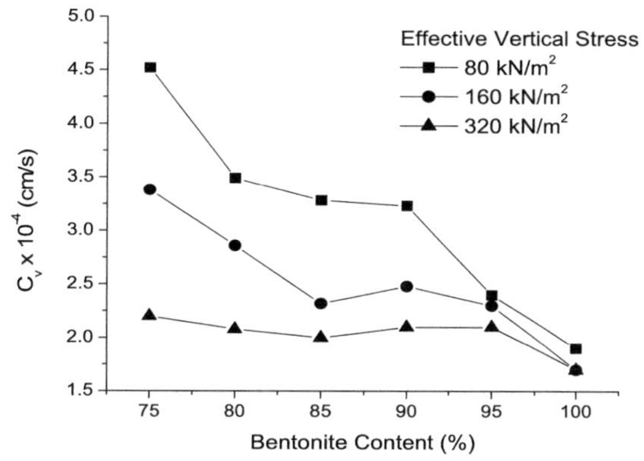

Fig. 3 Coefficient of consolidation versus bentonite content

(e) versus log of effective stress (σ'_v) plot, and its values for the different mixtures are shown in Fig. 4.

It is observed that C_C decreases with addition of sand content. The void ratio (e) versus log of effective stress (σ'_v) plot without including the expansion and recompression part is shown in Fig. 5.

It has been observed from the plot that with the addition of sand to bentonite, the void ratio has increased.

The permeability range of compacted clay liners should be very low. For many engineering projects, the permeability values usually lie between 10^{-9} and 10^{-7} cm/s.

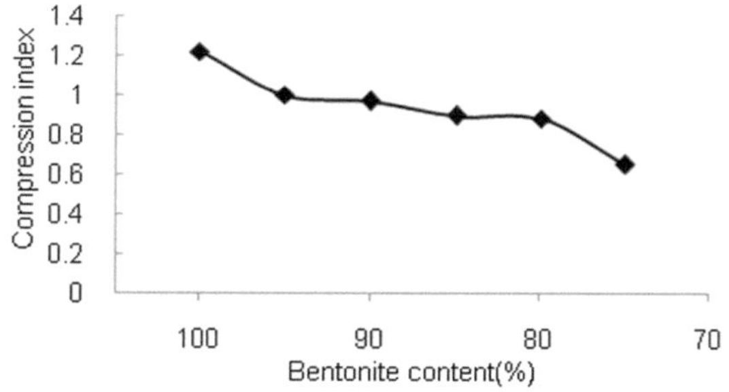

Fig. 4 Compression index versus bentonite content

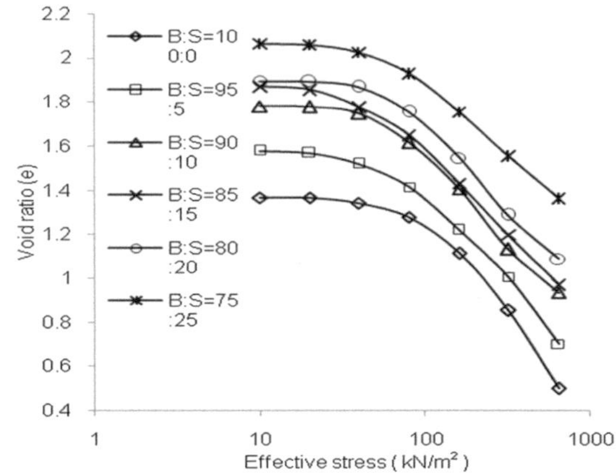

Fig. 5 Relationship between void ratio and effective stress (σ'_v) plot for the six samples

In this present study, the permeability behaviour of the bentonite–sand mixtures was studied in the laboratory using the oedometer test setups after full swelling was achieved.

Evaluation of permeability of bentonite–sand mixture as function of loading pressures on basis of experimental results by performing falling head permeability tests has been plotted graphically in Fig. 6.

It is observed for all the specimens that the permeability varies inversely with the loading pressures. The effect of effective vertical pressure on permeability is less significant once it crosses 320 kN/m^2. The coefficient of permeability (*k*) for pure bentonite lies between 3.7×10^{-8} and 1.78×10^{-8} cm/s, whereas for mixture obtained by adding 25% sand to 75% bentonite the value ranges from 6.3×10^{-8} and 3.97×10^{-8} cm/s. Here, it is observed the *k* values of the mixtures are in a narrow range. This is because the bentonite content is more than which can be accommodated within the voids of the sand fractions, and the permeability is controlled by the bentonite content alone. Permeability values were also calculated for various pressure increments for the various mixtures using the experimentally determined C_v and m_v values.

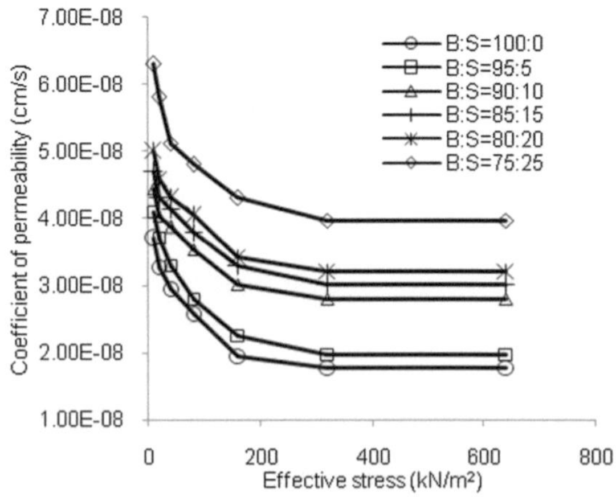

Fig. 6 Relationship between effective vertical stress and coefficient of permeability calculated experimentally

It was observed for each mixture $\log_{10} k$ was found to vary linearly with void ratio over the full range of pressure increments to which it was subjected to. Figure 7 shows the $e - \log_{10} k$ relationship for the six mixtures. The slope of the $e - \log_{10} k$ lines decreases with an increase in the percentage of bentonite.

The indirect evaluation of permeability from consolidation test data may be unreliable. However, in this study there is a good agreement between the experimental permeability with those determined theoretically from C_v and m_v. These are shown in Fig. 8a, b.

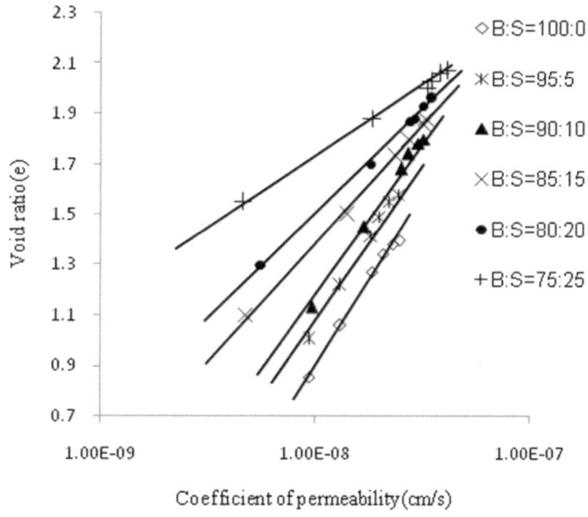

Fig. 7 Void ratio versus coefficient of permeability (k)

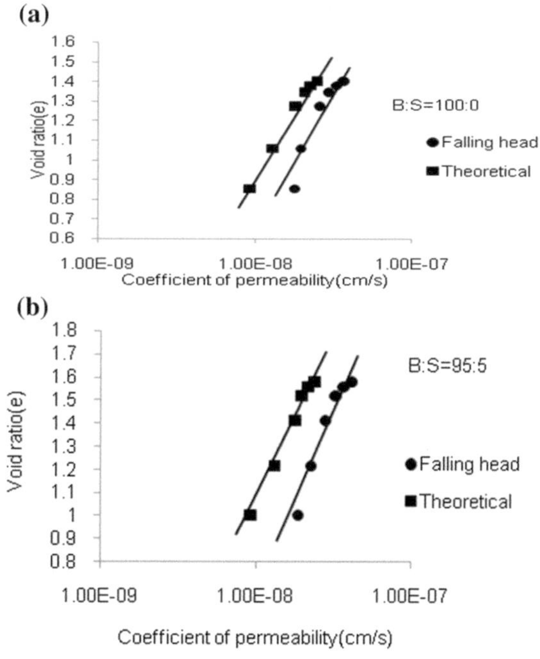

Fig. 8 **a** Comparison of permeability calculated experimentally and theoretically for pure bentonite and **b** comparison of permeability calculated experimentally and theoretically for B:S = 95:5

4 Conclusions

Liquid limit and plastic limit decreased with addition of sand to bentonite content. Coefficient of consolidation (C_v) showed a small increase with increase in sand content for low-stress ranges, whereas for high-stress range of 320 kN/m^2 it is found to remain constant. Compression index (C_C) decreased with increase in sand content. Addition of sand decreases the swelling pressure, total number of days taken for full swelling and amount of swelling percentage (%). The $e - \log \sigma'_v$ plot also shows that with the addition of sand to bentonite, the void ratio increases. The coefficient of permeability for pure bentonite is least and increases with addition of sand to bentonite. Moreover, there is a good agreement between the experimental permeability obtained through falling head permeability test with the theoretically calculated permeability values.

References

Cui, S. L., Zhang, H. Y., & Zhang, M. (2012). Swelling characteristics of compacted GMZ bentonite–sand mixtures as a buffer/backfill material in China. *Engineering Geology*, 141-142-65-73.

IS: 2720 (Part 5). (1985). *Method of test for soils: Determination of liquid and plastic limit*. New Delhi, India: Bureau of Indian Standards.

Mollins, L. H., Stewart, D. I., & Cousins, T. W. (1996). Predicting the properties of bentonite–sand mixtures. *Clay Minerals, 31,* 243–252.

Otoko, R. G., & Otoko, G. U. (2014). The permeability of ocean sand with bentonite. *International Journal of Engineering and Technology Research, 2*(1), 01–06. (ISSN; 2327-0349).

Sivapullaiah, P. V., Sridharan, A., & Stalin, V. K. (1996). Swelling behavior of soil bentonite mixtures. *Canadian Geotechnical Journal, 33,* 808–814.

Stewart, D. I., Studds, P. G., & Cousens, T. W. (2003). Factors controlling properties of bentonite enhanced sand. *Applied Clay Science, 23*(1–4), 97–110. (ISSN 0169-1317).

Effect of Plastic Fines on Shear Strength of Sands

N. Umaharathi, K. P. Bhargav Kumar and Balunaini Umashankar

Abstract Natural sands are rarely found as clean sands. They contain considerable amounts of plastic (clay) or non-plastic (silt) fines. If fines are present in sufficient quantity to separate the sand particles, soil behavior will be significantly controlled by the fines. The fabric of such a soil is referred to as floating fabric. Characterization of sand with the presence of fines is important in the design of geotechnical structures. This paper presents a detailed study focused on investigating the effect of plastic fines on the engineering properties of sand–fines mixture. Mixtures of sand and fines are prepared by mixing clean sand with plastic fines (clay). Indian Standard (IS) Grade II sand, widely known under the name of Ennore sand, is used during experimentation. This sand used is classified as poorly graded (SP) as per Unified Soil Classification System (USCS). Plastic fines are obtained locally, and the liquid limit (LL) and the plasticity index (PI) are found to be equal to 65 and 39%, respectively. The effect of plastic fines on shear strength of sands is studied using direct shear testing conducted in displacement-controlled mode. Tests are performed at 0.5 mm/min, and the samples are subjected to three normal stresses—50, 100, and 150 kPa. In addition to determining the shear strength, this study also focused on volumetric changes of specimens during shearing in a direct shear box. Tests are conducted on specimens of sand with various fines content ranging from 0 to 25%, while maintaining a constant total density of 1.7 g/cc throughout. Plots of shear stress versus horizontal displacement of lower box, and vertical displacement versus horizontal displacement are reported. The shear strength envelopes of sand specimens prepared at various percentages of fines are

N. Umaharathi · K. P. Bhargav Kumar (✉) · B. Umashankar
Department of Civil Engineering, Indian Institute of Technology Hyderabad, Kandi 502285, India
e-mail: ce15resch11003@iith.ac.in

N. Umaharathi
e-mail: ce13b1022@iith.ac.in

B. Umashankar
e-mail: buma@iith.ac.in

V. K. Stalin and M. Muttharam (eds.), *Geotechnical Characterisation and Geoenvironmental Engineering*, Lecture Notes in Civil Engineering 16,
https://doi.org/10.1007/978-981-13-0899-4_7

proposed. The test results showed that both friction angle and apparent cohesion decrease with increase in fines content in sand–fines mixture, thereby shifting the shear envelope downwards. In addition, there is a consistent decrease in peak shear strength with increase in fines content.

Keywords Plastic fines · Shear strength · Friction angle · Cohesion

1 Introduction

The mechanical behavior of clean sands was first investigated by Coloumb in 1776, and that of clays by Terzaghi in 1925. Studies on these soils continued over the years with researchers mainly focused on either clean sands or pure clays. Although these soil types define distinct boundaries of wide spectrum, they are seldom representative of nature soil deposits. Natural soil deposits are neither clean sands nor pure clays, but a combination of both in varying proportions. The characterization of sands containing fines is essential in geotechnical engineering practice. It has been shown that behavior of sands is greatly influenced by the fines content (Salgado et al. 2000), and mineralogy (Carraro et al. 2009). Salgado et al. (2000) conducted drained triaxial tests to investigate the effect of non-plastic fines on the shear strength of sands. It was found that for a given relative density and stress state, the addition of non-plastic fines increased the peak friction angle. Carraro et al. (2009) conducted drained triaxial tests to study shear strength and stiffness of sands containing plastic or non-plastic fines. The results showed that the addition of non-plastic fines to sands increased the peak friction angle of the sand and, on the other hand, the addition of plastic fines decreased the friction angle and imparted a less dilative response to the soil.

This study attempts to assess the mechanical behavior of 'sands containing plastic fines' or 'clayey sands.' Only sands containing small amounts (less than 25% by weight) of plastic fines were considered in the study. To this end, a series of direct shear tests on sands with varying amounts of clay (0–25%) was systemically conducted. The specific objectives of this study were to determine the effect of plastic fines on the shear strength of sands, and to characterize the volumetric behavior of such 'clayey sands.'

2 Overview

According to Unified Soil Classification System (USCS), clayey sands are soils that (i) contain more than 50% particles retained on the No. 200 sieve (75 μm); (ii) have 50% or more of the coarse fraction passing through No. 4 sieve (4.75 mm); and (iii) have more than 12% particles smaller than 75 μm, by weight. Sands with 5–12% fines require dual symbols in the USCS. Other classification systems establish different boundaries for the percentage of fines and particle size to classify sands containing fines. Henceforth, sand–fines mixtures used in this study are referred as 'clayey sands.'

Shearing requires that energy be supplied for (i) particles to overcome friction at the contacts so that relative motion can take place; (ii) particles are continuously rearranged for sustained shearing to take place; and (iii) particles climb over each other so that they can move past adjoining particles along the slip plane. This climbing action results in increase in volume (Salgado 2006).

Soil fabric: The fabric of soil in general has limited impact on the shear strength of soil, but it is of greater importance when fines are present. Moreover, if fines are present in sufficient quantity to separate the sand particles, the soil behavior becomes very different, controlled by fines. Such a fabric is referred to as floating fabric (Carraro et al. 2009). The transitional fines content is a key parameter that indicates whether the soil behaves as sand-dominated or fines-dominated, which is generally between 25 and 35% (Yang et al. 2006; Carraro et al. 2009). During the transition from non-floating to floating fabric, there would be a steep drop in the peak shear strength brought about by the increased interaction among the fines. The absence of such a drop in this study can be seen as an indication of non-floating conditions.

3 Experimental Program

3.1 Materials

Several clayey sand mixtures were reconstituted in the laboratory by mixing sand with clay, while maintaining a constant total density of 1.7 g/cc throughout. All the samples were in dry state with negligible moisture content. Indian Standard (IS) Grade II sand, popularly known as Ennore sand, was used. Figure 1 shows grain-size distribution curve of sand. It is classified as poorly graded sand (SP) with coefficient of uniformity (C_U) 2.17, according to Unified Soil Classification System (USCS). The shape of the sand particles was obtained using the scanning electron microscopy (SEM), and the particles were found to be sub-angular to angular (Fig. 2). The specific gravity of the sand was equal to 2.65.

Plastic fines (clay) were obtained from the surroundings of IIT Hyderabad laboratory situated in Yeddumailaram, Telangana. The plastic limit (PL) and liquid limit (LL) of clay were found to be equal to 39 and 25%, respectively. The mixtures of sands and plastic fines investigated in this study are 5, 15, and 25% clayey sands by weight. A 5% clayey sand sample would contain 5% clay and 95% sand by weight. Mixtures were prepared by first weighing the desired amount of sand and fines and then vigorously shaking in a closed container to obtain homogeneity.

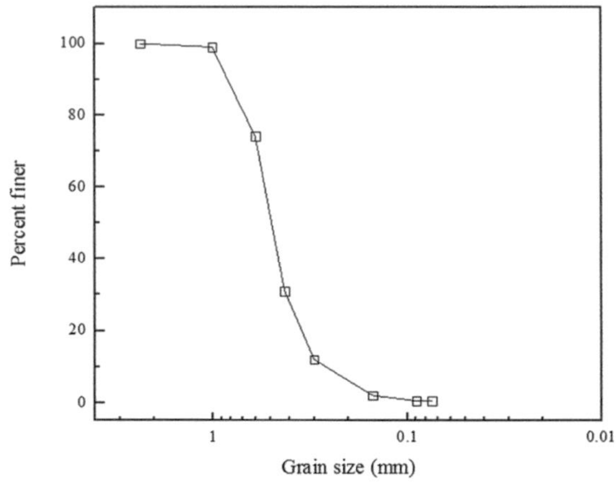

Fig. 1 Grain-size distribution curve of IS Grade II sand (Durga Prasad et al. 2014)

Fig. 2 Morphological features of sand particles using SEM (Durga Prasad et al. 2014)

3.2 Test Method

The effect of plastic fines on shear strength of sand was studied by conducting direct shear tests in displacement-controlled mode. The dimensions of the direct shear were equal to 60 mm × 60 mm (in plan). Tests were performed at a shear displacement rate 0.5 mm/min, and the samples were subjected to three normal stresses—50, 100, and 150 kPa. Plots of shear stress versus horizontal displacement to determine the peak shear strength (Figs. 3 and 4), and vertical displacement versus horizontal displacement to (volumetric behavior) were reported.

4 Results and Discussion

4.1 Sand Samples Without Fines

On observing the plot of shear stress versus horizontal displacement of a sand sample with no fines, it was observed that the resisting shear stress increases with shear displacement until it reaches a failure stress (indicating *peak-state shear strength* being reached). After failure stress is attained, the resisting shear stress gradually decreases until it finally reaches a constant value called the *critical-state shear strength*.

Upon observing the variation of the vertical displacement with respect to horizontal displacement of the lower box, the sample was found to be initially contractive in nature and dilative afterward (refer to Fig. 5). This is consistent with the fact that dense sand specimen first becomes denser as its particles occupy relatively stable position, and thereafter reaches an arrangement such that in order for shearing to continue particles must climb over one another and increase in volume occurs until the configuration is in equilibrium with the imposed stresses.

4.2 Sand Samples with Fines

4.2.1 Shear Strength

Tests conducted on mixtures of sand and fines (clayey sands) indicated that as the content of plastic fines increased, the peak shear strength mobilized from the sample decreased. Figure 4 shows that there is a steady decrease in the peak shear strength as the percentage of clay increased from 0 to 25% in the sand–fines mixture. These results suggest that presence of fines facilitates shearing with the clay particles acting as lubricants between the sand particles, enabling the sliding between sand grains. It is noted that in the absence of fines, the shearing takes place between sand grains alone. However, when fines are present, the shearing could take place between sand grains and fines or fines alone, which offer relatively less friction compared to angular and more irregular nature of sand grains (as shown in Fig. 2).

4.2.2 Volumetric Behavior

The volumetric response can be observed from plots showing vertical displacement versus horizontal displacement. As percentage of plastic fines increased, the dilative nature of clayey sand sample decreased. It can be seen from that the specimen behaves more contractive as the clay content increases (Fig. 5). The increase in volume is caused by the climbing action mentioned earlier. However, the presence of plastic fines seems to be adversely affecting climbing action, which in turn inhibits the dilative behavior of sand particles in mixtures.

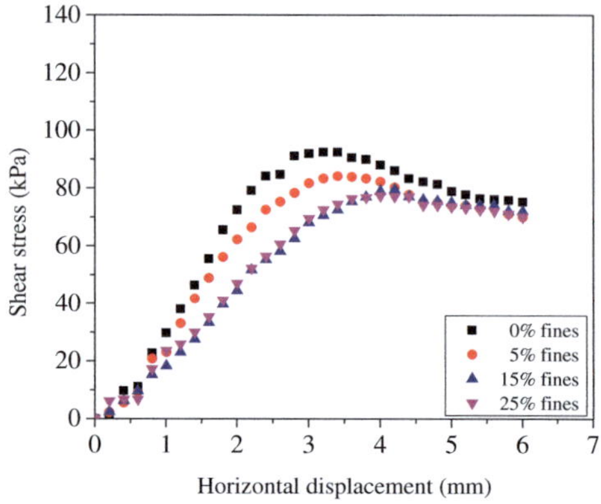

Fig. 3 Variation of shear stress and horizontal displacement for various fines content at normal stress equal to 100 kPa

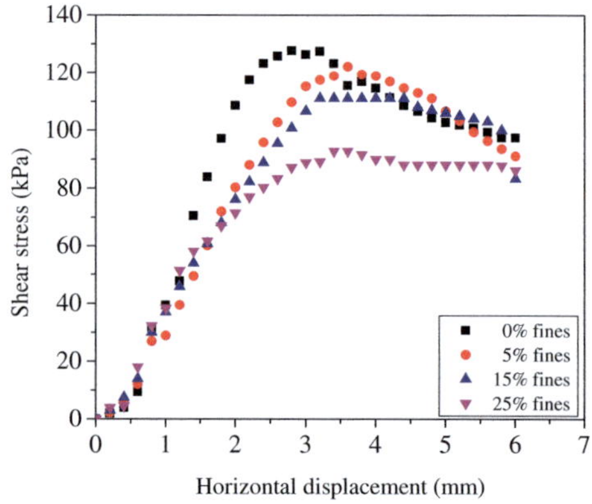

Fig. 4 Variation of shear stress and horizontal displacement for various fines content at normal stress equal to 150 kPa

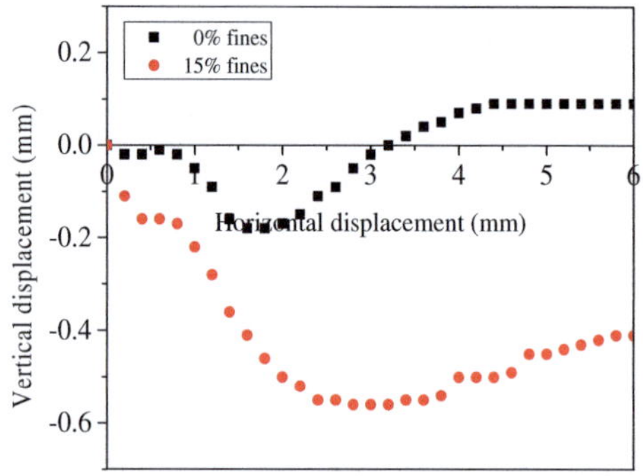

Fig. 5 Variation of vertical displacement of the sample with horizontal displacement of lower box for mixtures with 0 and 15% fines, at normal stress equal to 150 kPa

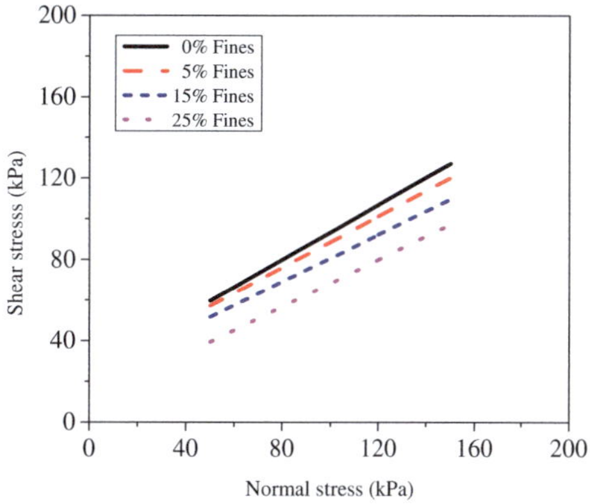

Fig. 6 Shear strength envelopes of different mixtures of sand and fines

4.2.3 Shear Strength Envelope

For designing any geotechnical structure, the shear stress on the failure plane is approximated as a linear function of normal stress. This envelope was plotted with the peak shear strength values obtained with normal stresses 50, 100, and 150 kPa. The values of apparent cohesion, c, and angle of shearing resistance, ϕ, are determined from intercept and slope of the equation Mohr–Coulomb shear strength envelope given by $\tau = c + \sigma \tan \phi$. It can be observed from Fig. 6 that with increase in percentage of plastic fines, the shear envelope shifts downwards indicating that both the friction angle (slope) and apparent cohesion (intercept) decrease. This seems to be a consequence of decrease in peak shear strength that is resulting in changes in strength envelope.

The above observations highlight the role played by fines in shaping the mechanical behavior of clayey sands.

5 Conclusions

An experimental study was undertaken to study the effect of plastic fines on shear strength and volumetric behavior of sands. To that end, care was taken to limit the floating fabric conditions, as mentioned before.

It was observed that

(1) The addition of plastic fines (clay) to the host sand decreases peak shear strength that can be mobilized from the sand sample. The addition of clay imparts a less dilative response to the sand.
(2) It appears that clay particles tend to lubricate the more angular and irregular surfaced contacts between sand grains, thereby facilitating shearing.

(3) As a consequence of above two results, a decrease in friction angle and apparent cohesion was observed, thereby shifting the shear strength envelope downwards.

References

Carraro, J., Antonio, H., Prezzi, M., & Salgado, R. (2009). Shear strength and stiffness of sands containing plastic or nonplastic fines. *Journal of Geotechnical and Geoenvironmental Engineering, 135*(9), 1167–1178.

Durga Prasad, B., Hari Prasad, C., & Umashankar, B. (2014). Shear strength of IS sand under various normal stress. In *Indian Geotechnical Conference (IGC-2014)*, Kakinada (pp. 109–113).

Prasad, A., & Pandey, B. (2013). Effect of fines on the mechanical behavior of sand. *International Journal of Structural and Civil Engineering Research, 2*(2), 41–47.

Salgado, R. (2006). *The engineering of foundations* (p. 181). McGraw-Hill Education.

Salgado, R., Bandini, P., & Karim, A. (2000). Shear strength and stiffness of silty sand. *Journal of Geotechnical and Geoenvironmental Engineering, 126*(4), 451–462. https://doi.org/10.1061/(asce)10900241.

Yang, S., Lacasse, S., & Sandven, R. (2006). Determination of the transitional fines content of mixtures of sand and non-plastic fines. *Geotechnical Testing Journal, 29*(2), 1–6.

Effect of Constant Energy Source on Coherence Function in Spectral Analysis of Surface Waves (SASW) Testing

Sayantan Chakraborty, Tejo V. Bheemasetti and Anand J. Puppala

Abstract The quality and acceptability of a waveform data collected during spectral analysis of surface waves (SASW) testing is judged based on the coherence function over the measured range of frequencies. However, many trials and repetitions are required during SASW testing to obtain data with acceptable coherence value (>0.95). This makes the test time-consuming, and in most cases, only small portion of the collected data that satisfies the acceptable coherence criteria can be used for analysis. In this research study, an attempt was made to study the effect of using an impact source of constant energy on the coherence function as compared to the use of traditional handheld hammers. Laboratory experimental studies were performed on sandy clay soil bed filled in a metal box of dimensions 1.5 m × 0.61 m × 0.45 m. Also, a series of field tests were performed to validate the applicability of the laboratory findings. In both laboratory and field testing, a 2.5 kg hammer was used with height of fall as variable parameter. Two sets of tests were performed. One with random height of fall that is similar to current practice and the other with fixed height of fall of 0.13 m to simulate impacts with same energy conditions. Test results depicted that unlike the conventional procedure of testing, the use of constant energy of impact leads to coherence close to 1 over a significantly large frequency bandwidth. This research highlights the effect of using constant and varying impact energy on coherence value over a wide range of frequencies obtained during SASW testing.

Keywords Spectral analysis of surface waves · Energy source
Coherence · Geophysical testing

S. Chakraborty (✉) · T. V. Bheemasetti · A. J. Puppala
Department of Civil Engineering, The University of Texas at Arlington,
Arlington, TX 76019, USA
e-mail: sayantan.chakraborty@mavs.uta.edu

T. V. Bheemasetti
e-mail: tejovikash.bheemasetti@mavs.uta.edu

A. J. Puppala
e-mail: anand@uta.edu

V. K. Stalin and M. Muttharam (eds.), *Geotechnical Characterisation and Geoenvironmental Engineering*, Lecture Notes in Civil Engineering 16,
https://doi.org/10.1007/978-981-13-0899-4_8

1 Introduction

Small strain shear modulus (G_{max}) with strain values less than 0.001% (Nazarian 1984; Nazarian and Stokoe 1985; Rix 1988; Joh 1996) is an important parameter used in design and analysis of geotechnical engineering projects. Some of the common uses of G_{max} include quality control after construction, slope stability, and landslide investigation, characterization of a geotechnical site based on stiffness properties and ground response analysis to earthquake loading (Joh 1996; Puppala et al. 2006; Bheemasetti et al. 2015). Over the years, several geophysical tests such as cross-hole test, down-hole test, seismic cone penetration testing, and spectral analysis of surface waves (SASW) test were developed to determine the shear modulus of a given site. This research study focuses on using SASW testing, and hence, a brief detail of it is discussed below.

SASW technique is a nondestructive and non-intrusive technique developed in early 1980s based on dispersive characteristics of surface waves (Heisey et al. 1981; Nazarian 1984; Nazarian and Stokoe 1985). SASW testing comprises three major tasks such as field test setup, generation of experimental dispersion curves, and performing inversion analysis to determine the shear wave velocity profile of a particular site (Nazarian 1984; Nazarian et al. 1988). The procedure involved in SASW testing is briefly discussed here; a detailed theory can be found elsewhere (Heisey et al. 1981). The testing process involves using hammers of different weights as sources of impact on the ground surface to produce Rayleigh waves of wide range of frequencies. These waves are received by a pair of vertical geophones placed at a predetermined distance from the source. The spacing between geophones is increased for every test in order to get information about deeper layers. Usually, the geophone spacing is doubled following common receiver midpoint geometry (Nazarian and Stokoe 1985; Joh 1996). The process is continued until data for preferred depth of interest is obtained. The global dispersion curve is obtained by combining the dispersion curves obtained for different geophone spacing. The inversion analysis is then performed on the global dispersion curve for determining the subsurface properties.

The quality of the experimental data is judged based on the value of coherence function. A coherence of 1 implies a perfect signal received by both the geophones. For practical purposes, a coherence value of 0.9–0.95 is considered acceptable (Nazarian and Stokoe 1985; Nazarian et al. 1988; Joh 1996). The frequency range with coherence value lower than the acceptable range is therefore discarded (masked) during analysis. Hence during the field test, it is imperative to accept only those tests which show coherence greater than 0.95 for the frequency bandwidth of importance.

One of the major challenges faced during the field test is to obtain the waveforms with coherence value close to 1 over the required frequency range. Several trials may be needed to obtain acceptable test results. In such cases, accepting the test data with a poor coherence may lead to erroneous dispersion curves and hence the actual shear wave velocity profile will be misinterpreted during analysis. In this research study, an attempt was made to study the effect of using a constant energy impact source instead of using the conventional test procedure using handheld hammers (which does not guarantee constant impact energy) on the coherence. The following sections provide details of the test procedure and analysis performed in this study.

2 Experimental Procedure

The main objective of the testing procedure adopted in this study is to investigate the effect of using an impact source with constant energy in lieu of a random energy source like a handheld hammer. In order to study this, tests were conducted on a control test section created in the laboratory using an iron box of 1.5 m length, 0.61 m wide, and 0.5 m deep filled with soil (Fig. 1).

Sandy clay soil was compacted in layers with approximately 16% moisture content. Due to the dimension constraint of the test box, the source to geophone spacing and distance between vertical geophones was restricted to 0.61 m (2 ft).

Two sets of experiments were performed such as (a) tests with varying height of fall (varying impact energy); (b) tests with constant height of fall (constant impact energy). Since the energy imparted by the impact is proportional to the height of fall, Case (a) represents the conventional case of using handheld hammers which does not have any control on the imparted energy. A random height of fall of a 2.5 kg drop hammer was used for this purpose, whereas Case (b) ensures constant impact energy every time the test is performed with a constant height of drop of 0.13 m. In order to verify the findings of the tests, a series of experiments were performed with both varying and constant energy of impact in a nearby field with receiver spacing of 0.61 m (2 ft), 1.22 m (4 ft), and 2.44 m (8 ft). For all the tests, 4 trials were conducted per test. The data obtained from the control section and field tests were analyzed to generate the experimentally obtained dispersion curve.

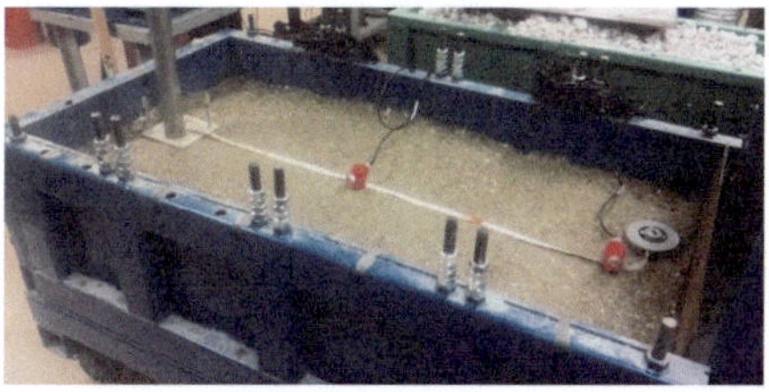

Fig. 1 Experimental setup for control section

3 Test Results and Analysis

This section presents the test results obtained from the two sets of experiments and the analysis performed for studying the effects of varying and constant energies. Figure 2 presents the waveform generated for the sets of experiments performed with varying impact energy. Figure 2a represents the coherence function, whereas Fig. 2b represents the wrapped phase data.

Figure 3 presents the waveforms generated for Case (b) studies, i.e., with constant impact energy. Figure 3a represents the coherence function, whereas Fig. 3b represents the wrapped phase. A high value of coherence data (close to 1) is required in order to perform further analysis. The data presented in Figs. 2b and 3b represents the wrapped phase data, which was used to generate the dispersion curve after masking the unacceptable sections.

A significant improvement in the coherence data can be observed from Figs. 2a to 3a when constant height of fall of hammer was used in all four trials in test. After masking the unwanted zones, effectively 12 cycles (as shown in Fig. 3b) can be used to produce the dispersion curve, whereas when random height of fall of hammer was used, an acceptable coherence value was obtained only for two cycles for a test as shown in Fig. 2a, b. The transfer function and wrapped phase are generated from the data obtained from average of four trials since averaging strengthens the signal after reducing the effect of background noise (Heisey et al. 1981). Therefore, non-uniformity in trials affects the averaged data obtained, especially for higher frequency since high-frequency waves attenuates faster. This leads to the reduction of coherence after a few cycles when the impact energy is not kept constant. Using the same impact energy for all the four trials leads to generation of almost identical set of waves. Hence, the final coherence data is found to be very close to one after averaging four identical trials. The wrapped phase data from both the tests (after masking) was used for generating the dispersion curves.

Figure 4 presents the dispersion curve profiles for the wrapped phase data obtained from constant and varying impact energies. Blue line in Fig. 4 represents

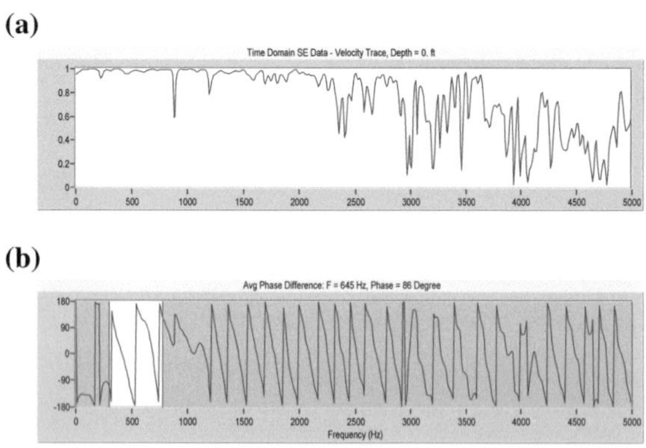

Fig. 2 **a** Coherence function and **b** wrapped phase for Case (a) tests on control section

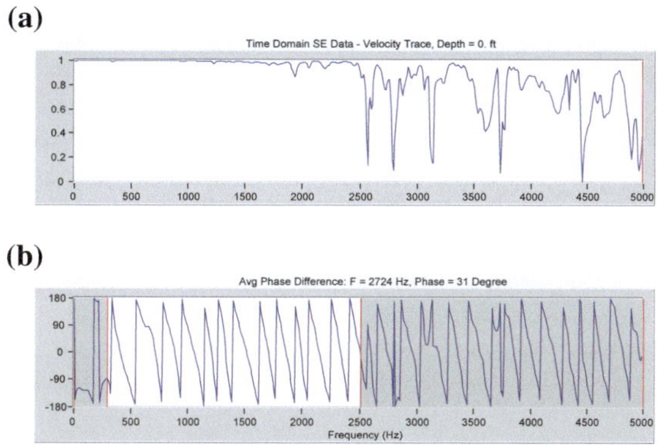

Fig. 3 **a** Coherence function and **b** wrapped phase for Case (b) tests on control section

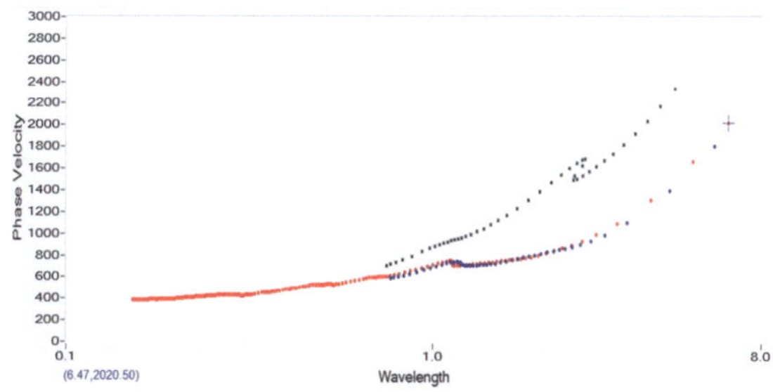

Fig. 4 Dispersion curve for tests conducted with different impact energy conditions on control section

the dispersion curve obtained from varying impact energy, whereas the red line represents the dispersion curve obtained for constant energy. It can be observed that the dispersion curve for the test with varying impact energy was found to match partially with that obtained for the constant energy. However for another test conducted with random height of fall of hammer, the coherence value for none of the cycles was found to reach the acceptable threshold value of 0.9–0.95. Nevertheless, when two cycles with maximum coherence was used to generate the dispersion curve, a significant amount of error was observed. This can be seen from the dispersion curve (shown in black) being significantly different from that shown in red and blue.

Hence, it can be inferred from the laboratory studies that the tests performed with the constant energy produced the waveform with coherence value close to 1 and more data can be used for the analysis for a larger frequency bandwidth. In order to validate the tests results from the laboratory, field tests were performed on a hydraulic dam located in Texas, USA. Tests were performed with receiver spacing of 0.61 m (2 ft), 1.22 m (4 ft), and 2.44 m (8 ft). The waveforms obtained from constant and varying impact energies showed similar observations as in laboratory testing. Figure 5 presents the test results obtained from the field investigations for 0.61 m (2 ft) receiver spacing.

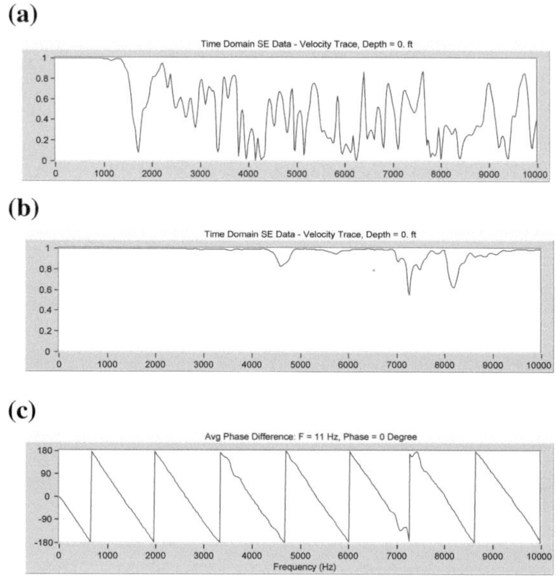

Fig. 5 **a** Coherence function for varying impact energy, **b** coherence function for constant impact energy, **c** wrapped phase for constant impact energy

Figure 5a represents the coherence function obtained with varying impact energy, whereas Fig. 5b, c represents the test results from constant impact energies. It was observed that, while conducting tests without maintaining constant impact energy, more trials were needed to obtain a good coherence. Even if coherence close to one was observed in a few cases, the high coherence value was observed to attenuate drastically after one cycle (Fig. 5a).

For higher receiver spacing of 1.22 m (4 ft) and 2.44 m (8 ft), a good coherence data was obtained when constant impact energy was used. But with increase in spacing, the coherence function was found to decrease beyond a few cycles (i.e., at high frequencies). This can be attributed to the fast attenuation of high-frequency shallow waves (Heisey et al. 1981). It was also observed that there is no significant impact of neglecting the low coherence data (high-frequency range) for higher geophone spacing (for, e.g., 8 ft) since information about shallow layers are obtained from the data obtained from lower receiver spacing (for, e.g., 2 ft or 4 ft).

4 Conclusion

The main objective of this research is to study the effect of using constant impact energy on the coherence function obtained in SASW testing. Based on the tests performed, a few significant conclusions are deduced:

- High coherence data (value close to one) was obtained in all the cases by using constant impact energy, whereas with varying impact energy, several trails were needed to perform for obtaining the data for a minimum threshold coherence value of 0.9. For performing SASW testing with constant impact energy, a drop hammer with a constant height of fall can be used.

- Laboratory and field test results depicted similar observations with respect to constant and varying impact energies. With constant impact energy, it was observed that the data with more frequency bandwidth can be used for dispersion curve analysis, whereas the frequency bandwidth was limited to only 2 cycles with varying impact energy. Also, the time and number of repetitions required during the test can be significantly reduced by using a constant impact energy source.
- High repeatability in the test results was observed by using the constant impact energy as the uncertainties associated with the height of fall was less, whereas with varying impact energies, less reliable waveforms were observed as the uncertainties associated with height of fall is different for each testing.

References

Bheemasetti, T., Puppala, A., Zou, H., Marshall, D., & Pedarla, A. (2015). Quality assurance method for assessment of stiffness of pipeline bedding material. *Journal of Materials in Civil Engineering*. https://doi.org/10.1061/(ASCE)MT.1943-5533.0001431,04015160.

Heisey, J. S., Stokoe, K. H., Hudson, W. R., & Meyer, A. H. (1981). *Determination of in situ shear wave velocities from spectral analysis of surface waves* (Master's thesis). University of Texas at Austin.

Joh, S. H. (1996). *Advances in the data interpretation technique for spectral-analysis-of-surface-waves (SASW) measurements* (Doctoral dissertation). University of Texas at Austin.

Nazarian, S. (1984). *In situ determination of elastic moduli of soil deposits and pavement systems by spectral-analysis-of-surface-waves method* (Doctoral dissertation). University of Texas at Austin.

Nazarian, S., & Stokoe, K. H. (1985). *In situ determination of elastic moduli of pavement systems by spectral-analysis-of-surface-waves method: Practical aspects.*

Nazarian, S., Stokoe, I. I., Kenneth, H., Briggs, R. C., & Rogers, R. (1988). Determination of pavement layer thicknesses and moduli by SASW method. *Transportation Research Record* (1196).

Puppala, A. J., Kadam, R., Madhyannapu, R., & Hoyos, L. R. (2006). Small-strain shear moduli of chemically stabilized sulfate-bearing cohesive soils. *Journal of Geotechnical and Geoenvironmental Engineering*, ASCE, 322–336.

Rix, G. (1988). *Experimental study of factors affecting the spectral-analysis-of-surface-waves method* (Doctoral dissertation). University of Texas at Austin.

Comparison of Properties of Cohesive Soils Along East Coast of India

Annam Madan Kumar and N. Kumar Pitchumani

Abstract The coastline of India is speckled with a large number of ports (major and minor). In connection with the construction of new ports and upgradation of existing ports, a lot of construction has taken place over a decade or two. This involved jetties, container yard, railway connectivity to existing railway network, deepening of berth pockets and channels, etc. On the geotechnical front, these involved reclamation, embankment construction, pile foundations, deep excavations, etc. The east coast of India is characterised by silty clay over a large distance from West Bengal to Tamil Nadu cutting across Odisha and Andhra Pradesh. A few kilometres inland from the coast, the region also comprises silty clay, probably residual deposits arising out of weathering of volcanic rocks. In this paper, laboratory test data from many of the sites are analysed to look for any behavioural pattern. The compression properties of the cohesive soils are looked into and compared with correlations available in the literature. It is seen that the compression characteristics do follow a definite trend although there is scatter. The aim of publishing this field data is to help designers in selecting parameters in the preliminary design stage before any detailed investigations are carried out. This could also help in estimating quantities for tender preparations with a fair amount of certainty on the primary consolidation settlements resulting from reclamation, embankment construction, and also the extent of ground improvement required, if any.

Keywords East coast · Compression index · Compression ratio
Design parameters · Ground improvement

A. M. Kumar
Keller India, 7th Floor, Centennial Square, Kodambakkam, Chennai 600024, India
e-mail: madankumar@kellerindia.com; madanannam@gmail.com

N. Kumar Pitchumani (✉)
AECOM India Pvt. Ltd., VBC Solitaire, T Nagar, Chennai 600017, India
e-mail: Nkumar.pitchumani@aecom.com; kpitchumani@gmail.com

V. K. Stalin and M. Muttharam (eds.), *Geotechnical Characterisation and Geoenvironmental Engineering*, Lecture Notes in Civil Engineering 16,
https://doi.org/10.1007/978-981-13-0899-4_9

1 Introduction

The East or Coromandel coast of India has seen a lot of activity in the past 2 decades in the form of major and minor ports, starting from Haldia to Tuticorin. Marine clay is encountered at most of the coastal locations. However at the mouths of rivers like Ganga, Mahanadi, Dhamra, Godavari, more of silty deposits of alluvial type are encountered.

During preliminary designs, lack of data inhibits designers to come up with relatively aggressive designs, especially in cohesive soils. The design assumptions of compressibility start with the use of Terzaghi's equation of compression index based on liquid limit which may not be applicable for Indian clays. This results in overestimation of primary consolidation settlements. This overestimation directly affects fill/reclamation quantities. Further, the contractor/bidder adds additional costs for risks resulting from variations in parameters. This risk is a result of insufficient data. On the whole, the cost of a project gets escalated unreasonably.

Various consultants and contractors have been working on projects along this coast, and a lot of data on the strata is available. To mitigate such cost escalations and to help clients/owners/contractors to arrive at a reasonably accurate estimation of primary consolidation settlements, data from various sites along the east coast are compiled in this paper and the compression characteristics are presented.

2 Description of Sites

Compressive soils from various places comprising of non-marine and marine sites as described in Table 1 were studied. The laboratory test data are taken from various consultancy projects carried out by the authors in the last 10–15 years. It is to be noted that the geotechnical investigations were conducted by various agencies. Some data that were found to be erroneous for various reasons have been discarded. Data are also collected from various published and unpublished literature. Data for Visakhapatnam are taken from Satyanarayana and Satyanarayana Reddy (2009) and Guru Vittal et al. (2006).

The soils at Dhamra are predominantly residual deposits of basalt from the Deccan. The data presented are along the railway alignment from Bhadrak to Dhamra. The silty clay deposits at Koyambedu are from the depot for Chennai Metro and are believed to be lacustrine deposits originated from shale. The deposits at Haldia and Paradip are alluvial in nature and are fluvial deposits from rivers.

Table 1 Sites considered for study

Non-marine sites	Dhamra (Odisha), Haldia (West Bengal) Koyambedu (Chennai, TN), Paradip (Odisha)
Marine sites	Visakhapatnam, Krishnapatnam, Bheemunipatnam (all in Andhra Pradesh)

The location of non-marine and marine sites is shown in Fig. 1. The classification of the soil for the non-marine and marine sites on the A line is shown in Figs. 2 and 3, respectively.

The plots depict that the non-marine deposits are silty clay of low, medium and high compressibility while the silty clay of marine deposits is of medium to high compressibility.

Fig. 1 Marine and non-marine site locations

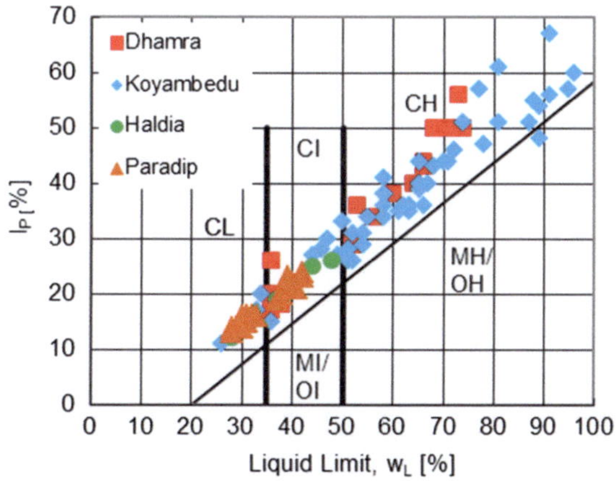

Fig. 2 Soil classification: non-marine sites

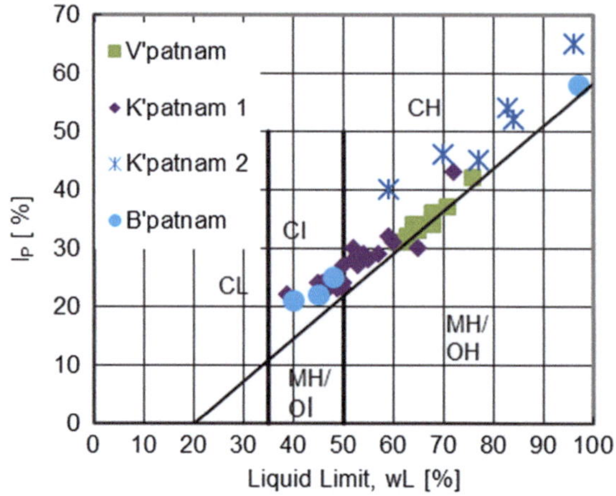

Fig. 3 Soil classification: marine sites

3 Soil Behaviour

Laboratory test results for the two groups of soils are plotted with various parameters as given in the literature.

The most commonly studied and compared parameter is the compression index (C_c) with index properties. A number of empirical equations are available in the literature (Satyanarayana and Satyanarayana Reddy 2009; Sridharan and Nagaraj 2000).

Lambe and Whitman (1969) and Cox (1970) have equations for the compression ratio CR = $C_c/(1 + e_0)$. From all these literature, it is clear that cohesive soils from different places show different variations. One equation from a particular site may not be suitable for another site.

Hence, the compression parameters are only plotted without fitting any trend line in this paper.

3.1 Compression Ratio

The compression ratio, CR, is plotted against liquid limit for the deposits of non-marine and marine sites in Figs. 4 and 5, respectively. Although there is scatter, there appears to be a trend. The marine deposits of different sites occupy a distinct spot on the plot.

The plot of CR with natural moisture content (NMC) is presented in Figs. 6 and 7 for the deposits of non-marine and marine sites. Here again there is scatter, nevertheless a trend is observed.

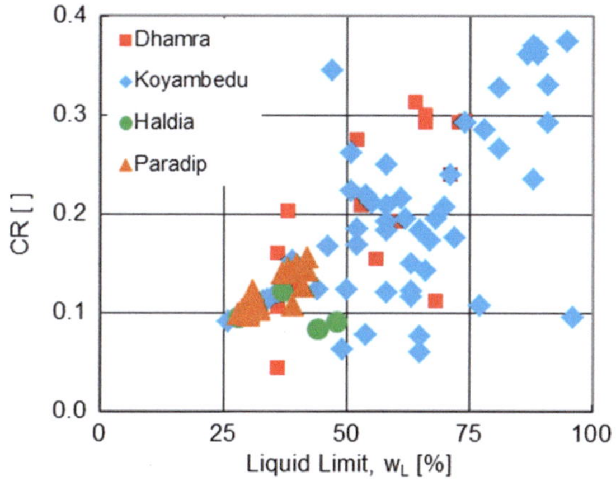

Fig. 4 Plot of compression ratio with liquid limit: non-marine sites

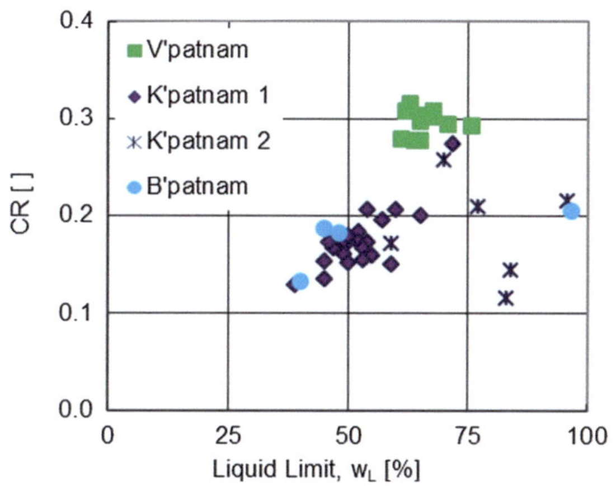

Fig. 5 Plot of compression ratio with liquid limit: marine sites

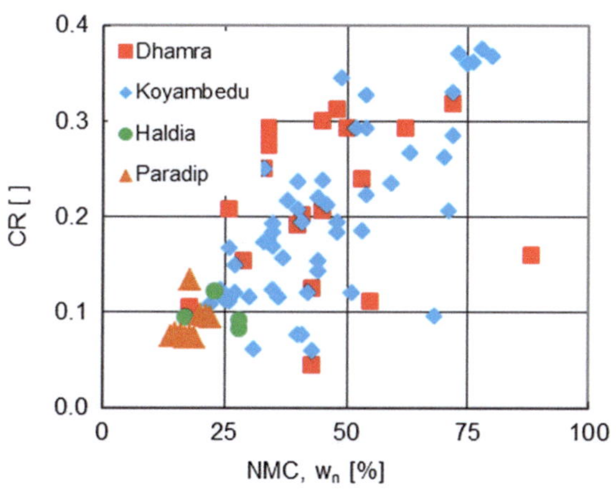

Fig. 6 Plot of compression ratio with natural moisture content: non-marine sites

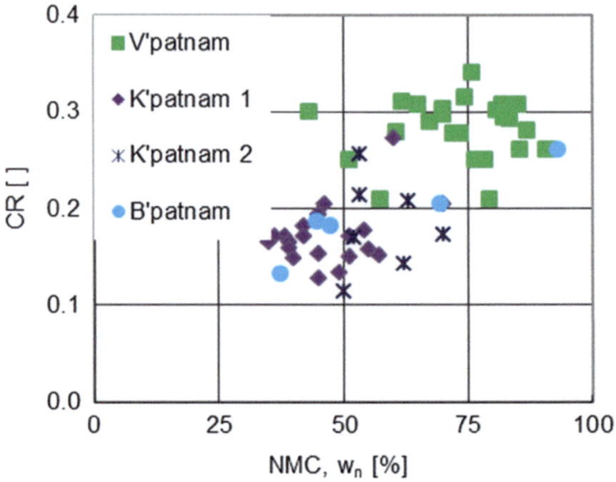

Fig. 7 Plot of compression ratio with natural moisture content: marine sites

Another important parameter is the consistency index I_c = (LL − NMC)/PI%. The plot of CR versus I_c is presented in Figs. 8 and 9, respectively, for the non-marine and marine sites.

It is seen that the non-marine and marine clays follow a similar pattern, and hence, plot of CR versus liquid limit and natural moisture content for all sites combined is presented in Figs. 10 and 11.

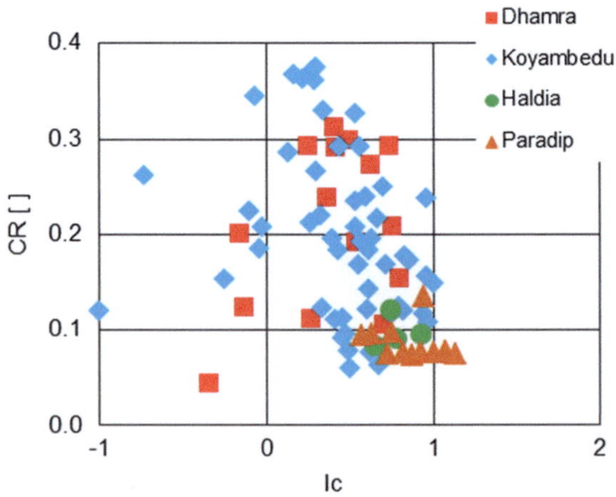

Fig. 8 Plot of compression ratio with consistency index: non-marine sites

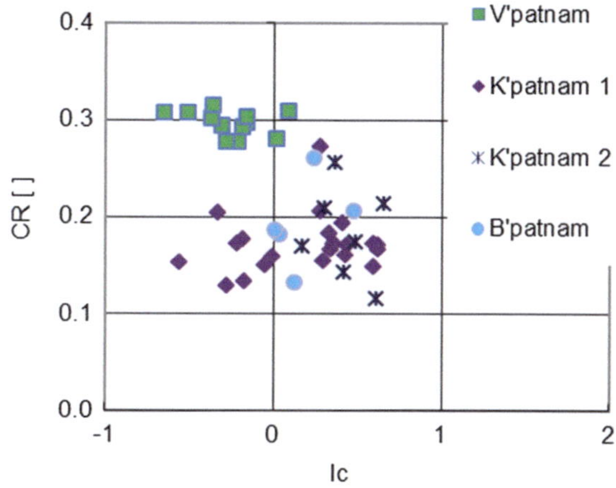

Fig. 9 Plot of compression ratio with consistency index: marine sites

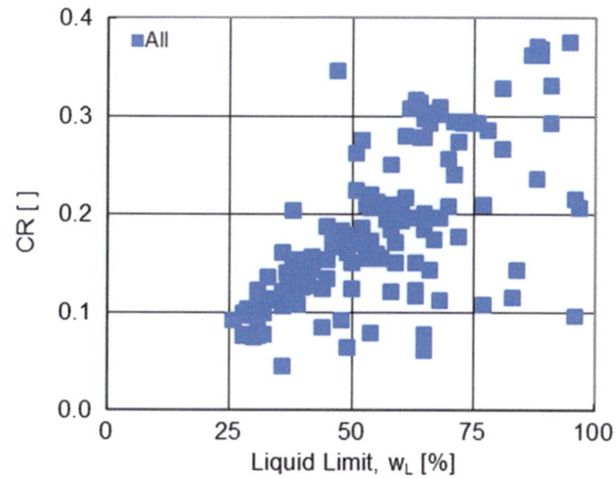

Fig. 10 Plot of compression ratio with liquid limit: all sites

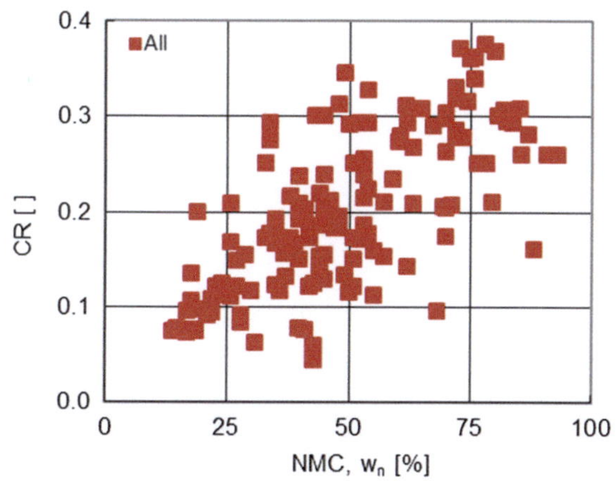

Fig. 11 Plot of compression ratio with natural moisture content: all sites

4 Conclusions

The aim of this paper is to present compression data from various sites along the east coast. The data presented are the compression ratio with various index properties generally published in the literature. The idea of presenting these data is to benefit the engineers to adopt more realistic parameters for designs either in the preliminary design stage or tender stage. This would help in arriving at a more pragmatic techno-commercial proposal for any project on soft soils along the east coast.

References

Cox, J. B. (1970). Shear strength characteristics of the recent marine clays in South East Asia. *Journal of the Southeast Asian Society of Soil Engineering, 1,* 1–28.

Guru Vittal, U. K., Subramanya Prasad, P., & Sudhir, M. (2006). Geotechnical properties of Visakhapatnam Marine Clay. In *Indian Geotechnical Conference*, 14–16, Chennai, India (pp. 329–330).

Lambe, T. W., & Whitman, R. V. (1969). *Soil mechanics*. New York: Wiley.

Satyanarayana, B, & Satyanarayana Reddy, C. N. V. (2009). Emperical correlations of compressibility of marine clays—a review. In *Indian Geotechnical Conference,* Guntur, India (pp. 65–70).

Sridharan, A., & Nagaraj, H. B. (2000). Compressibility behavior of remoulded, fine grained soils and correlations with index properties. *Canadian Geotechnical Journal, 37,* 712–722.

Effect of Pore Size Distribution on Unconfined Compressive Shear Strength

N. Saranya and D. N. Arnepalli

Abstract The unconfined compressive strength is one of the influencing parameters that are used for determining the in situ strength of soft, fine-grained soil deposits. Since many previous research works have highlighted the influence of pore fluid type, electrolyte concentration, pH and valence of the pore fluid on unconfined compressive shear strength. The present study has inferred the effect of pore size distribution (PSD) on unconfined compressive shear strength (UCS) of bentonite and kaolinite minerals. It is observed that the pore size distribution of bentonite is a bimodal distribution representing both interpore and intrapore, whereas the kaolinite mineral exhibits a unimodal distribution representing only the interpore. The interpore represents the water molecules bounded between soil aggregates, whereas the intrapore represents the water molecules bounded within the soil aggregated and on the clay surface. From the obtained UCS value of bentonite and kaolinite minerals, it can be inferred that the UCS strength variation of bentonite mineral is strongly influenced by the water molecules bounded on the clay surface or diffused double layer water, whereas the UCS strength of kaolinite minerals is controlled by the net attractive force between the clay particles. The present study demonstrated that the pore size distribution is also one of the parameters that strongly influences the unconfined compressive strength of the soil.

1 Introduction

The unconfined compressive strength is a simple and time-independent test which is generally carried out to characterize the in situ strength of soft, fine-grained soil deposits. In view of this, many research works were carried out to infer the

---

N. Saranya (✉) · D. N. Arnepalli
Department of Civil Engineering, Indian Institute of Technology Madras, Chennai 600036, India
e-mail: saranya.nithiyanandan@gmail.com

D. N. Arnepalli
e-mail: arnepalli@iitm.ac.in

V. K. Stalin and M. Muttharam (eds.), *Geotechnical Characterisation and Geoenvironmental Engineering*, Lecture Notes in Civil Engineering 16,
https://doi.org/10.1007/978-981-13-0899-4_10

influencing parameters that control the unconfined compressive strength variation. Du et al. (1999) analyzed the effect of dry unit weight, soil mineralogy, and soil fabric arrangement on the swelling-shrinkage and unconfined compressive strength characteristics. In addition, Sridharan and Prakash (1999) highlighted the reason behind the distinct undrained shear strength behavior of kaolinite and montmorillonite minerals. From the obtained results, it was concluded that the undrained shear strength of kaolinite minerals was strongly influenced by the net attractive force between the particles, whereas in case of montmorillonite minerals, the undrained shear strength was controlled by the viscous shear resistance of the diffused double layer (DDL) water.

Umesh et al. (2011) highlighted the effect of acid contamination on geotechnical properties of black cotton soil, and it has been concluded that the observed change in behavior of black cotton soil was mainly attributed by the reduction in double layer thickness, as well as the reduction in its moisture-holding capacity and loss of cohesion with increase in acid concentration. Heeralal et al. (2012) and Lin and Cerato (2012) inferred the effect of microstructural change on macrostructure behavior of clay minerals in a detailed way. Further, Sasanian and Newson (2013) characterized the effect of moisture content on soil fabric using mercury intrusion porosimetry (MIP) analysis, and from the obtained pore size distribution curve, the peaks associated with larger pore diameters were termed as inter-cluster pores, representing the water molecules bounded between soil aggregates. On the other hand, the peaks associated with smaller peaks were termed as intra-cluster pore which quantifies the water molecules bounded within the soil aggregate and on the clay surface.

From the literature review, it can be observed that all the macro-level behavior of clay minerals was strongly influenced by the corresponding change that occurs at its micro-level. Further, not many efforts were made to correlate the micro-level behavior with its macro-level changes; keeping this in view, an attempt was made to establish the relationship between the change in pore size distribution of clay minerals compacted at different moisture content with the corresponding change in its unconfined compressive strength value.

2 Experimental Investigation

The experimental investigation includes the selection of different clay minerals, characterising them for physical, chemical, mineralogical, and compaction characteristics. Further, the unconfined compressive strength of soils compacted at different water content was established; in addition, the pore size distribution of clay minerals was obtained using mercury intrusion porosimetry.

2.1 Material Selection and Characterisation

The present study has selected sodium bentonite (denoted as Na-BT) and kaolinite (denoted as KT) as representative samples for testing. The collected geomaterials were processed and characterised for their physical, mineralogical, and geotechnical properties by following the guidelines presented in the Indian standards. The obtained results are presented in Table 1.

2.2 Mineralogical Characteristics

The mineralogical compositions of selected geomaterials were obtained using X-ray diffraction (XRD) technique. From the established XRD pattern, the minerals present in the geomaterials were quantified using International Centre for Diffraction Database, and the same are detailed in Table 1.

2.3 Compaction Characteristics

The relationship between the moisture content and dry unit weight of the selected geomaterials was obtained using miniature proctor compaction apparatus. The

Table 1 Basic characteristic of geomaterials

Properties	Na-BT	KT
Physical characteristics		
Specific gravity (G)	2.72	2.67
Specific surface area (m^2/g)	338	47
Plasticity properties		
Liquid limit (%)	287	40
Plastic limit (%)	62	25
Particle size characteristics (percent by weight)		
Gravel size	0	0
Sand size	2	39
Silt size	40	28
Clay size	58	33
Classification of soil (as per IS: 1498-1970)	CH	CI
Mineralogical characteristics		
Minerals present	Kaolinite, montmorillonite, quartz	Kaolinite, quartz
Compaction characteristics		
γ_{dmax} (kN/m^3)	13.86	16.35
OMC (%)	29.1	20.35

selected geomaterials were mixed with different proportion of distilled water and it was compacted in the miniature compaction mold.

From the compacted soil mass and moisture content, the relationship between dry unit weight, $\gamma_{d,}$ and water content, w, has been established, and the same is shown in Fig. 1. The compaction characteristics such as the maximum dry unit weight and optimum moisture content value for the selected geomaterials were calculated, and the same are shown in Table 1.

From the obtained compaction characteristic curve, the sample compacted at different moisture content was selected for further testing. The details of molding moisture content and dry unit weight selected for unconfined compressive strength test and pore size distribution are detailed in Table 2.

2.4 Unconfined Compressive Strength

To obtain the unconfined compressive strength value, the geomaterials were initially mixed with a desired quantity of distilled water and stored in an airtight bag for 48 h to ensure uniform maturation of the soil. The prepared soil mass was

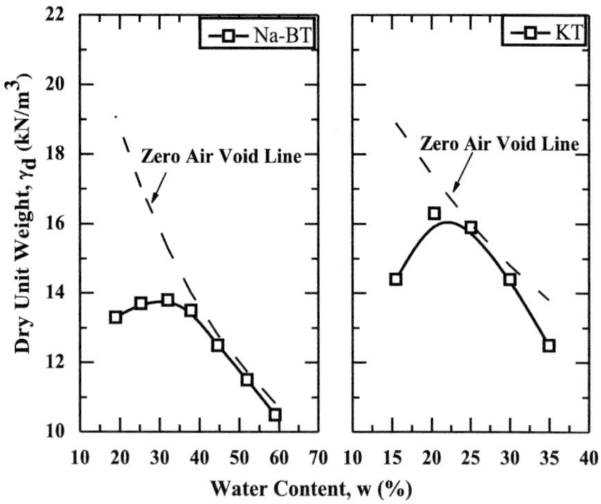

Fig. 1 Compaction curve of geomaterials

Table 2 Sample details for UCS and MIP analysis

Compaction state	Na-BT		KT	
	γ_d (kN/m^3)	w (%)	γ_d (kN/m^3)	w (%)
Dry of optimum	13.5	21.35	14.5	10.93
OMC	13.86	29.1	16.35	20.35
Wet of optimum	13.5	36.86	14.5	29.75

compacted to the desired density and it was tested in the triaxial test setup for its UCS value. The obtained stress–strain behavior of selected geomaterials is illustrated in Fig. 2. From the obtained stress–strain relationship, the unconfined compressive strength values were measured, and the same are depicted in Table 3.

2.5 Pore Size Distribution Characteristics

The pore size distribution (PSD) is one of the main influencing factors that control the engineering properties of geomaterials to a great extent. In view of this, the pore size distribution characteristic of the compacted geomaterials has been established using MIP technique. The geomaterials compacted for the desired density and water content were freeze dried and tested in MIP for their pore size distribution characteristics, and the same are presented in the form of Figs. 3 and 4. Further, the log-differential curve of Na-BT shown in Fig. 3 exhibits a bimodal pattern, which explains the type of the pores present in the soil; i.e., the pore size distribution of Na-BT is mainly contributed by two types of pores. The peak associated with large diameter pores (interpores) represents the water molecules bounded between soil aggregates, whereas the peak associated with smaller diameter pores (intrapores) is

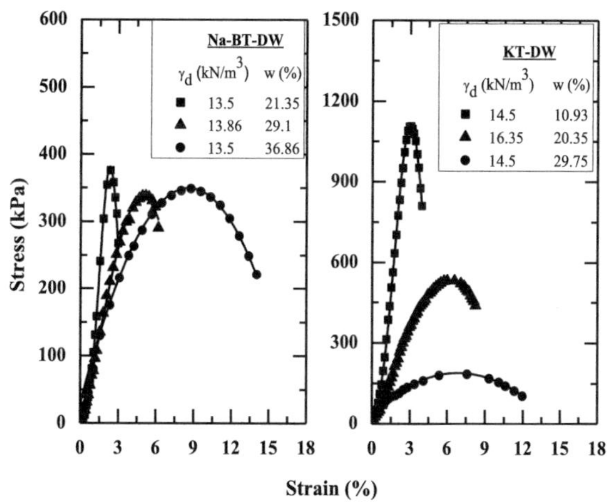

Fig. 2 Stress–strain variations of geomaterials

Table 3 UCS values of geomaterials

Geomaterial	γ_d (kN/m^3)	w (%)	UCS (kPa)
Na-BT	13.5	21.35	376
	13.86	29.1	338
	13.5	36.86	349
KT	14.5	10.9	1107
	16.35	20.35	531
	14.5	29.75	184

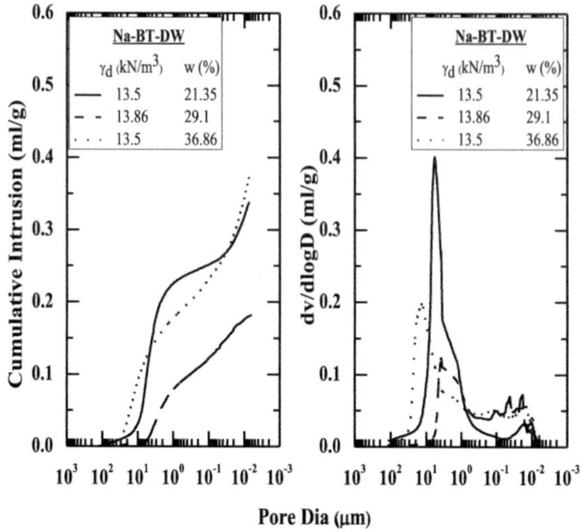

Fig. 3 PSD of Na-BT

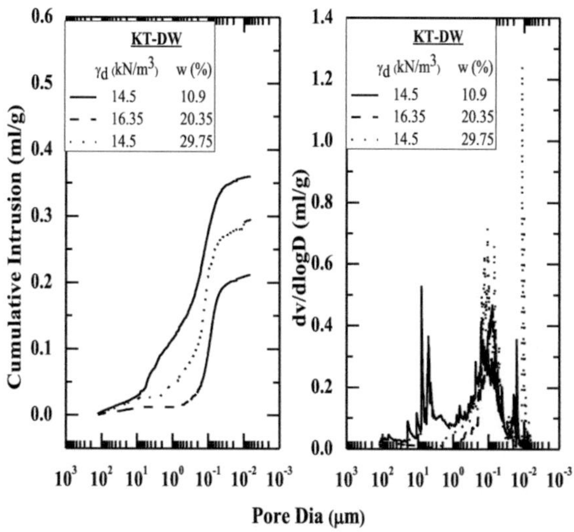

Fig. 4 PSD of KT

related to the water molecules bounded within the soil aggregated and on the clay surface (Romero et al. 1999). From the obtained pore size distribution and log-differential curve of KT mineral shown in Fig. 4, it was inferred that the KT mineral exhibits a unimodal pattern; i.e., the pore size distribution of KT mineral is mainly influenced by the interpore volume rather than the intrapore volume. As kaolinite mineral is 1:1 mineral having hydrogen bonding between the successive clay layers, the interlayer swelling of kaolinite mineral is significantly less (Mitchell 1976) due to which the formation of intrapores in kaolinite mineral is significantly very low when compared to that of sodium bentonite. The measured inter- and intrapore volumes of selected geomaterials are shown in Table 4.

Table 4 Pore volume distribution of geomaterials

Geomaterial	γ_d (kN/m^3)	*w* (%)	Interpore volume (ml/g)	Intrapore volume (ml/g)
Na-BT	13.5	21.35	0.25	0.088
	13.86	29.1	0.122	0.078
	13.5	36.86	0.231	0.142
KT	14.5	10.9	0.360	–
	16.35	20.35	0.211	–
	14.5	29.75	0.295	–

3 Conclusions

Based on the experimental results, the following observation has been made.

From the pore size distribution curve of sodium bentonite it was observed that the increase in water content from dry of optimum to optimum moisture content (OMC) the inter pores have shifted to the left corresponding to larger diameter pores, the observed shift in the pore diameter is due to the volume and diameter expansion of free water held in between the soil aggregates (inter pores). Further, the samples compacted at dry of optimum and wet of optimum exhibit similar particle size distribution pattern except that the samples compacted at wet of optimum have shown reduction in interpores when compared with the samples compacted at dry of optimum.

The log-differential curve of sodium bentonite exhibits a bimodal pattern; i.e., the pore size distribution of sodium bentonite is influenced by both inter- and intrapore volume. In case of kaolinite mineral, the pore size distribution is mainly contributed by the interpore volume, which can be visualized from its unimodal pattern.

From the measured inter- and intrapore volume of sodium bentonite, it was inferred that the sample compacted at wet side of optimum has higher intrapore volume when compared with sample compacted at other two compaction states. The obtained pore volume distribution dictates the increase in diffused double layer water formed around the clay surface when it is compacted at higher water content.

In case of kaolinite mineral, the formation of intrapore is negligible when compared with sodium bentonite, and this is due to the fact that the hydrogen bonding between the successive clay layers of kaolinite mineral resists the inter-layer swelling between the clay layers.

The UCS value of sodium bentonite compacted at wet side of optimum is higher when compared with soil compacted at optimum moisture content, whereas the UCS value of kaolinite mineral decreases with increase in water content. The observed change in UCS characteristics of sodium bentonite and kaolinite mineral is due to the fact that UCS strength variation of montmorillonite mineral is mainly contributed by the shear resistance offered by the diffused double layer water, whereas the UCS strength of kaolinite mineral is controlled by the net attractive force between the clay minerals (Sridharan and Prakash 1999).

References

Du, Y., Li, S., & Hayashi, S. (1999). Swelling-shrinkage properties and soil improvement of compacted expansive soil, Ning-Liang Highway, China. *Engineering Geology, 53,* 351–358.

Heeralal, M., Murty, V. R., Praveen, G. V., & Shankar, S. (2012). Influence of calcium chloride and sodium silicate on index and engineering properties of bentonite. In *International Conference on Chemical, Environmental science and Engineering (ICEEB'2012)* (pp. 52–57).

Lin, B., & Cerato, A. B. (2012). Prediction of expansive soil selling based on four micro-scale properties. *Bulletin of Engineering Geology and Environment, 71,* 71–78.

Mitchell, J. K. (1976). *Fundamentals of soil behavior* (2nd ed.). New York: Wiley.

Romero, E., Gens, A., & Lloret, A. (1999). Water permeability, water retention and micro-structure of unsaturated Boom clay. *Engineering Geology, 54,* 117–127.

Sasanian, S., & Newson, T. A. (2013). Use of mercury intrusion porosimetry for microstructural investigation of reconstituted clays at high water contents. *Engineering Geology, 158,* 15–22.

Sridharan, A., & Prakash, K. (1999). Mechanisms controlling the undrained shear strength behavior of clays. *Canadian Geotechnical Journal, 36,* 1030–1038.

Umesh, T. S., Sharma, H. D., Dinesh, S. V., Sivapullaiah, P. V., & Basim, S. C. (2011). Physico-chemical changes in soil due to sulphuric acid contamination. In *Proceedings of Indian Geotechnical Conference* (Paper no. L-320, pp. 765–768).

Synthesis of Fly Ash-GGBS-Blended Geopolymer Composits

Subhashree Samantasinghar and Suresh Prasad Singh

Abstract Fly ash and blast-furnace slag are well-known industrial by-products and are used to manufacture blended cements and concrete. Fly ash has been used as pozzolanic material to improve the physical, mechanical as well as chemical properties of the cements and concrete, whereas blast-furnace slag cements are characterized by their low heat of hydration and high sulphate and sea water resistance. On the other hand, addition of blast-furnace slag to fly ash may have substantial influence on the strength development of geopolymer binder when cured under ambient temperature condition. This paper presents the compressive strength of geopolymer binders synthesized from Class-F fly ash (FA) blended with ground granulated blast-furnace slag (GGBS) by optimizing the influential parameters. The chemical activation of fly ash–slag mixtures with different concentrations of sodium hydroxide (NaOH) for varying solution to solid ratio has been made. The test results showed that the development of compressive strength is directly concomitant to concentration of NaOH solution. Moreover, as the slag content in the mixture increases, the compressive strength increases. Hence, an inorganic polymer can be synthesized from fly ash–slag mixture by activating with appropriate amount of NaOH which can be used as cementitious material.

Keywords Fly ash · GGBS · Alkali concentration · Solution to solid ratio
Compressive strength

S. Samantasinghar (✉) · S. P. Singh
Department of Civil Engineering, National Institute of Technology Rourkela,
Rourkela 769008, Odisha, India
e-mail: shree.singhar@gmail.com

S. P. Singh
e-mail: spsingh@nitrkl.ac.in

V. K. Stalin and M. Muttharam (eds.), *Geotechnical Characterisation and Geoenvironmental Engineering*, Lecture Notes in Civil Engineering 16,
https://doi.org/10.1007/978-981-13-0899-4_11

1 Introduction

Geopolymer binders based on industrial by-product materials such as fly ash and slag can play a vital role in the context of sustainability and environmental issues. Approximately, 5% of global CO_2 emissions originate from the manufacturing of Portland cement. On the other hand, cements manufactured from industrial by-products such as slag has been shown to release up to 80% less greenhouse gas emissions, and there are 80–90% less greenhouse gas emissions in case of fly ash. Therefore, ground-granulated blast-furnace slag or fly ash would significantly reduce the CO_2 emission of geopolymer binder production.

Geopolymer is an alternative binder in which an alkali-activated alumino-silicate material is used as the binder instead of the traditional cement binder. Thus, the traditional binder based on cement or cement and other pozzolanic materials is replaced by the alkali-activated inorganic binder in geopolymer.

The major processes involve the reaction between an alumino-silicate source such as fly ash, metakaolin or blast-furnace slag and an alkaline solution which leads to final hardening of the matrix by exclusion of excess water and the growth of an inorganic polymer. Previous studies indicated that the reaction of the selected pozzolanic material is the most significant factor in producing a mechanically sound binder via the geopolymerization process. The chemical reaction and the rate of strength development of geopolymer are influenced by several factors based on chemical compositions of the source materials, alkaline activators and curing condition.

2 Literature Review

Geopolymerization mechanism includes the dissolution of aluminium and silicon in highly alkali medium, transportation of the dissolved species, and formation a three-dimensional network of alumino-silicate structures (De Silva et al. 2007). Geopolymer binders usually consist of reactive solids which contain silica and alumina and an alkaline solution. When reactive solids react with alkaline solution, it builds an alumino-silicate network which is amorphous to partially crystalline nature (Duxson et al. 2007a, b).

Puertas et al. (2000) synthesized fly ash–slag geopolymer. The ratio of fly ash to slag and the activator concentration always result to be significant factor. The influence of curing temperature in the development of the strength of the pastes is lower than the contribution due to other factors. At 28 days of reaction, the mixture of 50% fly ash and 50% slag activated with 10 M NaOH and cured at 25 °C developed a compressive strength of about 50 MPa. Deb et al. (2014) blended ground-granulated blast-furnace slag with class-F fly ash and experimented on compressive strength and stated that strength development of the geopolymer concrete mixtures slowed down after the age of 28 days and continued to increase at slower rates until 180 days of age and the compressive strength increases with the increase of slag content in the mixtures. Rajini and Rao (2014) studied the effect of fly ash and slag on mechanical properties of geopolymer concrete, and the results reveal that the compressive strength increases with slag content and curing period. Nath and Sarker (2014) experimented on the effect of slag on compressive strength of fly ash geopolymer and concluded that variation of the amount of alkaline activator affects the compressive strength of the mixtures. An increase in the activator solution content from 35% to up to 45% gradually reduced the strength of geopolymer concrete and mortar. Marjanovic et al. (2015) tested the compressive strength of blast furnace slag–fly ash-blended mortars. The strength is highly depended on the blend composition, activator concentration and water/binder ratio. Wardhono et al. (2015) investigated the strength of alkali-activated slag–fly ash (AASF) mortar blends. AASF specimen test results suggest that the hydration reaction of slag and the polymerization reaction of fly ash could occur separately or simultaneously. The simultaneous reaction is the most likely with the GGBS reaction activating the fly ash, enabling it to react at room temperature.

3 Experimental Programme

3.1 Materials

The raw materials used for this research work are fly ash, ground-granulated blast-furnace slag and sodium hydroxide. The fly ash was collected from captive power plant, and granulated blast-furnace slag was collected from slag granulation plant of the Rourkela Steel Plant (RSP), Sundargarh, Odisha. The materials were dried in oven to remove the water present in raw material. The fly ash had grayish white colour. The slag was ground in a ball mill to increase the fineness to enhance the reactivity of the material. The materials have been sieved through 75 μm. The fineness value of the slag was 410 m^2/kg. NaOH flakes with 98% purity were used as alkaline activator and obtained from Loba chemie, Mumbai. The chemical compositions for the raw materials are given in Table 1.

The XRD patterns and SEM image of fly ash and slag are shown in Figs. 1 and 2, respectively. The predominant constituents in fly ash are silicon dioxide and aluminium oxide. From the XRD test result of slag, it is observed that the slag is glassy material with some alumina and silica compounds. The microstructure for fly ash reveals that most of the particles are spherical structure with few irregular particles, whereas slag contains rough and angular-shaped particles.

Table 1 Chemical composition of fly ash and slag

Composition (%)	Fly ash	Slag
MgO	1.7	9.52
Al_2O_3	28.1	21.06
SiO_2	53.6	30.82
K_2O	1.97	1.04
P_2O_5	1.72	–
CaO	2.65	32.02
Fe_2O_3	1.8	1.37
Na_2O	0.5	0.088
MnO	0.3	0.14
TiO_2	0.85	1.04
SO_3	–	0.66
Loss on ignition	6.5	1.81

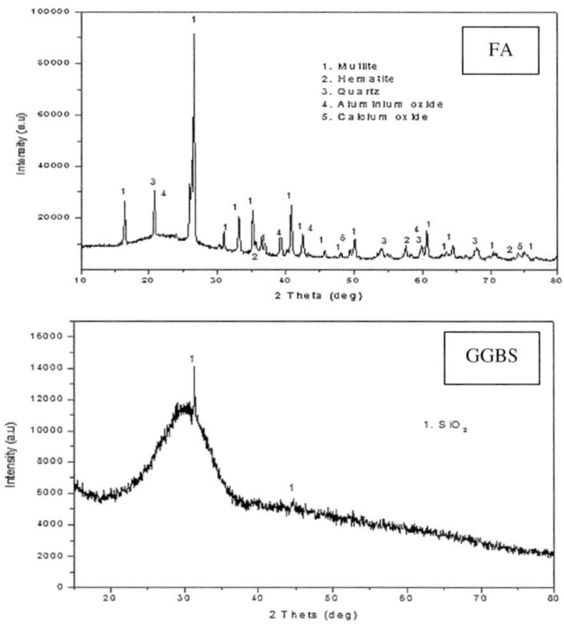

Fig. 1 XRD patterns of fly ash and GGBS

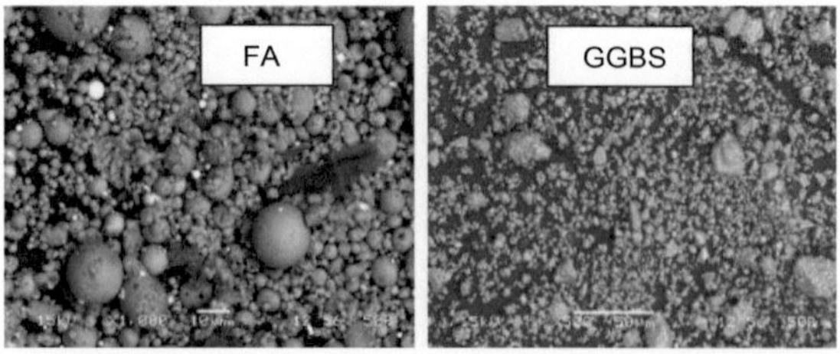

Fig. 2 SEM images of fly ash and GGBS

3.2 *Specimen Preparation and Curing*

3.2.1 Activator Solution

The sodium hydroxide solution was prepared by dissolving the flakes in water. The mass of NaOH solids in a solution varied depending on the concentration of the solution expressed in terms of molarity (M). For instance, NaOH solution with a concentration of 1 M consisted of $1 \times 40 = 400$ g of NaOH solids (in flake or pellet form) per litre of the solution, where 40 is the molecular weight of NaOH. The NaOH solution was prepared before 24 h of sample preparation to ensure proper dissolution of the sodium hydroxide flakes. The concentrations of the sodium hydroxide solution used were 2, 4, 8, 12 and 16 M.

3.2.2 Preparation of Fresh Mortar

A total 6 fly ash–slag mixtures were prepared by varying fly ash content of the mixture from 0 to 100% at 20% intervals. The NaOH solutions of different concentrations are added to the solid fly ash–slag mixtures by varying the solution to solid ratios of 0.25, 0.3, 0.35, 0.4 and 0.45 ml/g. The solids and solutions were mixed for at least for 1 min to achieve homogeneity.

3.2.3 Preparation of Specimen

For each binder, cylindrical specimens with a 2:1 aspect ratio were prepared. The fresh paste of NaOH-activated fly ash-slag mixture was rapidly poured into cylindrical PVC moulds and vibrated for 2 min on vibrating table to remove the air voids. The moulds were sealed from the atmosphere, and after 24 h, the specimens were demoulded. Immediately after demoulding, the test specimens were covered with plastic film to minimize the water evaporation during curing.

3.3 *Unconfined Compressive Strength*

The unconfined compressive strength (UCS) tests of the specimens were conducted to determine the compressive strength of the binder. For each mix proportion and each curing period, three identical specimens were prepared and the average of the strengths was reported as the compressive strength of the mix. Specimens are tested to determine the average compressive strength at the ages of 7 and 28 days. All specimens were prepared, cured and tested under ambient temperature.

4 Results and Discussion

4.1 *Effect of Alkali Concentration*

The variation of 28 days unconfined compressive strength with varying molar concentrations of NaOH solution keeping the solution to solid ratio of 0.35 ml/g as constant is shown in Fig. 3. From the strength results obtained, it is seen that the alkali concentration played a vital role in strength achievement. As the NaOH concentration increases, the strength increases significantly up to 8 M beyond that a declination in strength is observed. The development in strength is mainly governed by the amount of leachable alumino-silcates which is high for 8 M NaOH concentration. But, at a higher concentration (when the NaOH concentration increases from 8 to 16 M) of NaOH excess leaching of silica obstructs further leaching and geopolymerization process. Hence, a reduced strength is obtained.

4.2 *Effect of Solution to Solid Ratio*

Figure 4 shows variations in 28 days unconfined compressive strength with activator solution to binder solid ratio for 8 M NaOH concentration. It is observed that solution to solid ratio in the paste affects the compressive strength up to some extent. The strength is maximum for a particular solution to solid ratio for all the mixtures. Beyond this ratio, strength improvement gets hampered. This may be due to excess Na^+ ions (at higher solution to solid ratio) in the matrix which disturbs the geopolymer structure. As the polymer structure comprises of SiO_4 and AlO_4 tetrahedral anions, and the positive charges of the Na^+ ions compensates the negative charge of Al^{3+}, an excess of Na^+ ions hinders the framework. For most of the fly ash rich mixtures, a particular solution to solid mixture of 0.3 gives higher strength and the same is obtained for 0.35 in slag rich mixtures.

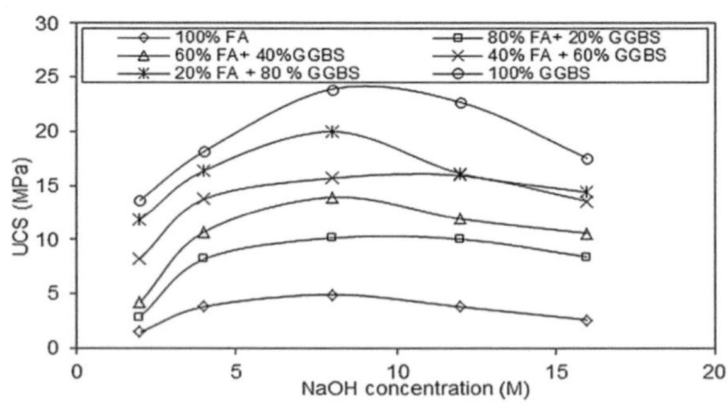

Fig. 3 Variation of 28 days UCS value with NaOH concentration

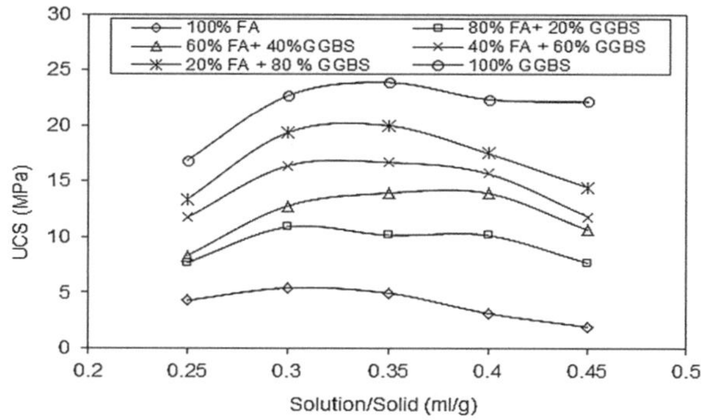

Fig. 4 Variation of 28 days UCS value with solution/solid ratio

4.3 *Effect of GGBS*

The compressive strength for the geopolymer binders are also significantly affected by the slag content in the fly ash-slag mixture. Figure 5 presents the 28 days unconfined compressive strength for all the fly ash–slag mixture with varying slag content for 8 M NaOH. The increased slag content in the mixture resulted in increased strength. This is for the reason that the high reactive contents of alumino-silicates in slag are primarily responsible for increased strength. Furthermore, oxides of calcium present in slag play vital role in strength improvement.

4.4 *Effect of Curing Period*

The development of unconfined compressive strength at the age of 7 and 28 days is shown in Fig. 6 where the NaOH concentration and solution to solid ratio of 8 M and 0.35 ml/g, respectively, are kept constant. Curing period is a parameter which

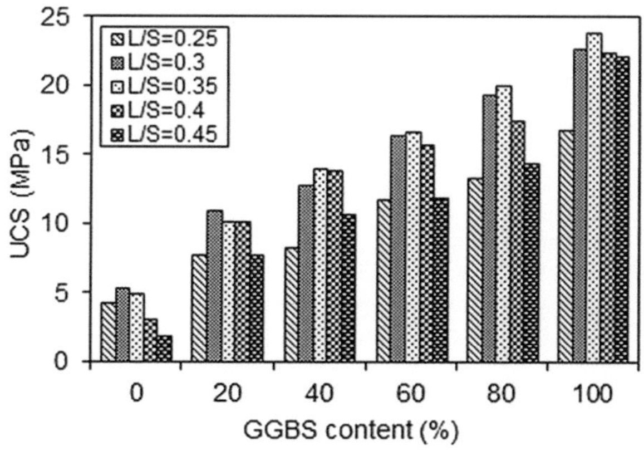

Fig. 5 Variation of 28 days UCS value with GGBS content

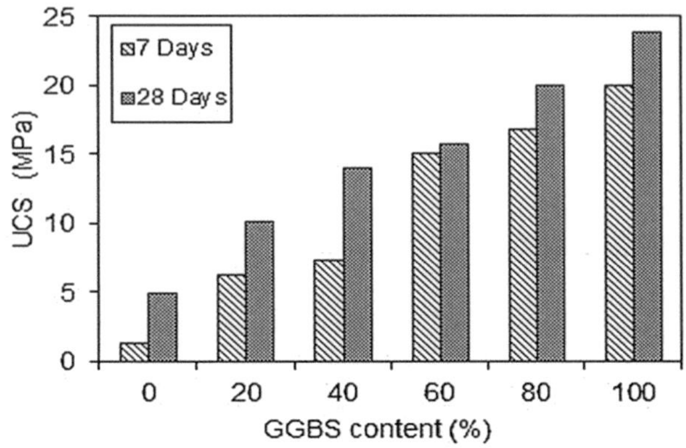

Fig. 6 Development of UCS value with curing period

influences the strength development performance. Most of the slag-rich specimens gained 85% of 28 days strength at the age of 7 days.

5 Conclusions

The following conclusions can be drawn from this experimental study:

1. The compressive strength increases up to 8 M NaOH concentration after which showed a declination in strength.
2. For most of the fly ash rich mixtures, maximum compressive strength is achieved at lower solution to solid ratio compared to slag-rich mixtures.
3. The compressive strength of geopolymer mortar increases as the GGBS content is increased.
4. The strength improves with increased curing period from 7 to 28 days.
5. The use of locally available waste materials such as GGBS and FA could be used for development of sustainable binding material.

References

Deb, P. S., Nath, P., & Sarker, P. K. (2014). The effects of ground granulated blast-furnace slag blending with fly ash and activator content on the workability and strength properties of geopolymer concrete cured at ambient temperature. *Materials and Design, 62,* 32–39.

De Silva, P., Sagoe-Crenstil, K., & Sirivivatnanon, V. (2007). Kinetics of geopolymerization: role of Al_2O_3 and SiO_2. *Cement and Concrete Research, 37*(4), 512–518.

Duxson, P., Fernández-Jiménez, A., Provis, J. L., Lukey, G. C., Palomo, A., & Van Deventer, J. S. J. (2007a). Geopolymer technology: the current state of the art. *Journal of Materials Science, 42*(9), 2917–2933.

Duxson, P. S. W. M., Mallicoat, S. W., Lukey, G. C., Kriven, W. M., & Van Deventer, J. S. J. (2007b). The effect of alkali and Si/Al ratio on the development of mechanical properties of metakaolin-based geopolymers. *Colloids and Surfaces A: Physicochemical and Engineering Aspects, 292*(1), 8–20.

Marjanovic, N., Komljenovic, M., Bascarevic, Z., Nikolic, V., & Petrovic, R. (2015). Physical–mechanical and microstructural properties of alkali-activated fly ash–blast furnace slag blends. *Ceramics International, 41*(1), 1421–1435.

Nath, P., & Sarker, P. K. (2014). Effect of GGBFS on setting, workability and early strength properties of fly ash geopolymer concrete cured in ambient condition. *Construction and Building Materials, 66,* 163–171.

Puertas, F., Martıinez-Ramıirez, S., Alonso, S., & Vazquez, T. (2000). Alkali-activated fly ash/slag cements: strength behaviour and hydration products. *Cement and Concrete Research, 30*(10), 1625–1632.

Rajini, B., & Narasimha Rao, A. V. (2014). Mechanical properties of geopolymer concrete with fly ash and GGBS as source materials. *International Journal of Innovative Research in Science Engineering and Technology, 3*(9), 15944–15953.

Wardhono, A., Law, D. W., & Strano, A. (2015). The strength of alkali-activated slag/fly ash mortar blends at ambient temperature. *Procedia Engineering, 125,* 650–656.

Local Strain Measurements in Triaxial Tests Using On-Sample Transducers

Shiv Shankar Kumar, A. Murali Krishna and Arindam Dey

Abstract Conventional triaxial tests mostly use external LVDTs attached to the actuator of automated triaxial system so as to measure the strain. However, these LVDTs measure the external strain applied and not the strain developed in the sample during shearing. This paper presents the use of on-sample transducer to measure localized strain in a soil sampled subjected to triaxial tests. Specimens prepared at two relative densities (30 and 90%) and three effective confining pressures (50, 100 and 150 kPa) were tested with displacement rates of 0.005 and 1.2 mm/min. It has been observed that the slower rate of loading poses significantly higher secant stiffness of soil. The use of on-sample transducers provides a wide range of strains, i.e., from low strain ($\sim 1 \times 10^{-3}\%$) to high strain (>1%), and can be used to evaluate the modulus reduction curve over the investigated wide range of strain. The capability of on-sample transducers can be effectively harnessed to evaluate the maximum shear modulus of a soil.

Keywords Local strain · Triaxial tests · On-sample transducer
Cohesionless soil

1 Introduction

The response of soils subjected to static and dynamic loading are highly dependent on the strains induced in the soils. The range of these strains caused by various sources such as earthquake, machine vibration, and traffic can be categorized into

S. S. Kumar (✉) · A. Murali Krishna · A. Dey
Department of Civil Engineering, Indian Institute of Technology Guwahati, Guwahati 781039, Assam, India
e-mail: k.shiv@iitg.ernet.in

A. Murali Krishna
e-mail: amurali@iitg.ernet.in

A. Dey
e-mail: arindamdeyiitk@gmail.com

V. K. Stalin and M. Muttharam (eds.), *Geotechnical Characterisation and Geoenvironmental Engineering*, Lecture Notes in Civil Engineering 16,
https://doi.org/10.1007/978-981-13-0899-4_12

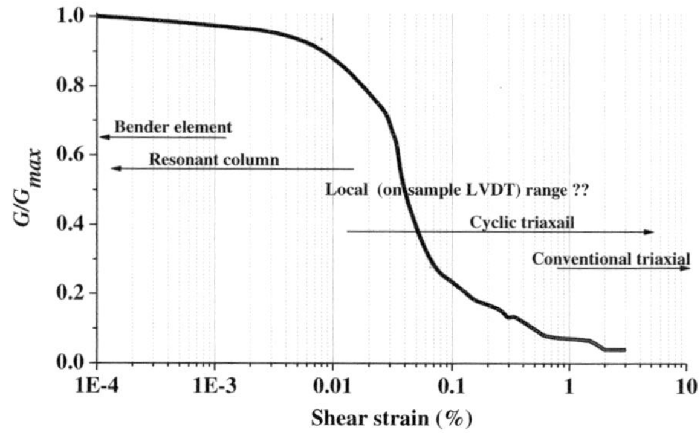

Fig. 1 Typical range of strain properties measured in laboratory (after Likitlersuang et al. 2013)

two sections, namely low strain (i.e., from 0.0001 to 0.001%) and high strain (i.e., from 0.01 to 0.1%). The variations of soil stiffness over such a wide range of shear strain are generally represented by the modulus (stiffness) degradation curve (Fig. 1).

In soil dynamics and earthquake engineering, the small strain shear modulus ($G_{\max}$), secant modulus (G) and the damping ratio (D) are important parameters in soil characterization (Likitlersuang et al. 2013). The small strain shear modulus of the soil is very important parameter to normalize the secant modulus obtained at different shear strains. To obtain the initial shear modulus, researchers have used resonant column and bender element tests. Recently, researchers have used on-sample linear variable differential transducers (LVDTs) to measure the local strain mobilized on the soil specimens (Cuccovillo and Coop 1997; Santagata et al. 2005). Some studies on small strain stiffness of cohesive soil using on-sample LVDTs were reported, but the same on cohesionless soil are limited. In this study, the small strain shear modulus of cohesionless soil was determined in laboratory using on-sample LVDTs. The soil specimens prepared at relative densities (D_r) of 30 and 90% and confining pressures of 50, 100, 150 kPa were sheared at displacement rates of 0.005 and 1.2 mm/min.

2 Previous Studies

Burland and Symes (1982) initially introduced and used the electrolytic liquid levels as horizontal inclinometers in a triaxial test apparatus for the measurement of local displacements and strains on soil specimens. Further, Symes and Burland (1984), Jardine et al. (1984, 1985), Cuccovillo and Coop (1997), and Santagata et al. (2005) have used different types of on-sample LVDTs on clay and sandy soil triaxial specimens and reported undrained shear stiffness at a very small axial strain. The present study investigates the local strain (or, small strain) mobilized on the soil specimen during monotonic shearing.

3 Experimental Investigation

3.1 Soil Characteristics

Brahmaputra sand (BS) obtained from Guwahati region (Assam, India) has been used for the study. Particle size of BS (Fig. 2) was determined by conducting dry sieve analysis (IS: 2720, part-4) and classified them as poorly graded sand as per relevant standards (ASTM D2487; IS: 1498). It can be observed that the soil belongs to the category of severely liquefiable soils (Tsuchida 1970; Xenaki and Athanasopoulos 2003). Index properties of the soil such as minimum and maximum dry unit weight, specific gravity, coefficient of curvature (C_c), uniformity coefficient (C_u) were determined as per relevant standards (IS: 2720, part-3; IS: 2720, part-14) and were enlisted in Table 1.

3.2 Testing Apparatus

An automated pneumatic-controlled cyclic triaxial apparatus also facilitating monotonic triaxial tests, shown in Fig. 3, was used for the experimental investigations. It consists of a 100-kN-capacity loading frame fitted with an actuator (± 15 mm displacement range) operating over a frequency range of 0.01–10 Hz; a

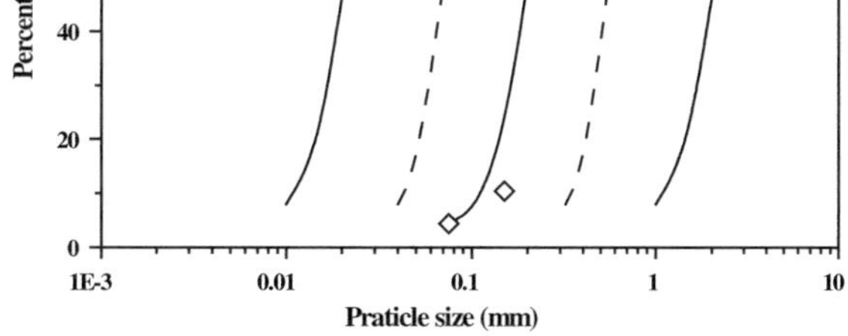

Fig. 2 Particle size distribution

Table 1 Physical properties of collected Brahmaputra sand

Unit weight (kN/m³)		Specific gravity	D_{10} (mm)	C_u	C_c
γ_{max}	γ_{min}				
16.84	13.85	2.7	0.13	1.47	1.09

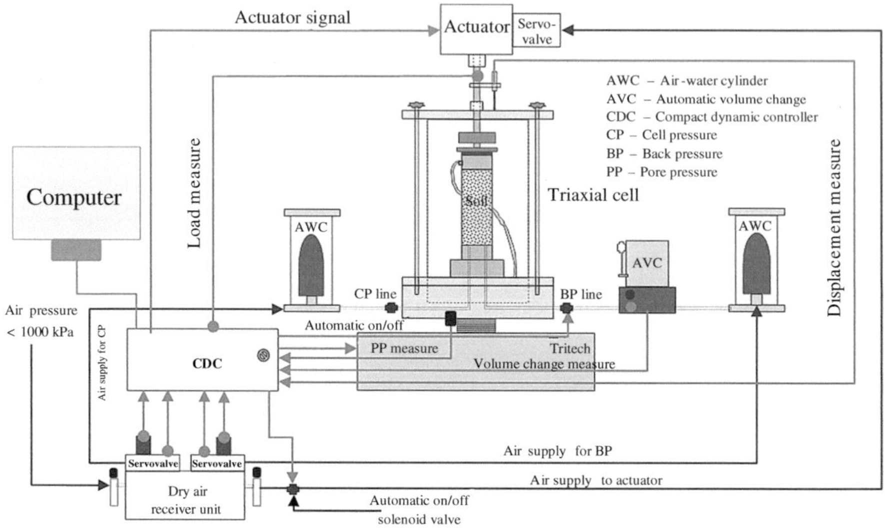

Fig. 3 Layout of cyclic triaxial test setup

triaxial cell having (2000 kPa capacity), and an air compressor having a maximum capacity of 800 kPa. During monotonic compression shear, the shearing was done with upward movement of Tritech while actuator was fixed. Instrumentations available with the apparatus are as follows: linear variable differential transducers (LVDTs) of measuring displacement 30 mm; one submersible load cell of capacity 25 kN; three pressure transducers of 1000 kPa capacity to measure cell pressure, back pressure and pore-water pressure, and one volume change measuring device. The testing is controlled by a compact dynamic controller (CDC) unit, which conveys the instructions provided by software and records the data by data logger.

3.3 Sample Preparation

Dry pluviation technique (ASTM D3999) was adopted to prepare the cylindrical specimens of BS having 70 mm diameter and 140 mm height. A nominal vacuum pressure of 15–20 kPa has been used to maintain verticality of the specimen (Ishihara et al. 1978). In order to achieve a quick saturation, carbon dioxide was flushed through the specimen, for 10–15 min, with a pressure lesser than the applied cell pressure (CP), followed by flushing with de-aired water. To attain the saturation, the CP and back pressure (BP) were then gradually increased in stages while maintaining an almost constant differential pressure of 10 kPa and checking the pore pressure parameter (B) after each CP increment. At a BP of 200 kPa, the

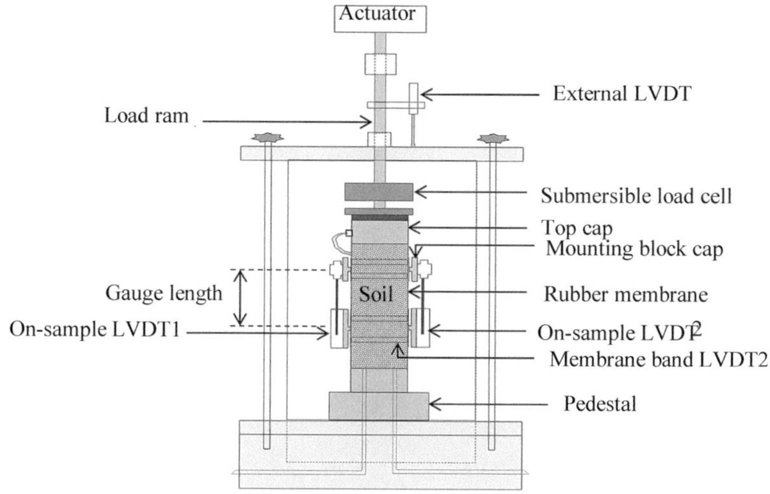

Fig. 4 Schematic diagram of the use of on-sample LVDTs

B value was obtained to be greater than 0.96 and then saturation process was terminated. The test specimens were consolidated to the targeted effective stress levels, before the application of monotonic loading.

3.4 On-Sample LVDTs

The local axial strain in a triaxial test specimen was measured using two linear variable displacement transducers (LVDTs), oriented vertically at the middle of the specimen as shown in Fig. 4. Each transducer was fixed on the specimens with two mounting blocks, which displaces relative to one another as the specimen deforms. The mounting block was fixed on the soil specimens with rubber band of the same stiffness as of membrane. The displacement recorded by transducers required to calculate the local axial strain uses the initial distance between mounting blocks, i.e., the gauge length, rather than the initial specimen length.

4 Results and Discussions

Static triaxial tests were performed on the sandy soil specimens to measure local strain using on-sample LVDTs. The specimens prepared at D_r = 30 and 90% were sheared at σ'_c = 50, 100, 150 kPa with displacement rate of 1.2 mm/min. For the sake of brevity, only the results for specimens with D_r = 90% are presented in this paper.

Local axial strain (based on the on-sample LVDTs) was evaluated and compared with the global axial strain (based on the external LVDT) as shown in Fig. 5. Figure 5a shows the variation of both global and local axial strain at D_r = 90% and at different confining pressures. It can be observed from Fig. 5a that the local axial strain varies nonlinearly in contrast to the global axial strain, the difference being approximately more than 50%. Figure 5a also hints that the axial strain applied at the top of the specimen does not get uniformly transferred throughout the length of the specimen. The stiffness evaluated at these local strains would be higher in comparison to the global axial strain, which can be represented as maximum stiffness of soil. Thus, of the measured local axial strain can be treated as small strain for the evaluation of maximum shear stiffness of the specimen. The local

Fig. 5 Variation of **a** global and local strain, **b** stress–strain response and **c** soil stiffness with on-sample and external LVDTs

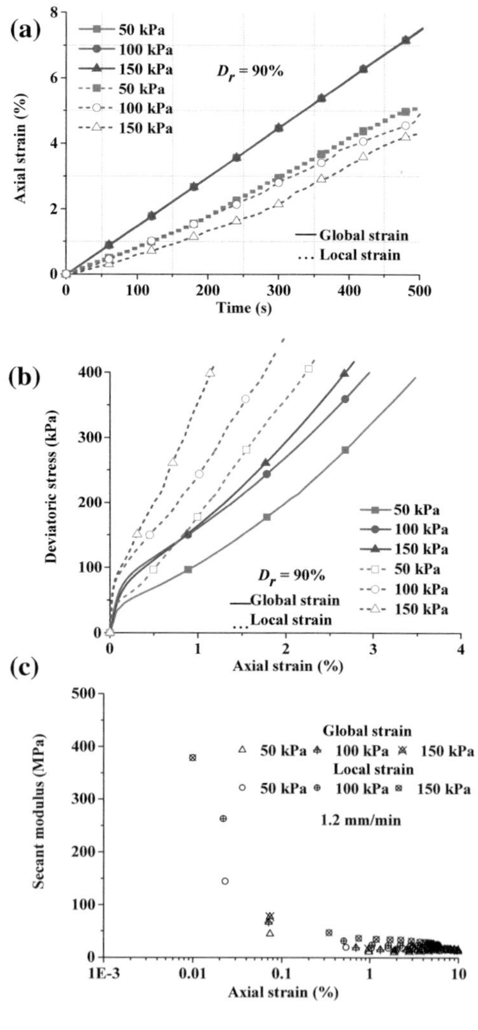

axial strain results have been evaluated based on the average of the records of both longitudinal deformation transducers LVDT-1 and LVDT-2, in order to eliminate any inaccuracy originating due to bedding error. The response of soil specimen during static compression shear has been depicted in Fig. 5b, which also indicates that the high deviatoric stress with on-sample (local strain) LVDTs results in higher initial soil stiffness than the global strain. Figure 5c shows secant shear modulus evaluated for different confining pressure at a displacement rate of 1.2 mm/min. It can be observed from Fig. 5c that the on-sample LVDTs capture relatively higher stiffness corresponding to local strain in comparison to the external LVDT.

Figure 6 presents the response of soil at two different displacement rates 0.005 and 1.2 mm/min. Figure 6a illustrates the stress–strain response of soil at two displacement rates, where significantly higher deviation of deviatoric stress was observed at lower displacement rate. It can also be observed from Fig. 6a that the deviatoric stress becomes higher from on-sample LVDT because of small local strain. Figure 6b depicts the secant modulus variation at two displacement rates which shows loading rate significantly affects the soil stiffness. Specimen sheared with low displacement rate shows higher shear stiffness using on-sample LVDTs, whereas the same showed by external LVDT is significantly lower value. Soil stiffness becomes higher at higher displacement rate (1.2 mm/min) which is reflected by external LVDT, whereas with on-sample LVDTs the same shows higher values at low displacement rate (0.005 mm/min). Hence, it can be stated that on-sample transducer helps in more in-depth of measurement and understanding of the stiffness of the soil specimens.

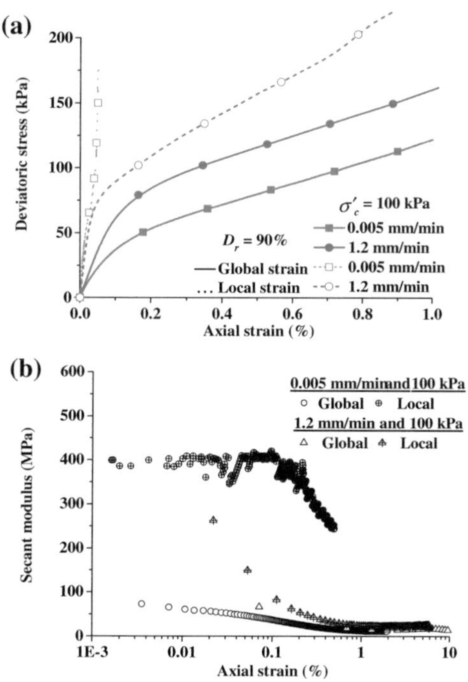

Fig. 6 Variation of **a** global and local stress–strain and **b** soil stiffness at two different displacement rates

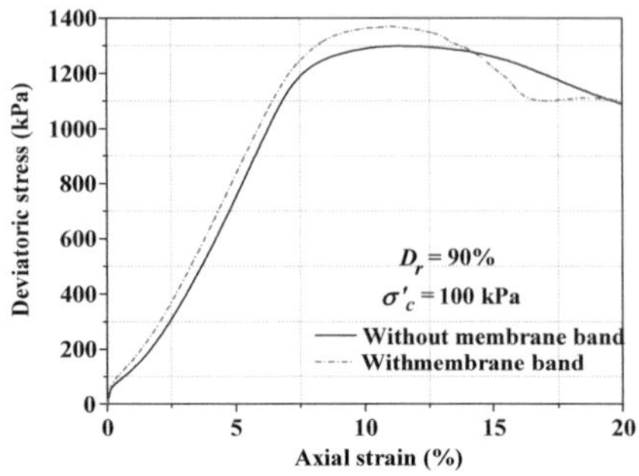

Fig. 7 Response of soil on the use of rubber strip

4.1 *Effect of Rubber Band Attachment*

Rubber band having identical stiffness as that of the rubber membrane was used for an effective attachment of the on-sample LVDTs with the soil specimen. In order to quantify the effect of the presence of rubber band on enhancing the stiffness of the soil specimen at the location of its attachment, a controlled test experiment was conducted in the absence of the rubber band attachment, wherein the on-sample LVDTs were directly attached on the rubber membrane surrounding the soil specimen. Figure 7 shows that the deviatoric stress increased marginally by 3% in the presence of the rubber strip (sample prepared at 90% relative density). The specimens prepared at lower relative density and tested at lower confining stress may depict greater effect of rubber band attachment on the stress–strain response in comparison to the specimen prepared at higher relative density and higher confining stress which has to be further investigated.

5 Conclusions

From the present study, it can be concluded that the on-sample transducers (LVDTs) play a significant role to eliminate the bedding error and to capture the small strain in term of local strain in the soil specimen. The use of on-sample LVDTs provide a wide range of strains, i.e., from small strain ($\sim 1 \times 10^{-3}$ %) to high strain over wide range of strain. The evaluated local strain (small strain) can be used to find out the maximum shear modulus. Membrane band used to hold the LVDTs reveal marginal effect on the stress response at higher relative density and confining pressure. Secant stiffness of soil is significantly affected by the rate of loading. Slower displacement rate of shearing exhibits significantly higher secant stiffness of soil specimens.

References

Burland, J. B., & Symes, M. (1982). A simple axial displacement gauge for use in the triaxial apparatus. *Geotechnique, 32*(1), 62–65.

Cuccovillo, T., & Coop, M. R. (1997). The measurement of local axial strains in triaxial tests using LVDTs. *Geotechnique, 47*(1), 167–171.

Ishihara, K., Silver, L. M., & Kitagawa, H. (1978). Cyclic strength of unsaturated sands obtained by large diameter sampling. *Soils and Foundations, 18*, 61–76.

Jardine, R. J., Symes, M. J., & Burland, J. B. (1984). 'The measurement of soil stiffness in the triaxial apparatus. *Geotechnique, 34*(3), 323–340.

Jardine et al. (1985). Field and laboratory measurements of soil stiffness. *Proceedings of 11th International Conference on Soil Mechanics and Foundation Engineering*, San Francisco.

Likitlersuang, et al. (2013). Small strain stiffness and stiffness degradation curve of Bangkok clays. *Soils and Foundations, 53*(4), 498–509.

Santagata, et al. (2005). Factors affecting the initial stiffness of cohesive soils. *Journal of Geotechnical and Geoenvironmental Engineering, 131*(4), 430–441.

Symes, M., & Burland, J. (1984). Determination of local displacements on soil samples. *Geotechnical Testing Journal, 7*(2), 49–59.

Tsuchida, H. (1970). Prediction and counter measure against the liquefaction in sand deposits. *Proceedings of the Seminar in the Port and Harbour Research Institute*, Ministry of Transport (1–33).

Xenaki, V. C., & Athanasopoulos, G. A. (2003). Liquefaction resistance of sand-mixtures: An experimental investigation of the effect of fines. *Soil Dynamics and Earthquake Engineering, 21*(3), 1–12.

Geotechnical Characterization of Hillslope Soils of Guwahati Region

Chiranjib Prasad Sarma, A. Murali Krishna and Arindam Dey

Abstract Hillslopes within the city of Guwahati consist of geological stratification that characterize progressive stages of residual weathering, which can be categorized as basal rock, decomposed granitic rock and corestones, saprolitic and lateritic residual overburden. Saprolite formation is the layer of residual soil derived from isovolumetric weathering of the bedrock. Undisturbed saprolite formation looks very compact and retains much of the parent rock structure and fabric, but actually is extremely porous due to washing out of the finer particles. Such soils are very friable and once disturbed are exceedingly susceptible to landslides. Several rainfall-induced landslide occurrences have been reported in the hillslopes of Guwahati region. Characterization of these soils is important for proper analysis of hillslope stability of this region. Laboratory test has been performed on disturbed and undisturbed soil samples to assess the geotechnical characteristics. Test to determine grain size distribution, specific gravity, Atterberg's limits, in situ dry density, shear strength parameters, and permeability were conducted. Hillslopes within Guwahati region consist of residual soils in unsaturated condition. The soil–water characteristic curve, which is an important characteristic of unsaturated soil, is determined and compared with that obtained from empirical methods using grain size distribution and Atterberg's limits provided in the literature. Unconsolidated undrained triaxial test was conducted on statically compacted samples remoulded to in situ density at different water content to understand the effect of degree of saturation and moisture content on the shear strength of the soil. Such detailed characterization would provide the requisite understanding for an efficient analysis of the rainfall-induced natural hillslope failure in this region.

C. P. Sarma (✉) · A. Murali Krishna · A. Dey
Department of Civil Engineering, Indian Institute of Technology Guwahati, Guwahati 781039, Assam, India
e-mail: s.chiranjib@iitg.ernet.in

A. Murali Krishna
e-mail: amurali@iitg.ernet.in

A. Dey
e-mail: arindam.dey@iitg.ernet.in

V. K. Stalin and M. Muttharam (eds.), *Geotechnical Characterisation and Geoenvironmental Engineering*, Lecture Notes in Civil Engineering 16,
https://doi.org/10.1007/978-981-13-0899-4_13

Keywords Geotechnical characterization · Soil–water characteristic curve
Laterite · Saprolite · Landslides

1 Introduction

Numerous geotechnical slope stability models are developed which are exceedingly capable of providing detailed description of potential instability taking into account the changing environmental and climatic conditions and thus establish threshold values of the triggering rainfall. However, before being able to conduct such an analysis, proper characterization of the behaviour of the unsaturated residual soils is essential. A proper understanding of the geology and the geotechnical characterization is absolutely necessary to develop a meaningful model for analysis of the phenomena.

The residual soils of the hills within the Guwahati municipality region are derived through physical and chemical weathering process on the parent rock, viz. granite, gneiss, and porphyritic granite, which form the constituent basal stratum of the hills of this region (Maswood 1981). The influence of rainfall in initiating landslides in this soil slopes is widely documented. Infiltration of rainwater increases the water content and thus decreases the matric suction, thereby raising the unit weight and reducing the shear strength of the natural hillslope soils. To assess the propensity to rainfall-induced landslide, a comprehensive understanding of changes in water content and matric suction in response to rainfall infiltration is essential (Selby 1993; Wesley 2010). In this context, severe limitations are observed as far as the literature pertaining to characterizing the unsaturated soil behaviour of this region is concerned. Hillslopes within the city of Guwahati consist of residual soils, in unsaturated condition (Das and Saikia 2010, 2011); therefore, characterizing the unsaturated soil parameters is essential for thoroughly assessing the stability of the natural hillslopes. The study attempts to provide a detailed geotechnical characterization that would provide the requisite understanding for an efficient analysis of the rainfall-induced natural hillslope failure in this region.

2 Hillslope Soil Characterization

Laboratory test has been performed on disturbed and undisturbed soil samples to assess the geotechnical characteristics. Test to determine grain size distribution, specific gravity, Atterberg's limits, in situ dry density, shear strength parameters, and permeability was conducted. Figure 1 gives the images of the cut slope from where the samples have been collected. Two distinct layers of soil can be easily identified from the images. Irregular zones of contrasting weathering, weathered corestones, etc., can be observed from Fig. 1. Figure 2 gives the cross section of the undisturbed soil samples collected at different sites.

Fig. 1 IITG cut slope site depicting two layers of soil and weathered corestones (top layer is referred as soil 1 while the bottom layer as soil 2)

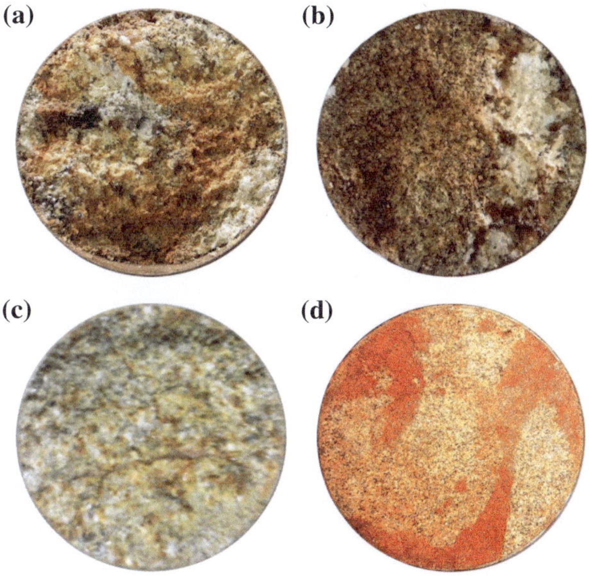

Fig. 2 Cross section of the undisturbed soil samples

Variations in soil structure at the material scale can be easily observed in the samples. Weathering decreases with depth; the uppermost layers (0–2 m) of the soil can be considered as true or mature residual soils, i.e., lateritic soils. Underlying the lateritic soils is the saprolitic formation. This soil layer retains much of the parent rock structure and fabric but with a much lower density. The saprolite formation is very porous and friable in nature and can easily crumble from the slightest disturbance. In equatorial climates of heavy rainfall, saprolitic soil formation can reach

Table 1 Index properties and Atterberg's limits for the two different types of soils

Soil characteristics	Experimental results—soil 1	Experimental results—soil 2
Referred as	RSC_EXP1	PGSS_EXP2
Specific gravity	2.68	2.68
In situ dry density (gcm^{-3})	1.50	1.57
Liquid limit (%)	47	35
Plastic limit (%)	27	Non—Plastic
Coeff. of uniformity	18	12.12
Coeff. of curvature	2.1	2.546
Fines content (%)	77.8	36.75
Porosity	0.44	0.41
In situ degree of saturation (%)	95	47.79
Saturated permeability (ms^{-1})	10^{-7}	10^{-5}

depths of tens of metres (Selby 1993). Table 1 gives the test results for index properties and Atterberg's Limits for the two different types of soils.

Concluding from field investigation, two types of commonly found overburden residual soils constituting the hillslopes around Guwahati can be reported. The top laterite formation of reddish silty clay varying in thickness from few centimetres to few meters, underlain by a saprolite formation of pale yellowish poorly graded silty sand. Hillslopes within the city of Guwahati consist of geological stratification, that characterize progressive stages of residual weathering, typical of tropical and sub-tropical climatic conditions with very high amount of annual precipitation and excessive fluctuation of the ground water table, which can be categorized as basal rock, decomposed granitic rock and corestones, saprolitic and lateritic residual overburden.

2.1 *Determination of Shear Strength Parameters*

Triaxial tests were conducted on undisturbed samples as well as remoulded samples, compacted to in situ dry density for evaluating the shear strength characteristics of the hillslope soils. The tests were conducted at effective confining stress of 50, 100, and 200 kPa. Strain rate of 20%/hr, i.e., 0.4667 mm/min, was found sufficiently low to prevent any excess pore pressure generation in the consolidated drained test and hence was adopted for all further test.

Figure 3 gives the p'-q plot for the consolidated drained triaxial test, whereas Fig. 4 gives the p'-q plot for the consolidated undrained triaxial test. From Fig. 4, it can be observed that the soil behaved as normally consolidated even at low confining stress of 50 kPa. Similar results were observed for all the other samples.

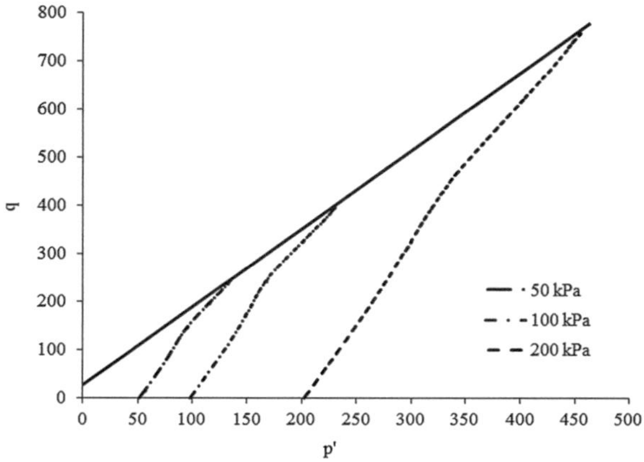

Fig. 3 p'-q plot of consolidated drained triaxial test of IITG cut slope site sample

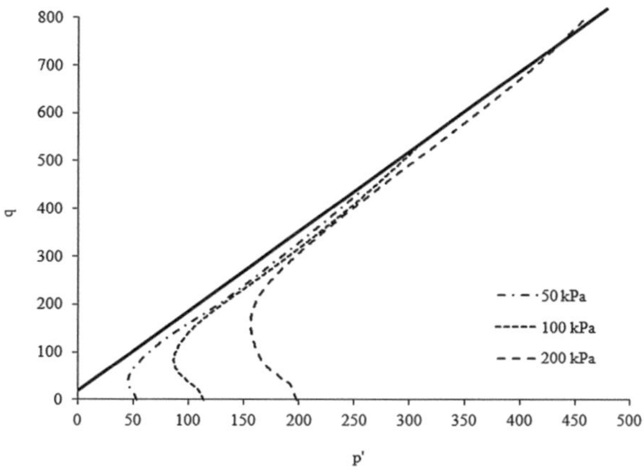

Fig. 4 p'-q plot of consolidated undrained triaxial test of IITG cut slope site sample

Table 2 gives the summary of the test procedure conducted and the corresponding cohesion and angle of internal friction evaluated from the test results.

Table 2 Triaxial test procedure and cohesion and angle of internal friction

Sample No.:/Site	Triaxial test	c (kPa)	φ' (°)
IITG cut slope site 1	Consolidated drained	16.73	30.38
IITG cut slope site 1	Consolidated undrained	11.87	30.9
IITG cut slope site 2	Consolidated undrained	14.53	30.63
IITG Lothia Gate cut slope	Consolidated undrained	15.31	31.24
IITG cut slope site 1 (remoulded to in situ density)	Consolidated undrained	10.42	30.4

2.2 Soil–Water Characteristics (SWCC)

Soil–water characteristic curve is the relationship between the water content and matric suction. SWCC can be applied for quantifying unsaturated shear strength and derivation of unsaturated hydraulic conductivity. The SWCC for the soil was determined using dielectric water potential sensor, (MPS–1) combined with volumetric water content sensor (EC–5). The experimentally obtained SWCC was then compared with the SWCC estimated using the modified Kovac's method (Aubertin et al. 2003) from the grain size distribution and liquid limit of the soil.

Figure 5 gives the SWCC for silty sand (PGSS_EXP2) along with that estimated using grain size distribution and liquid limit data. It can be observed from Fig. 5 that the estimation procedure is ineffective in predicting the SWCC for this type of soil.

2.3 Influence of Moisture Content on the Shear Strength

The test procedure consists of determining the shear strength of statically compacted partially saturated soil specimens. Soil samples were mixed with water in proper weight ratio so that after compacting the soil samples attain the required degree of saturation and dry density. The soil samples are so prepared such that the dry density of the soil specimen turns out to be approximately 1.57 gcm^{-3}, while the water content is varied so that the specimens attain the desired degree of saturation (equal to 40, 50, 60% and so on up to 90%). The samples were then kept properly packed in airtight plastic pouches and left in desiccators for 48 h to attain equilibrium distribution of water content. The soil samples were then statically compacted to specimens of 38 mm diameter and 76 mm length using constant volume mould. The specimens were again kept in desiccator for 24 h. The samples

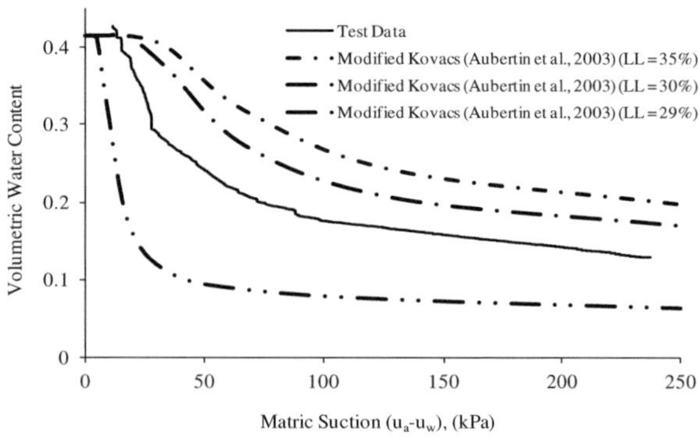

Fig. 5 Soil–water characteristic curves for the saprolitic soil (PGSS_EXP2)

were then setup in the triaxial test equipment. Conventional unconsolidated undrained triaxial test procedures were followed to shear the soil specimens. The samples were sheared at minimal cell pressure of 25 kPa. The stress–strain behaviour is presented in Fig. 6. The peak deviatoric stresses are then plotted in Fig. 7 against the degree of saturation. It can be observed from Figs. 6 and 7 that with decrease in water content the peak deviatoric stress increases. With decrease in the water content, the stiffness of the soil also increases.

Fig. 6 Deviatoric stress versus strain (%) for partially saturated soil sample

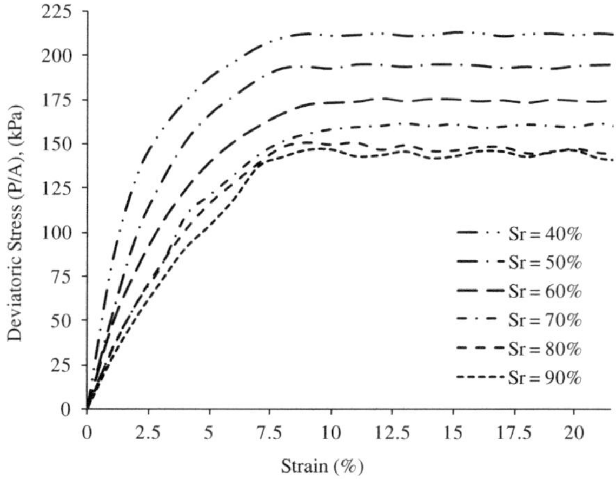

Fig. 7 Increase in peak deviatoric stress with decrease in degree of saturation (%)

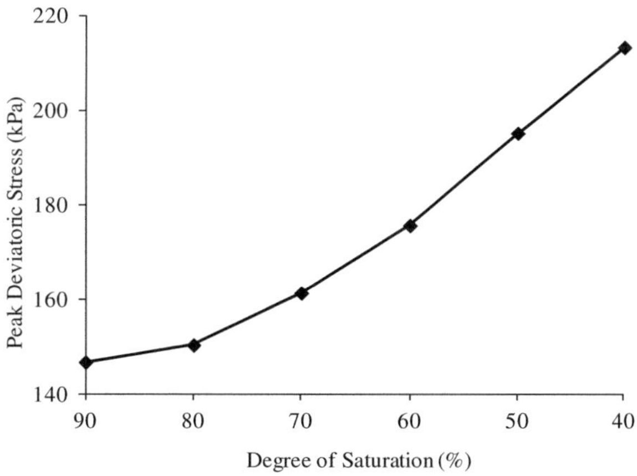

3 Conclusions

Two types of soil can be identified in the hillslopes of this region. The uppermost layers, lateritic in nature, varying in thickness from few centimetres to few 1–2 m can be considered as mature residual soils and are characterized by high plasticity index, higher percentage content of clay size particles, and low permeability values. Underlying the lateritic soils are the saprolitic soils which can be classified as poorly graded silty sand.

Saprolite formation is isovolumetrically weathered from rocks and retains much of the parent rock structure and fabric but with lower density. Such soils are very porous and friable in nature and can easily disintegrate from the slightest disturbance. However, in in situ conditions, such soils appear to be very stiff giving a feeling of stability, which actually is not existent. In equatorial climates of heavy rainfall, saprolitic soil formation can reach depths of tens of metres.

Though considerable differences are observed at material scale of the soil samples, however, variations in the shear strength were not very significant. The angle of internal friction was around 30°–31°, while the cohesion was varying from $\sim$10 to 17 kPa.

The SWCC of the soil could not be estimated applying the modified Kovac's method from the grain size distribution and liquid limit. Thus, for analysis purpose, experimentally obtained SWCC should be used as far as possible.

Moisture content had a significant effect on the shear strength of the soil. With degree of saturation decreasing, increase in the peak deviatoric stress was observed. The stiffness of the soil also increased with decrease in the water content.

References

Aubertin, M., Mbonimpa, M., Bussière, B., & Chapuis, R. P. (2003). A model to predict the water retention curve from basic geotechnical properties. *Canadian Geotechnical Journal, 40*(6), 1104–1122.

Das, U. K., & Saikia, B.D. (2010). Shear strength of unsaturated residual soils of the Hills in Guwahati. *Proceedings of Indian Geotechnical Conference, GEOtrendz* (679–682).

Das, U.K., & Saikia, B.D. (2011). Evaluation of a prediction model for shear strength of unsaturated soils. *Proceedings of Indian Geotechnical Conference*, Kochi (643–646).

Maswood, Md. (1981). Granite gneisses around Guwahati, Assam, *Journal of Geological & Mineralogical Society of India, 53*(3, 4), 115–124.

Selby, M. J. (1993). *Hillslope Materials and Processes*. New York: Oxford University Press Inc.

Wesley, L. D. (2010), *Geotechnical Engineering in Residual Soils*. Wiley.

Effect of Box Size on Dilative Behaviour of Sand in Direct Shear Test

S. R. Mohapatra, S. R. Mishra, S. Nithin, K. Rajagopal and Jitendra Sharma

Abstract In this paper, an attempt is made to analyse the dilative behaviour of dense sand at two different sizes of the direct shear box, i.e., small (60 mm × 60 mm × 30 mm) and large (305 mm × 305 mm × 140 mm). A three-dimensional numerical model is developed using the FLAC^{3D} software to analyse the size effect on dilative behaviour of dense sand along the top and the shear plane of the box at 15 kPa normal pressure. It is observed that the vertical deformation of soil on top plane increases linearly with horizontal displacement, whereas on shear plane, the vertical deformation remains constant after yielding of sand. It is also found that there is greater movement of sand particles at the front and the back of the box for the large shear box compared with that for the small shear box.

Keywords Direct shear · Numerical modelling · FLAC · Dilation Mohr–Coulomb

Abbreviations and Notations

B	Width of the shear box
DSB	Direct shear box
H	Height of the shear box

S. R. Mohapatra · S. R. Mishra (✉) · S. Nithin · K. Rajagopal
Department of Civil Engineering, Indian Institute of Technology Madras, Chennai, India
e-mail: srm.cet@gmail.com

S. R. Mohapatra
e-mail: siluuce@gmail.com

S. Nithin
e-mail: nithinsudersanan@gmail.com

K. Rajagopal
e-mail: gopalkr@iitm.ac.in

J. Sharma
Department of Civil Engineering, York University, Toronto, Canada
e-mail: jit.sharma@lassonde.yorku.ca

V. K. Stalin and M. Muttharam (eds.), *Geotechnical Characterisation and Geoenvironmental Engineering*, Lecture Notes in Civil Engineering 16,
https://doi.org/10.1007/978-981-13-0899-4_14

L	Length of the shear box
MC	Mohr–Coulomb
2D	Two dimensional
3D	Three-dimensional
D_{50}	Diameter corresponding to percentage finer than 50%
ψ	Dilation angle
ϕ	Peak friction angle
ϕ_{cv}	Friction angle at 40-mm horizontal displacement
σ_n	Normal pressure
τ	Peak shear stress
τ_{40}	Shear–stress at 40-mm horizontal displacement

1 Introduction

Granular soils are primarily used as a backfill material for mechanically stabilized earth wall and reinforced soil slopes. A direct shear box (DSB) is commonly used in most of the geotechnical laboratories to measure the shear strength parameters (cohesion interpcept c and friction angle ϕ) of granular soils (Jewell and Wroth 1987; Jewell 1989; Thronton and Zhang 2003; Cui and O'Sullivan 2006; Zhang and Thornton 2007; Indraratna et al. 2014). It consists of an upper box and a lower box. Usually, the upper box is restrained and the lower box moves relative to upper box. The shear force is measured using a proving ring or load cell (Mohapatra et al. 2014, 2016). DSBs of various shapes (square and circular) and sizes (small, medium and large) are used according to the maximum particle size of the soil (Shibuya et al. 1997; Lings and Dietz 2004; Bareither et al. 2007). According to AASHTO T236-08-UL (AASHTO 2008) or ASTM D3080 (ASTM 2011), the width of the DSB must be 10 times the maximum particle size and the initial specimen thickness must be 6 times the maximum particle size of the soil. To satisfy the above criteria, mostly three different sizes of DSB, i.e., small (60 mm × 60 mm), medium (150 mm × 150 mm) and large (305 mm × 305 mm), are normally used in a typical geotechnical laboratory.

Various studies are reported in the literature to compare the shear strength parameters of the granular soil (mostly sand) measured using DSBs of different sizes. Bareither et al. (2008) carried out laboratory experiments to understand the size effect of DSB on measured shear strength of sand backfill. They concluded that friction angles measured using small and large DSBs are essentially same.

In a DSB, the vertical movement of the top cap is measured using a dial gauge or a LVDT for the estimation of angle of dilation. The vertical displacement versus the horizontal displacement plot does not give the entire picture of dilative behaviour of the soil, which may vary from the top cap to the shear plane. Various studies are reported in the literature to understand the dilation behaviour of granular soil (Newland and Allely 1957; Bolton 1986; Houlsby 1991; Chakraborty and Salgado 2010). Very limited studies have been conducted to understand the effect of box size on dilative behaviour of sand in a direct shear test.

In the present study, an attempt is made to analyse the dilative behaviour of dense sand using two different sizes of DSB, i.e., small and large. A three-dimensional (3D) numerical model is developed using the finite difference software FLAC^{3D} (Itasca 2005) to analyse the size effect on dilative behaviour of dense sand under low normal pressure (σ_n= 15 kPa). The behaviour is analysed along the top boundary and the shear plane. Additionally, the movement of soil particles along the front and back boundaries of the DSB are also analysed and compared for the small and the large DSB.

2 Numerical Modelling Procedure

Three-dimensional finite difference grid models were created in FLAC^{3D} for the small and the large DSBs. Rigid walls of the DSB were not modelled explicitly. Instead boundary conditions as described below were used to simulate the conditions during direct shear tests. Roller boundary conditions were given to four vertical sides of top and bottom part of the sample, similar to the test condition. All the grid points on four vertical faces of bottom box and the bottom surface of the box were given equal horizontal displacement to simulate shearing of soil during the direct shear test. Potts et al. (1987) used similar approach for two-dimensional finite element modelling of the direct shear test. Size of the model was similar to the sample size inside the DSB, i.e., 60 mm × 60 mm × 30 mm for the small and 305 mm × 305 mm × 140 mm for the large DSB. To improve the solution accuracy, very fine grid meshes are used along the shear plane (Fig. 1). In the current model, an equal horizontal velocity was applied to all the nodes on the bottom box and the model is allowed to run for certain time steps to achieve the required horizontal displacement. The unbalanced forces generated due to the displacement of the bottom box were summed up and divided by the plan area of the shear box to determine the shear stress corresponding to different horizontal displacements.

Elastic-perfectly plastic Mohr–Coulomb (MC) model was used to model the dense sand. Properties of sand used for the modelling is given in Table 1. Grain size distribution of sand is given in Fig. 2. The input parameters ϕ and ψ for sand are obtained from laboratory experiments (Mohapatra et al. 2016). From the shear

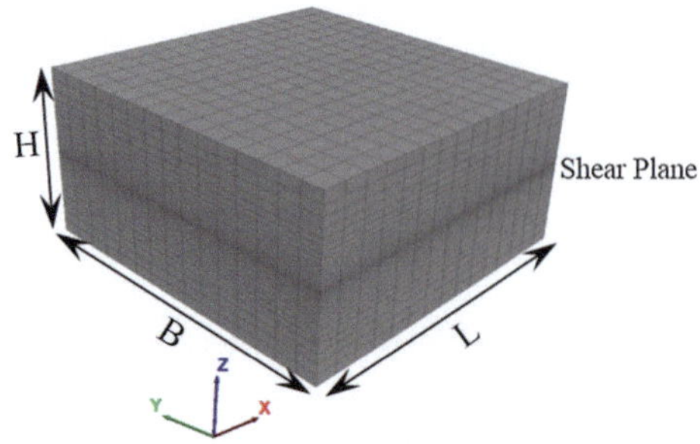

Fig. 1 Isometric view of the direct shear model

Table 1 Properties of sand used in the modelling

Shear box type	Peak friction angle; ϕ (°)	Dilation angle; ψ (°)
Large	53	16
Small	48	5

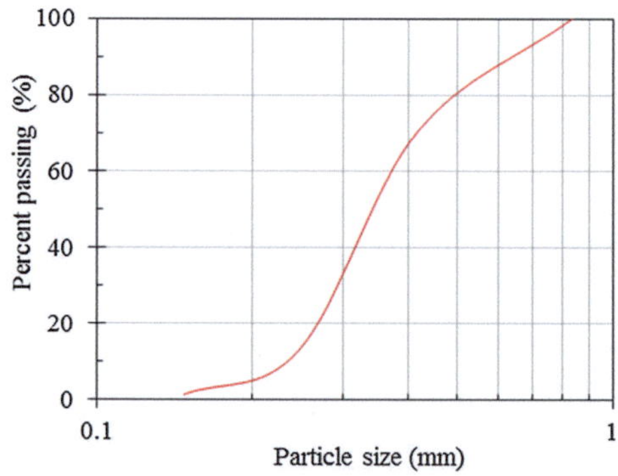

Fig. 2 Grain size distribution of sand

stress versus horizontal displacement plot (Fig. 3), friction angle at peak (ϕ) and 40-mm horizontal displacement (ϕ_{cv}) are obtained using Eqs. (1) and (2).

$$\phi = \tan^{-1}\frac{\tau}{\sigma_n} \tag{1}$$

$$\phi_{cv} = \tan^{-1}\frac{\tau_{cv}}{\sigma_n} \tag{2}$$

where τ and τ_{cv} are the shear stresses mobilized at peak and at 40-mm horizontal displacement, respectively, and σ_n is the applied normal pressure. The dilation angle (ψ) of the soil was obtained from Eq. (3) proposed by Bolton (1986)

$$\phi = \phi_{cv} + 0.8\psi \tag{3}$$

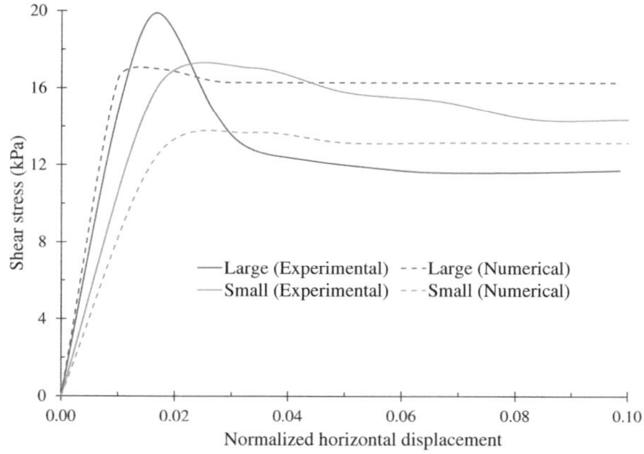

Fig. 3 Shear stress versus normalized horizontal displacement

3 Results and Discussion

The results from the numerical simulation are validated with the laboratory tests, and some other relevant data are also extracted from the numerical modelling thereafter. Figure 3 shows the validation using MC model with the shear stress versus normalized horizontal displacement plot obtained from the experiments. The horizontal displacement is normalized with the length of the box. As MC model is an elastic-perfectly plastic model, the post-failure strain softening of granular material cannot be modelled. However, the model can be seen to assume the failure stress states appropriately.

As there was no provision to measure the vertical displacement during the experiments, the relevant data are extracted from the numerical simulation and are shown in Figs. 4 and 5. The values are normalized with the respective dimensions of the box. Generally in the laboratory, only the top box movements are measured but the same extent of dilation may not be occurring at the shear plane which is

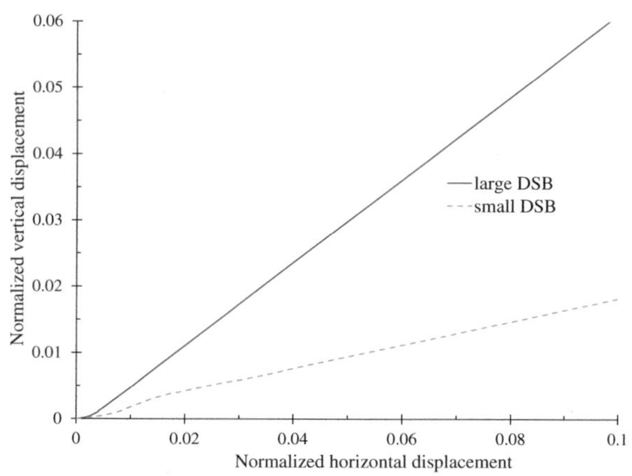

Fig. 4 Comparison of vertical displacement at top plane

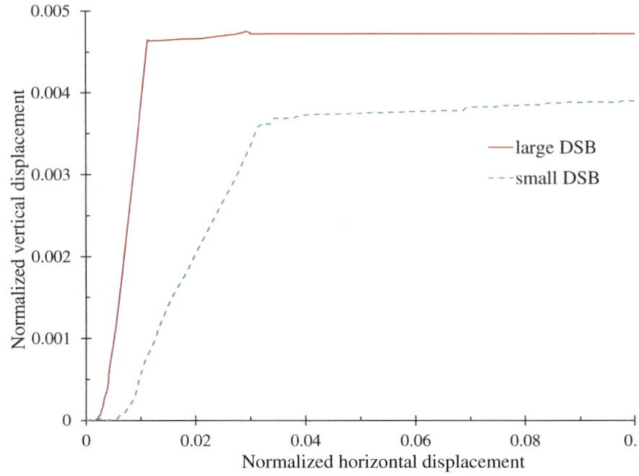

Fig. 5 Comparison of vertical displacement at shear plane

actually the plane of interest. The shear behaviour is studied at the shear plane, whereas the dilation behaviour is observed at the top boundary. The 3D model helps in determining the spatial variation throughout the box rather than just on a particular plane as in the case of a 2D model.

Figures 4 and 5 show the variation of vertical displacement at two planes, (a) top plane and (b) shear plane of the DSB. From the figures, it can be observed that the vertical displacement of the top box increases at a constant rate throughout the test, whereas in shear plane, it remains constant after the yielding. This behaviour is observed for both sizes of the DSBs.

Figure 6 shows the contours of vertical displacement in case of large and small DSB. From the figure, it can be observed that soil grains at the back side of the

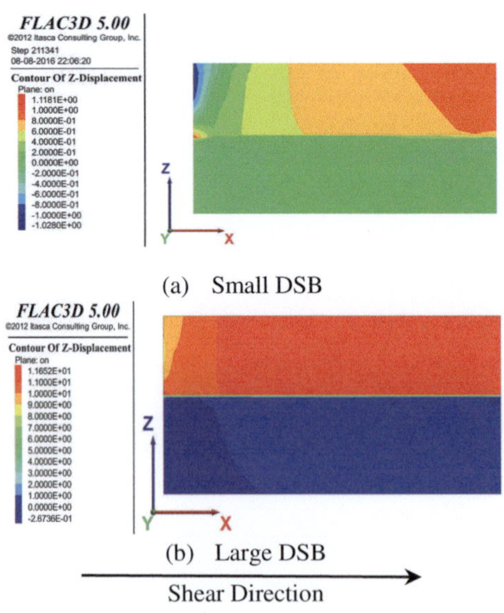

Fig. 6 Contours of vertical displacement after shearing

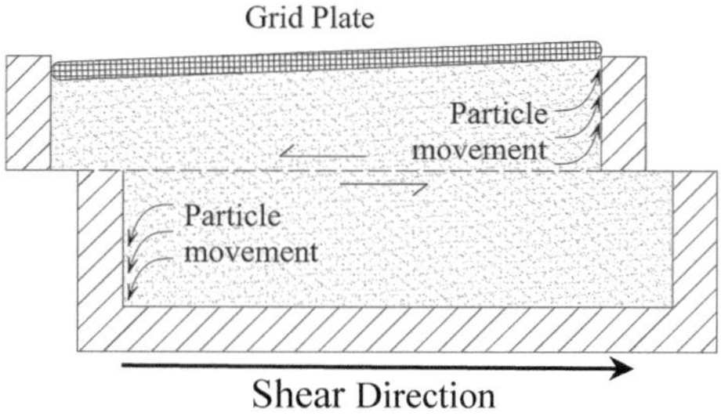

Fig. 7 Particle movement during shearing in large DSB

bottom box undergo downward movement in case of large DSB, whereas in case of small DSB, the particle movement is restricted to the top box. The relative magnitude of vertical displacement is higher in case of large DSB. This particle movement in large DSB results in the particle-to-particle force concentrations that are transferred to the particle–box interface, resulting in increased measured shear resistance (Fig. 7).

In the small DSB, particle–box interaction was insignificant compared to that in the large DSB due to greater number of particles moving within the shear band. Liu (2006) reported similar observation from the results obtained from discrete element modelling of direct shear test of dense sand.

4 Conclusion

From the above study, the following conclusions can be derived

- Dilation angle of sand measured from small and large DSB is not same, and significant variation in dilation angle is observed due to the difference in particle movement in small and large shear box.
- Vertical displacement measured at top of the box during laboratory experiment does not give entire picture of dilative behaviour of sand inside the shear box.
- Higher shear forces are measured in large DSB due to larger number of particle movement.

References

AASHTO. (2008). *Standard method of test for direct shear test of soil under consolidated drained conditions*. Washington, D.C.: T236-08-UL American Association of State Highway and Transportation Officials.

ASTM. (2011). *ASTM D3080/D3080M: Standard test method for direct shear test for soils under consolidated drained conditions*. West Conshohocken, Pennsylvania, USA: American Society for Testing and Materials.

Bareither, C. A., Benson, C. H., & Edil, T. B. (2007). Reproducibility of direct shear tests conducted on granular backfill materials. *ASTM Geotechnical Testing Journal, 31*(1).
Bareither, C. A., Benson, C. H., & Edil, T. B. (2008). Comparison of shear strength of sand backfills measured in small-scale and large-scale direct shear tests. *Canadian Geotechnical Journal, 45,* 1224–1236.
Bolton, M. D. (1986). The strength and dilatancy of sand. *Géotechnique, 36*(1), 65–78.
Chakraborty, T., & Salgado, R. (2010). Dilatancy and shear strength of sand at low confining pressures. *ASCE Journal of Geotechnical and Geoenvironmental Engineering, 136*(3), 527–532.
Cui, L., & O'Sullivan, C. (2006). Exploring the macro- and micro-scale response of an idealized granular material in the direct shear apparatus. *Géotechnique, 56*(7), 455–468.
Houlsby, G. T. (1991). *How the dilatancy of soils affect their behaviour*. Report No. OUEL-1888/91. Oxford, UK: Oxford University Engineering Laboratory.
Indraratna, B., Ngo, N. T., Rujikiatkamjorn, C., & Vinod, J. S. (2014). Behavior of fresh and fouled railway ballast subjected to direct shear testing: Discrete element simulation. *International Journal of Geomechanics, 14*(1), 34–44.
Itasca. (2005). *FLAC^{3D} 5.00 user's manual*. Minneapolis, USA: Itasca Consulting Group Inc.
Jewell, R. A. (1989). Direct shear tests on sand. *Géotechnique, 39*(2), 309–322.
Jewell, R. A., & Wroth, C. P. (1987). Direct shear tests on reinforced sand. *Géotechnique, 37*(1), 53–68.
Lings, M. L., & Dietz, M. S. (2004). An improved direct shear apparatus for sand. *Géotechnique, 54*(4), 245–256.
Liu, H. S. (2006). Simulating a direct shear box test by DEM. *Canadian Geotechnical Journal, 43,* 165–178.
Mohapatra, S. R., Rajagopal, K., & Sharma, J. S. (2014). Analysis of geotextile-reinforced stone columns subjected to lateral loading. In *Proceedings of the 10th International Conference on Geosynthetics*, Berlin, Germany.
Mohapatra, S. R., Rajagopal, K., & Sharma, J. S. (2016). Large direct shear load test on geosynthetic encased granular columns. *Geotextiles & Geomembranes, 44*(3), 396–405.
Newland, P. L., & Allely, B. H. (1957). Volume changes in drained triaxial tests on granular materials. *Géotechnique, 7*(1), 17–34.
Ozer, C., & Arshiya, A. (2015). Dilatancy and friction angles based on in situ soil conditions. *ASCE Journal of Geotechnical and Geoenvironmental Engineering, 141.*
Potts, D. M., Dounias, G. T., & Vaughan, P. R. (1987). *Géotechnique, 37*(1), 11–23.
Shibuya, S., Mitachi, T., & Tamate, S. (1997). Interpretation of direct shear box testing of sands as quasi-simple shear. *Géotechnique, 47*(4), 769–790.
Thronton, C., & Zhang, L. (2003). Numerical simulations of the direct shear test. *Chemical Engineering & Technology, 26*(2), 153–156.
Zhang, L., & Thornton, C. (2007). A numerical examination of the direct shear test. *Géotechnique, 57*(4), 343–354.

Part II
Geoenvironmental Engineering and Behaviour of Unsaturated Soils

Influence of Initial Conditions on Undrained Response of Soft Clays

Dipty Sarin Isaac, Kodi Rangaswamy and S. Chandrakaran

Abstract Soft soils are found along east and west coast of Kerala. They exhibit unusual engineering properties like high natural water content, high compressibility, low permeability and low shear strength. Availability of land for the development of commercial, housing, industrial and transportation, infrastructure, etc., is getting scarce nowadays. This necessitated the use of land, which has weak strata, wherein the geotechnical engineers are challenged by the presence of different problematic soils with varied engineering characteristics. Not much work has been done to understand the behaviour of these soils. Hence, it is necessary to study the geotechnical behaviour of such soils including the effect of various parameters. As far as Kerala is considered, soft clays are predominant in Kuttanad and greater parts of Cochin. Research works carried on Kuttanad clay and Cochin marine clay mainly focused on compaction, compressibility and UCC strength. Clayey soils are impervious and do not permit drainage of water. There will be a cumulative build-up of pore water leading to increase in pore water pressure and reduction in shear strength, finally leading to increase in settlement and deformation. Not much works have been done on laboratory modelling of the soft clays of Kerala and its behaviour under the influence of varying parameters. Soil properties vary from point to point due to the variations of field parameters. The objective of this paper is to investigate the effect of various parameters influencing undrained behaviour of Kuttanad clay. Confining stress and density were studied. For this purpose,

D. S. Isaac (✉)
Department of Civil Engineering, Saintgits College of Engineering,
Pathamuttom, Kottayam 686532, India
e-mail: dipty.sarin@gmail.com

K. Rangaswamy · S. Chandrakaran
National Institute of Technology Calicut, Kozhikode 673601, Kerala, India
e-mail: ranga@nitc.ac.in

S. Chandrakaran
e-mail: chandra@nitc.ac.in

V. K. Stalin and M. Muttharam (eds.), *Geotechnical Characterisation and Geoenvironmental Engineering*, Lecture Notes in Civil Engineering 16,
https://doi.org/10.1007/978-981-13-0899-4_15

consolidated undrained triaxial test was performed with pore pressure measurements. Typical deviator stress–strain graphs, pore pressure versus strain graphs, and stress paths were drawn for all the tests conducted.

Keywords Density · Confining stress · Deviator stress · Pore pressure

1 Introduction

Soft clay is defined as a disturbed cohesive soil whose water content is higher than its liquid limit; such materials display extremely low-yield stresses and represent difficult construction conditions. Soft clay deposits are widespread, and they present special problems. By definition, soft clays are of high natural water content, low shear strength, high compressibility, low permeability and low shear strength, and many are sensitive, in that their strength is reduced by disturbances. Foundation failures in soft clay are comparatively common, and surface loading, e.g. in the form of embankments, inevitably results in large settlements. The present study aims to investigate on the parameters influencing undrained behaviour of soft clay under triaxial loading conditions. Not much works have been done on laboratory modelling of the soft clays of Kerala and its behaviour under the influence of varying parameters. Soil properties vary from point to point due to the variations of field soil parameters. So, it is highly essential to study the behaviour of clay under varying parameters. Kuttanad clays are dark brown-coloured medium sensitive alluvial deposits spread over the Kuttanad region in the state of Kerala in India. These clays are characterized by high compressibility, low shear strength and high percentage of organic matter, which are unfavourable from the geotechnical engineering point of view. A large number of embankment failures and foundation failures have been reported in this soil due to its poor shear strength and compressibility characteristics. It is proposed to make an attempt on such clay soils to investigate on the undrained behaviour of soils prior to improvement by adopting suitable measures.

2 Literature Review

Saranya and Muttharam (2013) carried out studies to determine the effect of consolidation stress on strength of lime stabilized soils. The various laboratory tests such as vane shear, unconfined compressive strength and triaxial tests were conducted. It was found that the undrained cohesion and the angle of internal friction are increases with increase in lime content and curing periods. However for a given lime content and curing period, the undrained cohesion increases with an increase in consolidation stress and angle of internal friction decreases with an increase in consolidation stress. Sachan and Penumadu (2007) studied the influence of confining pressure on clay specimens with dispersed and flocculated micro-fabric. In this research, the normally consolidated kaolin clay shows its dilative nature during shearing for dispersed micro-fabric and contractive nature for flocculated micro-fabric. The dispersed micro-fabric showed relatively sudden failure response at low strain levels followed by significant softening when compared to flocculated micro-fabric due to its volume change behaviour. Elsayed and Swan (2010) introduced a new preparation technique, termed volume control (VC), which attempts to achieve a pre-determined, pre-shear relative density (Drc) for granular soil samples used in triaxial tests. Korichi et al. (2013) studied the effects of compression on porous texture of clay powder. Application of uniaxial pressing pressures (31.84 at 127.38 MPa) of powder clay produces materials with increased density and reduction of porosity. Caridad et al. (2014) conducted experimental measurements of thermal conductivity (λ) and the density (ρ) of several clay pastes used in pelotherapy (peloids). A clear correlation was established between the density and the thermal conductivity of the peloids.

3 Experimental Programme

Clay sample selected for the present study was collected from Champakkulam (Kuttanad region) from 2 to 3 m depth as per the procedure described in IS: 2132 (Part 2)—1986. Water table was very close to ground surface at the time of sampling. The soil sample was brought to the laboratory without any loss of water content and it was stored under humid condition. Disturbed samples were also collected. Soil samples were tested for Atterberg limits and grain size distribution characteristics. Specific gravity was calculated from known values of shrinkage limit. Grain size analysis was done by hydrometer method. The weight of wet sample which will contain 50 g dry soil is estimated based on water content. All the tests were conducted according to IS code procedures. The index properties determined are shown in Table 1.

Table 1 Index and strength properties

Property	Value
Index property	
Natural water content (%)	130
Specific gravity	2.53
Liquid limit (%)	135
Plastic limit (%)	55
Shrinkage limit (%)	25
Plasticity index	79
Clay size (%)	50
Strength property	
Maximum dry density (kN/m^3)	14
Optimum moisture content (%)	31
UCS (kPa) (at w_c = 60%)	15.2
Organic content (%)	12

4 Triaxial Compression

Among all the shear tests, triaxial compression is more preferable for accurate research work. There is a complete control over the drainage conditions. Among the different drainage conditions, consolidated undrained test conditions are suitable for pore pressure measurements. The test set-up comprised of a standard triaxial test apparatus with facilities for measuring pore water pressure. A proving ring of 2.5 kN capacity was calibrated and was used for measuring axial loads. A dial gauge of 0.01 mm sensitivity was used to measure the axial displacements. It contains pressure control system to control the cell and back pressure applications. Electronic kit device is attached to monitor the displacement as well as axial loads applied to the sample. Cylindrical specimens were prepared in the mould of 38 mm dia. and 76 mm long at the various densities of 12, 13 and 14 kN/m^3. Samples' air dried to 60% water content was used for the convenience of sample preparation. Samples were saturated by immersion method. For each specimens, tested back pressure is increased in steps and corresponding increase in pore pressure is noted to check B parameter for saturation (B approximately 0.96). Specimens were tested under consolidated undrained condition and subjected to confining pressures of 50, 100 and 150 kPa. After application of each confining stress, axial load is applied.

5 Results and Discussion

5.1 Effect of Density

The effect of densities on undrained response in terms of stress–strain relations and pore pressure variations at particular confining pressure of 50, 100and 150 kPa is shown in Figs. 1, 2 and 3, respectively. Figures 1, 2 and 3 are indicating that the deviator stresses are increasing with increase in the density irrespective of all pressures. The increase of deviator stress is due to decrease of pore water pressure with increase in the density. The effective stress paths of samples confined at the pressures of 50, 100 and 150 kPa are shown in Fig. 4. It illustrated that the clay samples moulded at low densities possesses high contraction behaviour and exhibits less resistance to shear loads when to compared to that at high densities irrespective of all confining pressures. It also observed that the reduction in mean effective normal stress is increasing with decrease in density of clay samples.

5.2 *Effect of Confining Stress*

The effect of confining pressure is to be expressed in terms of normalized stress ratios as the test samples are subjected to different pressures. The deviator stress

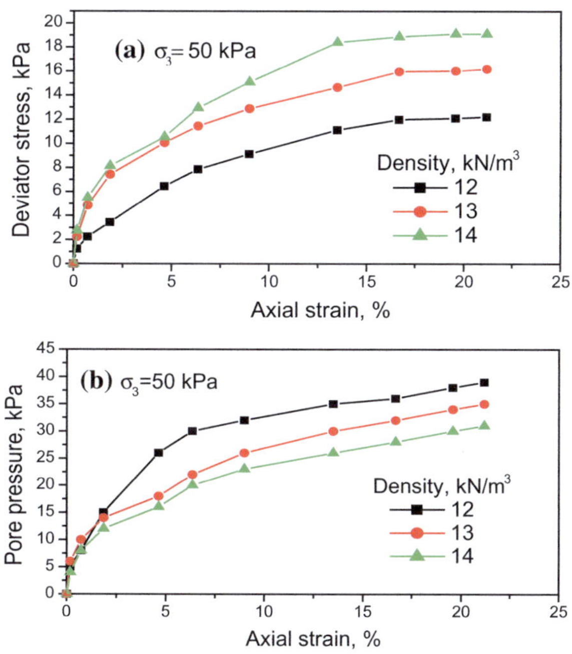

Fig. 1 Effect of density on **a** stress–strain relations and **b** pore pressure variation of Kuttanad clay sample (σ_3 = 50 kPa)

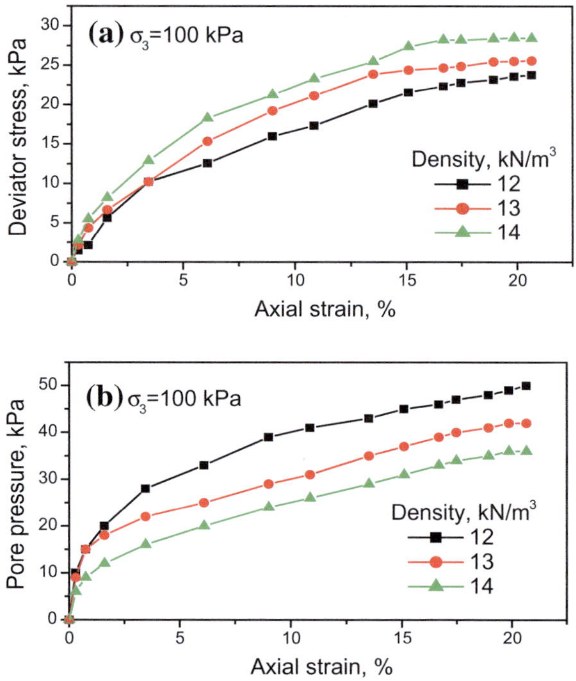

Fig. 2 Effect of density on **a** stress–strain relations and **b** pore pressure variation of Kuttanad clay sample (σ_3 = 100 kPa)

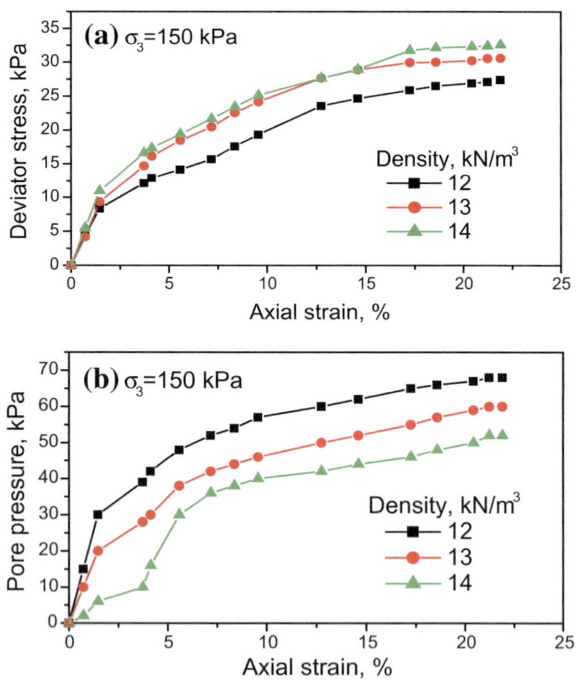

Fig. 3 Effect of density on **a** stress–strain relations and **b** pore pressure variation of Kuttanad clay sample (σ_3 = 150 kPa)

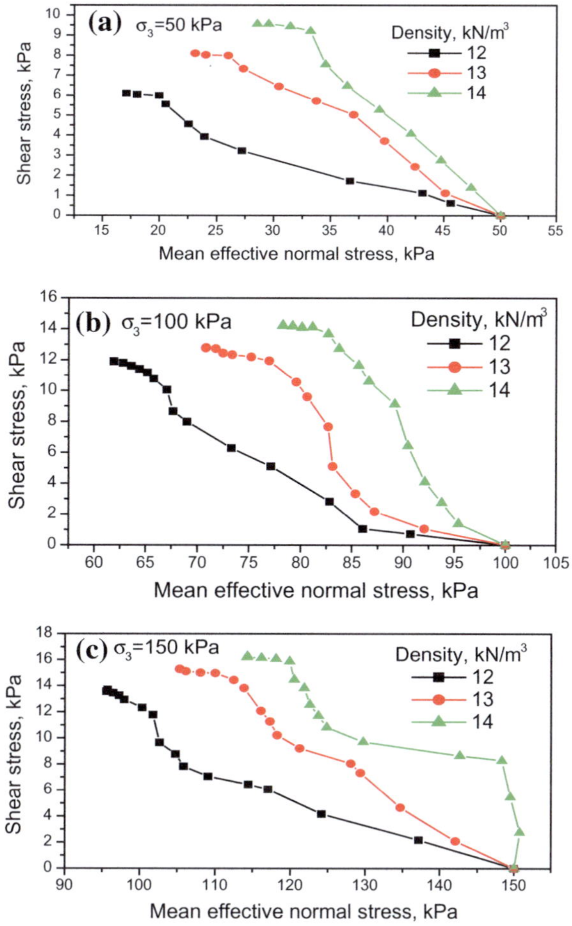

Fig. 4 Effect of density on stress paths of samples at confining pressures of **a** 50 kPa, **b** 100 kPa and **c** 150 kPa

ratios are increasing with decrease in the confining pressure irrespective of density. The deviator stress ratio is defined as ratio of deviator stress to applied confining stress. The increase of deviator stress ratio is due to decrease of pore water pressure with decrease the confining pressure (Fig. 5). The increase of deviator stress ratio is due to decrease of pore water pressure with decrease in the confining pressure of sample.

It indicates that the samples confined at low pressures are more resistant to static shear stress ratios (static liquefaction).

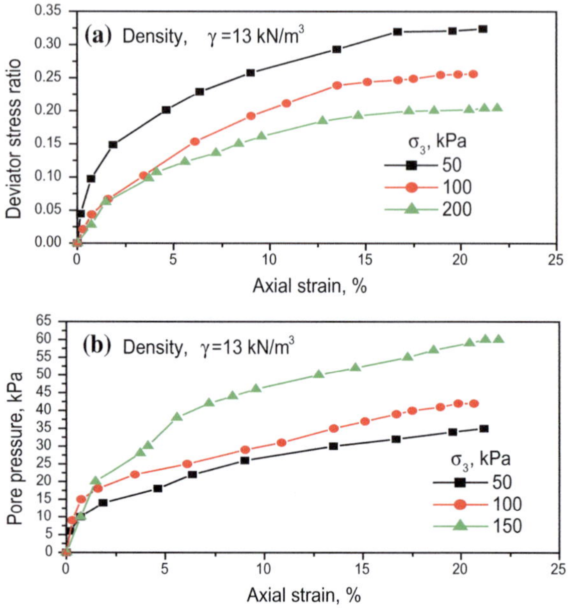

Fig. 5 Effect of confining pressure on **a** stress ratio–strain relations, and **b** pore pressure variation of Kuttanad clay sample at the density of 13 kN/m^3

6 Conclusions

Peak deviator stresses at failure strain level are increasing with increase in the density of samples, but the pore pressures decrease with increase in the densities irrespective of all confining pressures surrounded by the sample. Clay samples moulded at low densities possess high contraction behaviour and exhibit less resistance to shear loads when to compared at high densities irrespective of all confining pressures. The reduction in mean effective normal stress is increasing with decrease in the density of clay samples. Effect of confining pressure is to be expressed in terms of normalised stress ratios due to the test samples are subjected to different pressures. Peak deviator stress ratios are decreasing with increase in the confining pressures, but the pore pressures increase with increase in pressures irrespective of all density of samples. In contrary, the peak deviator stresses are increasing with increase in the confining pressures.

References

Caridad, V., de Zárate, J. M. O., Khayet, M., & Legidob, J. L. (2014). Thermal conductivity and density of clay pastes at various water contents for pelotherapy use. *Applied Clay Science, 93–94,* 23–27.

Elsayed, A. A., & Swan, C. W. (2010) Controlling preshear relative density in triaxial tests and its effects on undrained behaviour of sand. In *Geoflorida* (pp. 2581–2590).

Korichi, S., Keddam, M., & Bensmaili, A. (2013). Effects of compression on porous texture of clay powder: Application to uranium diffusion. *Chemical Engineering Research and Design CHERD-1409*.

Sachan, A., & Penumadu, D. (2007). Effect of microfabric on shear behaviour of Kaolin clay. *Journal of Geotechnical and Geoenvironmental Engineering, 33*(3), 306–318.

Saranya, K., & Muttharam, M. (2013). Consolidation stress effect on strength of lime stabilized soil. *International Journal of Engineering Research and Applications, 3*(5), 1515–1519.

A Methodology to Determine Water Vapour Diffusion Characteristics of Geomaterials

K. R. Arsha Lekshmi and D. N. Arnepalli

Abstract Water vapour diffusion is a predominant moisture transport mechanism in an unsaturated soil due to the presence of either thermal gradient or concentration gradient. In such soils, as water content decreases, the continuity of liquid films is lost, and as a result, water movement is mainly in the form of water vapour. The knowledge of water vapour migration characteristics of geomaterials is important in the performance evaluation of underground buried services, compacted/geosynthetic clay liners, geothermal energy utilization, thermally enhanced clean-up of contaminated sites and the containment facilities for disposal of high-level nuclear waste. In view of this, the present study describes a methodology to investigate the water vapour diffusion through geomaterials under isothermal conditions. Further, the study evaluates the effect of particle size and compaction state on water vapour diffusion characteristics of the soils.

Keywords Water vapour diffusion · Geomaterials · Containment facility Geothermal energy

1 Introduction

Recently, due to the innumerable applications in the field of geotechnical and geoenvironmental areas, the diffusion of water vapour through geomaterials has been a topic of paramount interest. An in-depth knowledge of water vapour migration characteristics of geomaterials is imperative for the performance evaluation of underground buried services, compacted/geosynthetic clay liners, geothermal energy utilization, thermally enhanced clean-up of contaminated sites and the containment facilities for disposal of high-level nuclear waste (Cleall et al. 2013).

K. R. Arsha Lekshmi (✉) · D. N. Arnepalli
Department of Civil Engineering, Indian Institute of Technology Madras, Chennai 600036, India
e-mail: arsha.lekshmi@gmail.com

D. N. Arnepalli
e-mail: arnepalli@iitm.ac.in

V. K. Stalin and M. Muttharam (eds.), *Geotechnical Characterisation and Geoenvironmental Engineering*, Lecture Notes in Civil Engineering 16,
https://doi.org/10.1007/978-981-13-0899-4_16

In unsaturated soils, moisture transfer is primarily in the form of vapour and liquid states, which in turn depends on the moisture content. For acutely dry unsaturated soil, the liquid film becomes discontinuous, and as a result, water in the pores gets transported mainly in the form of water vapour. The impetus behind the vapour transport is due to the gradient in vapour pressure. The factors such as temperature, moisture and salt concentration fluxes can impose a vapour pressure gradient. The water vapour movement occurs from an area of moist soil with a high vapour pressure to a dry soil region where the vapour pressure is low. Under isothermal conditions, moisture movement happens without the presence of a temperature gradient; i.e., it migrates as a result of moisture gradient.

The movement of water vapour in soils especially under isothermal conditions has been studied experimentally by many researchers (Hanks 1958; Rose 1963; Jackson 1964). Evgin and Svec (1988) conducted studies on compacted silty soil to determine the water vapour diffusion coefficients under isothermal conditions. The effect of porosity and temperature on the water vapour diffusion properties of dry silty clay of different aggregate sizes has been thoroughly investigated (Jabro 2009). It has been observed from this study that the rate of water vapour diffusion through dry silty clay is enhanced with the increase of temperature. On the other hand, the rate of diffusion through the dry soil is decreased when the porosity soil is reduced due to the increased sorption of water by the soil.

The moisture movement in soil under a thermal gradient has been extensively investigated by many researchers (Gurr et al. 1952; Cary 1964; Evgin and Svec 1988; Yong et al. 1997; Villar et al. 1996, Cleall et al. 2013). It has been concluded from the studies that in an unsaturated soil moisture migrates in the form of vapour by diffusion, along a thermal gradient, from a region of higher temperature to a region of lower temperature. The vapour gets condensed when it reaches a relatively colder region, thereby causing a local increase in water content. This impending increase of water content in the lower-temperature region results in the back flow of water in its liquid state towards the higher-temperature region.

In view of the above, this article presents an experimental methodology to determine the water vapour diffusion coefficient under isothermal conditions. Various samples of different grain size were considered. The diffusion coefficients are calculated using the experimentally obtained vapour flux values. The effect of grain size and compaction state on the diffusion characteristics is also evaluated.

2 Theory

Quantitative description of steady vapour flux, q_v (kg/m^2 s), can be represented by Fick's first law (Philip and de Vries 1957; Cassel et al. 1969).

For isothermal conditions, q_v is given by:

$$q_v = -D_V\left(\frac{\partial \rho_V}{\partial x}\right) \quad (1)$$

where q_v is the vapour flux (kg/m^2 s), D_v is the diffusion coefficient (m^2/s), ρ_v is the water vapour density (kg/m^3), and x is the distance (m).

Presuming that the equation of state for an ideal gas is valid considering the state of vapour, the concentration of vapour is given as a function of pressure and temperature. After modification of Fick's law, the diffusion coefficient can be written as

$$D_V = \left(\frac{m}{At}\right) \times \left(\frac{dRT}{M_w S(R_1 - R_2)}\right) \quad (2)$$

where m is the amount of water vapour diffused through the sample (kg), A is the sample area (m^2), d is the thickness of the sample (m), t is the period of time corresponding to the transport of water vapour (s), R is the universal gas constant (8.3143 J/mol/K), T is the absolute temperature (K), M_w is the molar mass of water (0.018 kg/mol), $R_1 - R_2$ is the difference in relative humidity under and above the specimen surface, and S is the saturation vapour pressure corresponding to the absolute temperature (kPa).

3 Material Selection and Methodology

The study is carried out by using various sizes of sand grains obtained by sieving medium-grade sand. The sand samples are air dried before performing the experiments. The grain sizes of sand considered for the study are 2–1 mm, 1 mm–600 μm and 600–425 μm.

Figure 1 shows the grain size distribution of the medium-grade sand. The grain size distribution of the medium-grade sand is conducted by dry sieve analysis according to the guidelines given by ASTM D422 (1994).

The specific gravity of the sand sample is found to be 2.65. For this purpose, a helium gas pycnometer (Ultra Pycnometer, Quantachrome Instruments, USA) has been employed.

Figure 2 depicts the schematic view of water vapour diffusion apparatus. The sand sample is compacted to a desired density in a PVC pipe of diameter 36 mm and then attached to an airtight container. A non-sorbent cloth mesh screen is mounted at the bottom end of the PVC pipe to support the sand. The thickness of the soil specimen is 300 mm. The water vapour density inside the container is controlled by means of calcium chloride desiccant placed inside the container. The desiccant maintains the relative humidity value very close to zero inside the chamber. The relative humidity inside the humidity chamber is measured with

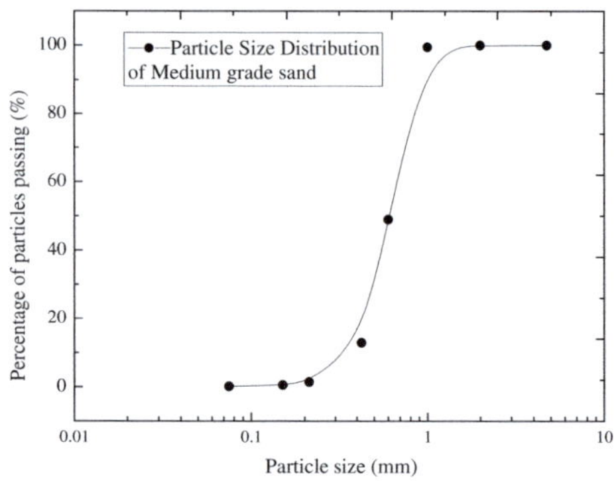

Fig. 1 Grain size distribution of medium-grade sand

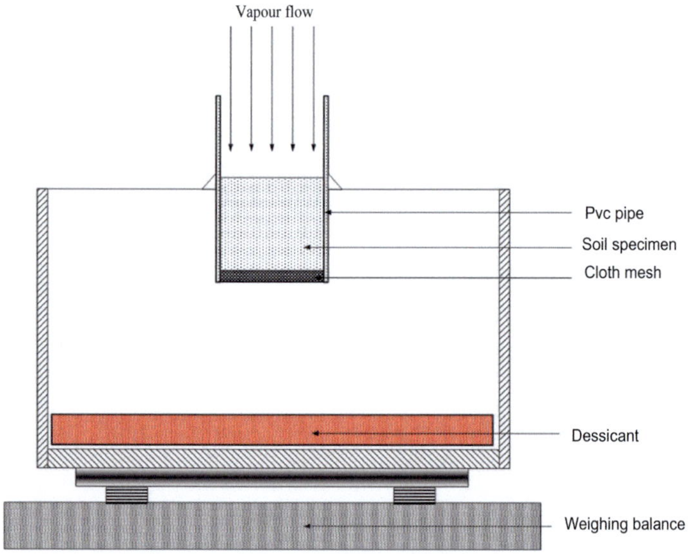

Fig. 2 Experimental set-up to study water vapour diffusion

the help of relative humidity sensors (Arduino DHT 22). The air and moisture tightness is ensured by sealing the container on all sides using silicon-based polymer. The entire set-up is then placed in a temperature- and humidity-controlled chamber and maintained the temperature of 25 °C and a relative humidity of 70%.

Under isothermal conditions, the vapour movement occurs due to the presence of concentration gradient. The test is conducted by varying the relative humidity across the specimen keeping the temperature constant. The difference in vapour pressure between the top and bottom of the specimen causes the moisture flow through it from humidity chamber to the container resulting in the increase in weight of desiccant calcium chloride. The rate of moisture movement can be measured by weighing the change in weight of the desiccant periodically.

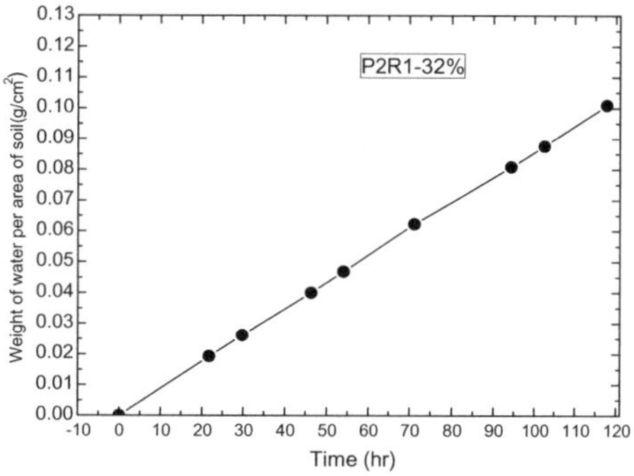

Fig. 3 Moisture flux of soil specimen P2R1

The weights are recorded with the help of a high-precision weighing balance having the precision of 1 μg. The test is terminated when a steady-state flow of vapour is attained. Further, experiments are carried for different particle gradations over a range of dry unit weights. Figure 3 shows the gain in water vapour by the desiccant per unit area of sample per unit time for a soil specimen of 32% porosity and passing through 2 mm and retaining on 1 mm (designated as P2 R1). Similar plots can be made for all samples. The slope of this graph yields the vapour flux. The value of vapour flux is substituted in the Eq. (2) to determine the water vapour diffusion coefficient. The saturation vapour pressure at 25 °C is taken as 3.169 kPa (Lide 2006).

4 Results and Discussions

Figure 4 indicated the variation of diffusion coefficients with porosity for different gradations of soil. In the plot, P2R1 and P1R600 represent the soil passing through 2 mm sieve and retaining on 1 mm and passing through 1 mm sieve and retaining on 600 μm, respectively. The porosity is varied over a wide range from 32 to 40%. It is evident that as the porosity increases the diffusion coefficient increases irrespective of the grain size. This is because of the fact that as the structure becomes more pervious, the space for the seamless movement of vapour increases. Furthermore, as the particle size increases, the pore diameter eventually increases causing a higher diffusion coefficient with gradation.

The main limitation of this experiment is the longer test duration which can result in poor reliability and reproducibility. The experiment needs to continue several weeks, and this can lead to problem such as inconsistent environmental conditions and material degradation. A slight change in relative humidity can substantially affect the vapour movement. An error in measuring the change in weight of the desiccant can also affect the accuracy of the method. This methodology is not desirable for soils with expansive clay minerals in them such as bentonite. For such soils, a separate methodology needs to be developed.

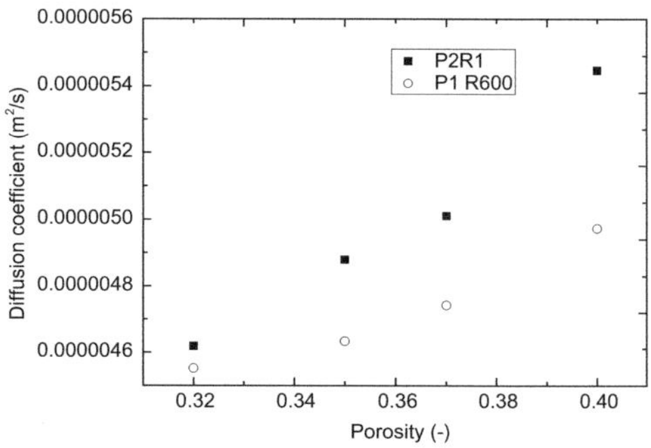

Fig. 4 Variation of water vapour diffusion coefficient with porosity

5 Conclusions

An experimental set-up to carry out the water vapour diffusion studies under isothermal conditions has been developed. The study was carried out for different aggregates of sizes 2–1 mm, 1 mm–600 μm and 600–425 μm. The diffusion coefficient of each soil sample has been calculated based on the experimental results, and the variation of diffusion coefficients with respect to porosity has been established. From the experimental studies, it is found that the water vapour diffusion coefficient is dependent on the porosity and the soil gradation.

References

ASTM D422 (1994). *Standard test method for particle size analysis of soils. Annual Book of ASTM Standards.* West Conshohocken, PA, USA: ASTM International.

Cary, J. W. (1964). An evaporation experiment and its irreversible thermodynamics. *International Journal of Heat and Mass Transfer, 7,* 531–538.

Cassel, D. K., Nielson, D. R., & Biggar, W. (1969). Soil water movement in response to temperature gradients. *Soil Science Society of America Journal, 33*(4), 493–500.

Cleall, P. J., Singh, R. M., & Thomas, H. R. (2013). Vapour transfer in unsaturated compacted bentonite. *Geotechnique, 63*(11), 957–964.

Evgin, E., & Svec, O. J. (1988). Heat and moisture transfer characteristics of compacted MacKenzie silt. *Geotechnical Testing Journal, 11*(2), 92–99.

Gurr, C. G., Marshall, T. J., & Hutton, J. T. (1952). Movement of water in soil due to a temperature gradient. *Journal of Soil Science, 74,* 335–345.

Hanks, R. J. (1958). Water vapour transfer in dry soil. *Soil Science Society of America Journal, 22,* 392–394.

Jabro, J. D. (2009). Water vapor diffusion through soil as affected by temperature and aggregate size. *Transport in Porous Media, 77*(3), 417–428.

Jackson, R. D. (1964). 'Water vapor diffusion in relatively dry soil: I. Theoretical considerations and sorption experiments. *Soil Science Society of America Journal, 28,* 172–176.

Lide, D. R. (Ed.). (2006). *Handbook of chemistry and physics* (87th ed.). New York: CRC Press.

Philip, J. R., & Vries, D. A. (1957). Moisture movement in porous materials under temperature gradient. *American Geophysics Union, 38*(2), 222–231.

Rose, D. A. (1963). Water movement in porous materials: Part 2—The separation of the components of water movement. *British Journal of Applied Physics, 14,* 491–496.

Villar, M. V., Cuevas, J., & Martin, P. L. (1996). Effects of heat, water flow interaction on compacted bentonite: Preliminary results. *Engineering Geology, 41,* 257–267.

Yong, R. N., Mohamed, A. M. O., Shooshpasha, I., & Onofrei, C. (1997). Hydro-thermal performance of unsaturated bentonite-sand buffer material. *Engineering Geology, 47*(4), 351–365.

Atterberg Limits of Sand-Bentonite Mixes and the Influence of Sand Composition

V. Srikanth and Anil Kumar Mishra

Abstract Sand-bentonite mixtures are assuming greater importance as geotechnical barrier materials in waste disposal schemes around the world, and like with any other geotechnical soil structure, a comprehensive understanding on the engineering behavior of various components used in the making of a geotechnical barrier becomes a necessity. A review of the literature indicated a host of variables influencing the engineering behavior of a barrier to various degrees, including and not limited to mineralogy of the soil, size, and shape of particles, their interactions among themselves, interactions with water, interactions with any other pore fluid that may exist in the soil, pH, temperature, cation exchange capacity, and quantity of cations in the soil pore fluid. Upon scrutiny, it has been observed that most of the studies conducted were primarily focusing on the role of bentonite in the sand-bentonite mixture, and studies highlighting the contributions of sand particle size, sand gradation, etc., were very few even though sand forms the major component in most of the sand-bentonite mixtures being employed around the world. This study sheds some light on the plasticity characteristics of sand-bentonite mixtures with varying bentonite proportions and sand compositions. In the current study, locally available sand has been washed thoroughly and sieved for different particle sizes (fine sand and medium sand) and two commercially available bentonites were procured. Fine sand-bentonite (FS-B), medium sand-bentonite (MS-B), and fine sand-medium sand-bentonite (FS-MS-B) mixtures were made with bentonite proportion varying from 10 to 50% by dry weight in the mix and were tested for Atterberg limits. Liquid limit results indicated that, for a given bentonite content, FS-B, MS-B, and FS-MS-B mixes exhibited different liquid limits and sand composition has a little influence on the liquid limits. Shrinkage limit results indicated that sand composition has a notable influence on the shrinkage characteristics of the sand-bentonite mixtures.

V. Srikanth (✉) · A. K. Mishra
Department of Civil Engineering, Indian Institute of Technology Guwahati, Guwahati 781039, India
e-mail: v.srikanth@iitg.ernet.in

A. K. Mishra
e-mail: anilmishra@iitg.ac.in

V. K. Stalin and M. Muttharam (eds.), *Geotechnical Characterisation and Geoenvironmental Engineering*, Lecture Notes in Civil Engineering 16,
https://doi.org/10.1007/978-981-13-0899-4_17

Keywords Atterberg limits · Sand-bentonite mixtures · Sand gradation
Bentonite content

1 Introduction

Consistency limits proposed by Atterberg were initially introduced into soil mechanics as means to classify soils (Wroth and Wood 1978). Atterberg limits are perhaps the oldest and most commonly used tests in the field of soil engineering (Sridharan and Venkatappa Rao 1975). As time progressed, attempts were made to correlate these Atterberg limits with other engineering characteristics of soils so as to provide soil engineers on site with a better assessment of the soils being encountered (Wroth and Wood 1978; Sivapullaiah et al. 2000; Skempton and Jones 1944; Sridharan and Nagaraj 2004). With sand-bentonite mixtures gaining importance as geotechnical barrier materials around the world, any information leading to a better understanding of these complex composite materials is a welcoming step.

A liquid limit test is essentially a measure of viscous resistance or shear strength of a soil that is so soft, and it approaches the liquid state (Sowers et al. 1959). Liquid limit of a sand-bentonite mixture generally depends upon the clay concentration and the type of clay mineral present in the mixture. Apart from the above-mentioned criterion, Sivapullaiah and Sridharan (1985) observed that the interactions between clay and sand particles also influence the plasticity characteristics, though no quantification of the same has been provided. Skempton and Jones (1944) indicated that liquid limit is particularly sensitive to particle-size distribution of a soil. This study is an attempt to understand the extent to which the particle-size distribution of sand can really influence the Atterberg limits of sand-bentonite mixtures.

2 Materials and Methods

To study the influence of sand content, sand particle size, and sand proportioning on the plasticity characteristics of sand-bentonite mixtures, a locally available sand was used in this study. Sand was sieved into two groups, i.e., fine sand (FS) (0.075–0.425 mm) and medium sand (MS) (0.425–2 mm), using a sieve shaker. Further, to study the influence of the bentonite quality on sand-bentonite mixtures, two commercially available bentonites with different swelling properties, namely bentonite-1 (B1) and bentonite-2 (B2), were selected for this study. Bentonite-1 is moderately swelling, while bentonite-2 is a high swelling bentonite. Geotechnical characteristics of bentonites used in the study are presented in Table 1. Liquid limit tests were performed on B1 and B2 using Casagrande liquid limit test apparatus.

Table 1 Characteristics of bentonites used in the study

Properties	Bentonite-1	Bentonite-2
Particle-size distribution		
Clay (%)	52.2	68.0
Silt (%)	43.4	28.3
Sand (%)	4.4	3.7
Atterberg limits		
Liquid limit (%)	346.0	609.0
Plastic limit (%)	33.0	41.5
Shrinkage limit (%)	23.0	10.4

Sand-bentonite mixtures used in the study were prepared by mixing sand and bentonite, with bentonite proportion varying from 10 to 50% by dry weight of the mixture. The study has been conducted in two phases; in phase one, fine sand-bentonite and medium sand-bentonite mixtures were tested for Atterberg limits to check if there is any difference in the plasticity behavior exhibited when different types of sand are mixed with bentonites. After confirming the differences, in phase two, fine sand-medium sand-bentonite mixtures were tested for Atterberg limits. Sand used in phase two of the study was prepared by mixing fine sand and medium sand in proportions varying from 30:70 to 70:30. Since the sand-bentonite mixtures contained a higher amount of non-plastic soil, liquid limit of the mixtures was determined using cone penetration method as suggested by Sivapullaiah and Sridharan (1985).

3 Results and Discussion

Influence of bentonite content, sand type, clay mineralogy, etc., on liquid limit of sand-bentonite mixtures is presented in Fig. 1. Liquid limit was found to be increasing with increasing bentonite content. It can be seen that, for any given bentonite content, fine sand-bentonite mixtures were exhibiting higher liquid limit values compared to medium sand-bentonite mixtures (Srikanth and Mishra 2016). Between fine sand-bentonite and medium sand-bentonite mixtures, differences in liquid limits exhibited are higher in case of mixtures with bentonite content greater than 30%. Bentonite-2 is higher swelling bentonite, and subsequently, fine sand-bentonite-2 and medium sand-bentonite-2 mixtures exhibited higher liquid limits.

Role of sand proportioning on liquid limit of sand-bentonite mixtures can be seen in Figs. 2 and 3. For any given bentonite content, variation in liquid limits is small (5% at the most) and for most cases observed values were lying in between those exhibited by fine sand-bentonite mixtures and medium sand-bentonite mixtures.

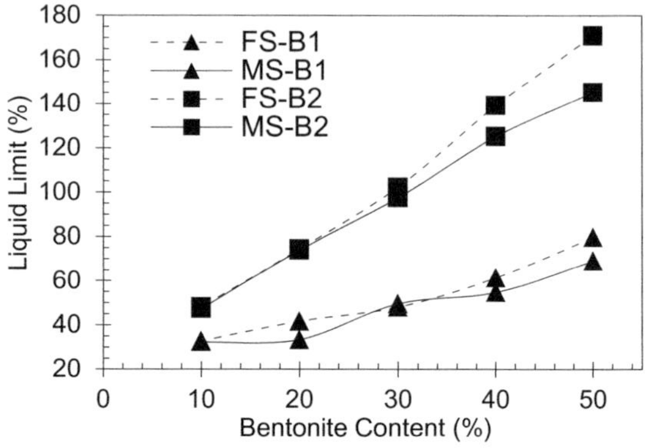

Fig. 1 Variation in liquid limit of fine sand-bentonite and medium sand-bentonite mixtures with bentonite content

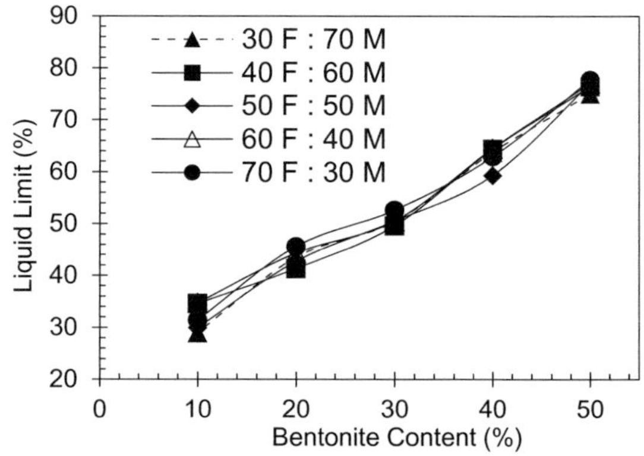

Fig. 2 Variation in liquid limit of fine sand-medium sand-bentonite-1 mixtures with bentonite content

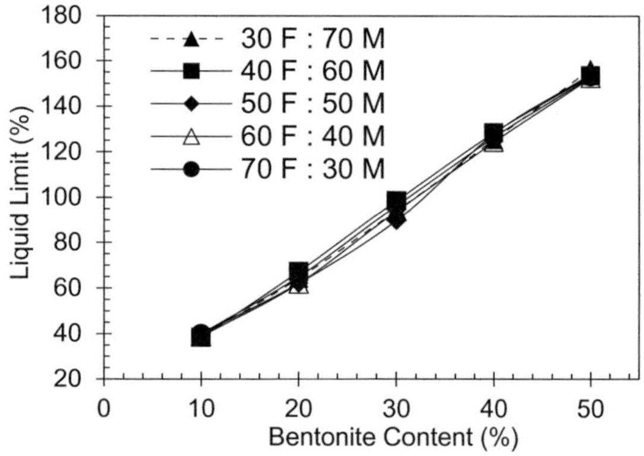

Fig. 3 Variation in liquid limit of fine sand-medium sand-bentonite-2 mixtures with bentonite content

Shrinkage limit results exhibited by fine sand-bentonite and medium sand-bentonite mixtures are presented in Fig. 4. Shrinkage limit was found to be decreasing with increasing bentonite content, and similar results were reported by

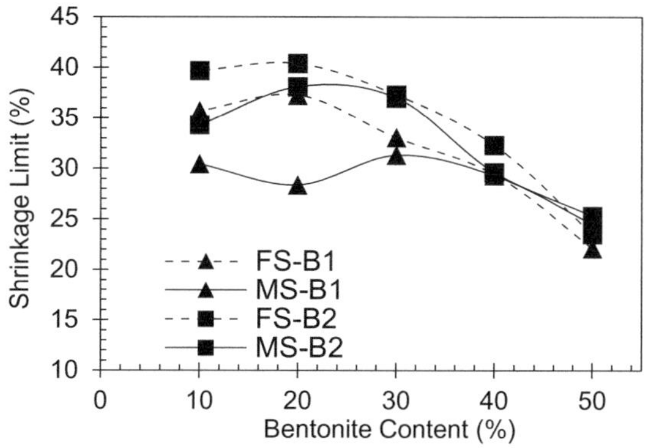

Fig. 4 Variation in shrinkage limit of fine sand-bentonite and medium sand-bentonite mixtures with bentonite content

Jong and Warkentin (1965). For the same bentonite content in the mixture, fine sand-bentonite mixtures exhibited higher shrinkage limits. Sridharan and Prakash (2000) proposed that shrinkage limit is primarily dependent on the relative particle-size distribution of soils, unlike other consistency limits.

Shrinkage limit results exhibited by fine sand-medium sand-bentonite-1 and fine sand-medium sand-bentonite-2 mixtures are presented in Figs. 5 and 6. Because of sand proportioning, resulting in a different particle-size distribution curve for every other fine sand-medium sand-bentonite mixture, different shrinkage limit values were observed for any given bentonite content with the variation being in the range of 5–15%. Upon comparing Figs. 5 and 6, no particular trend in shrinkage limit could be identified and further analysis is being carried out. Comparing fine sand-bentonite, medium sand-bentonite, and fine sand-medium sand-bentonite mixtures revealed that shrinkage limit is reducing because of sand proportioning.

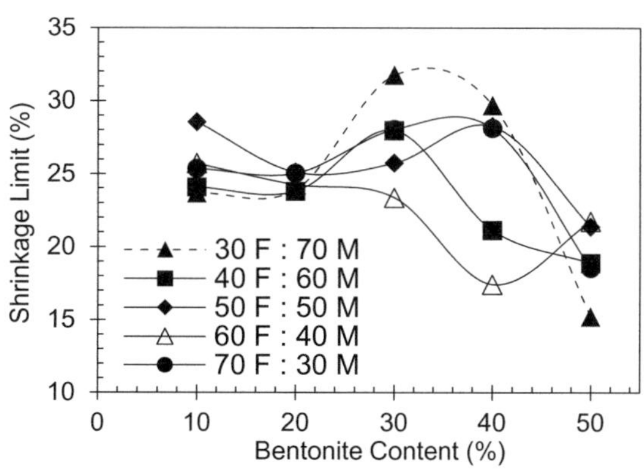

Fig. 5 Variation in shrinkage limit of fine sand-medium sand-bentonite-1 mixtures with bentonite content

Fig. 6 Variation in shrinkage limit of fine sand-medium sand-bentonite-2 mixtures with bentonite content

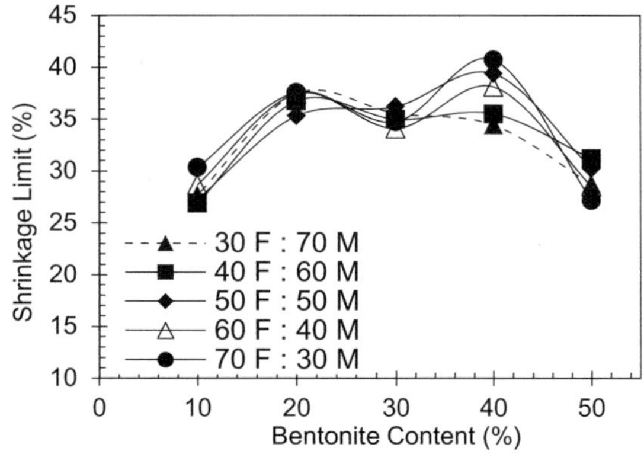

4 Conclusions

Geotechnical barriers form an integral part of engineered waste disposal schemes around the world. An improved understanding on the engineering behavior of various component members used in the making of a barrier may lead to a better and optimized design of the barrier system. Given the huge amounts of natural/ enhanced soil being employed in making the geotechnical barriers, any reduction in the raw material required would result in optimizing the necessary infrastructure for mining, processing, and transportation of raw material, thereby reducing project costs involved. Reduced project costs can help promote the concept of engineered waste disposal.

The investigation of Atterberg limits of fine sand-bentonite, medium sand-bentonite, and fine sand-medium sand-bentonite mixtures has resulted in the following understanding.

1. Liquid limit of a sand-bentonite mixture is influenced by the particle-size gradation of sand, though the influence is small particularly in the lower bentonite contents.
2. Quality of bentonite has a major influence on liquid limit along with bentonite content.
3. Shrinkage limit was found to be decreasing with increasing bentonite content for fine sand-bentonite and medium sand-bentonite mixtures.
4. Shrinkage limit was found to be decreasing with sand proportioning for a given bentonite content, though no particular trend could be identified, indicating the necessity for further studies in this regard.

References

Jong, E. D., & Warkentin, B. P. (1965). Shrinkage of soil samples with varying clay concentration. *Canadian Geotechnical Journal, 2*(1), 16–22.

Sivapullaiah, P. V., & Sridharan, A. (1985). Liquid limit of soil mixtures. *Geotechnical Testing Journal, 8*(3), 111–116.

Sivapullaiah, P. V., Sridharan, A., & Stalin, V. K. (2000). Hydraulic conductivity of bentonite-sand mixtures. *Canadian Geotechnical Journal, 37*(2), 406–413.

Skempton, A. W., & Jones, O. T. (1944). Notes on the compressibility of clays. *Quarterly Journal of the Geological Society, 100*(1–4), 119–135.

Sowers, G. F., Vesic, A., & Grandolfi, M. (1959). Penetration tests for liquid limit. *American Society for Testing and Materials, Special Technical Publication, 254,* 216–224.

Sridharan, A., & Nagaraj, H. B. (2004). Coefficient of consolidation and its correlation with index properties of remolded soils. *Geotechnical Testing Journal, 27*(5), 1–6.

Sridharan, A., & Prakash, K. (2000). Shrinkage limit of soil mixtures. *Geotechnical Testing Journal, 23*(1), 3–8.

Sridharan, A., & Venkatappa Rao, G. (1975). Mechanisms controlling the liquid limit of clays. In *Istanbul conference on soil mechanics and foundation engineering,* pp. 65–74.

Srikanth, V., & Mishra, A. K. (2016). A laboratory study on the geotechnical characteristics of sand-bentonite mixtures and the role of particle size of sand. *International Journal of Geosynthetics and Ground Engineering, 2*(1), 1–10.

Wroth, C. P., & Wood, D. M. (1978). The correlation of index properties with some basic engineering properties of soils. *Canadian Geotechnical Journal, 15*(2), 137–145.

Effect of Sample Thickness on Laboratory Determination of Gas Permeability of Buffer Material

S. S. Surya and D. N. Arnepalli

Abstract Permeability of gases through geomaterials are critical in areas like landfill cover design, buffer material characterization in deep geological repositories for radioactive waste containment, geosequestration of greenhouse gases, oil and gas recovery. In general, there are two types of laboratory methods for determination of gas permeability of geomaterials, such as steady-state and transient or pressure decay methods. In all these methods, the sample thickness that needs to be considered is important for accurate estimation of gas permeability characteristics. In view of this, the present study discusses the effect of sample thickness on the laboratory determination of gas permeation through geomaterials. A permeability apparatus has been developed for evaluating the gas permeability characteristics of geomaterials over a range of compaction state, based on the concept of pressure decay. Further, the experiments are conducted at different sample thicknesses along the standard Proctor compaction curve and the effect of sample thickness on gas permeability has been evaluated.

Keywords Gas permeability · Pressure decay test · Buffer materials
Deep geological repository

1 Introduction

An understanding of gas transport is important for evaluating movement of gaseous and semi-volatile radionuclides, such as ^{3}H, ^{14}C, and Rd from radioactive waste disposal facilities. It is also important to comprehend the mechanism of gas transport in unsaturated media for evaluation of soil aeration or movement of O_2

S. S. Surya (✉) · D. N. Arnepalli
Department of Civil Engineering, Indian Institute of Technology Madras,
Chennai 600036, India
e-mail: surlibra@gmail.com

D. N. Arnepalli
e-mail: arnepalli@iitm.ac.in

V. K. Stalin and M. Muttharam (eds.), *Geotechnical Characterisation and Geoenvironmental Engineering*, Lecture Notes in Civil Engineering 16,
https://doi.org/10.1007/978-981-13-0899-4_18

from the atmosphere to the soil and estimation of transport of volatile and semi-volatile organic compounds from contaminated sites through the unsaturated zone to the groundwater. Grout injection which is one of the most widely accepted soil stabilization technique also requires the measurement of permeability decrease after injection operation is proposed in order to quantify its improvement mechanism. The use of soil venting or soil vapor extraction as a technique for remediating contaminated sites has resulted in increased interest in gas transport through unsaturated zone. In many areas, the migration of gases from landfills such as methane formed by decomposition of biodegradable organic waste is important (Moore and Roulet 1991). Soil gas composition has also been used as a tool for mineral and petroleum exploration and for mapping organic contaminant plumes. The adverse health effects of radon and its decay products have led to evaluation of transport of radon in soils and into buildings. In order to assess all these issues, a thorough understanding of gas transport mechanism is required. Among the gas transport mechanisms, gas permeability is an important parameter to understand the transport characteristics. It depends on the geometry, connectivity of pores, and water content. Permeability variations indicate mechanical, hydraulic, and structural changes of porous material.

In general, either steady-state or transient method can be employed for the determining gas permeability characteristics of geomaterials. In steady-state method, the flow rate of gas from an external reservoir is required to be maintained a desired constant pressure during the experimentation. The steady-state method is time-taking, and slippage effect has a major role. However, in transient pulse technique, the test region is initially filled with the desired pressure and subsequent pressure decay is measured with time.

Transient pulse techniques (pressure decay tests) are usually used to determine permeability of low-permeable geomaterials, i.e., when the flow rates become so small that flow into the formation could be more accurately determined from the measured pressure drop in the test interval. Transient methods are less time-consuming, and it is not necessary to measure the flux because its value can be calculated from changes in pressure with time (Calogovic 1995; Billiote et al. 2008; Gardner et al. 2008; Fedor et al. 2008). Though many researchers have investigated gas permeability characteristics of geomaterials, the influence of sample thickness on gas permeability determination has not been explored. In view of this fact, the present study intends to investigate the influence of sample thickness on the determination of gas permeability characteristics of geomaterials, particularly buffer materials of deep geological repositories. Further, the effect of compaction state on permeation characteristics of buffer materials has been assessed.

1.1 Principle of Pressure Decay Method

To determine the permeability of gases through geomaterials, pressure decay method proposed by Billiotte et al. (2008) is employed in the present study. In this method, the sample is compacted in the sample holder and sandwiched between two reservoirs which are maintained at constant and uniform pressure at the beginning of the test. A sudden decline in pressure is then imposed in the downstream reservoir, and corresponding pressure changes in both the reservoirs are recorded. Figure 1 illustrates the schematic representation of principle of pressure decay method.

The pressure decay in terms of gas permeability through the sample during gas permeability tests, can be expressed by the following equation,

$$\frac{\partial\left(K \times \frac{\rho}{\mu} \times \frac{\partial P}{\partial x}\right)}{\partial x} = \frac{\beta n \rho \partial P}{\partial t} \quad (1)$$

By developing the equation of transient flow according to pressure the above relationship becomes,

$$\frac{\partial^2 P}{\partial x^2} + \frac{1}{K} \times \frac{\partial K}{\partial P} \times (\partial P/\partial x)^2 = \frac{\beta n \mu}{K} \times \frac{\partial P}{\partial t} \quad (2)$$

where K is effective gas permeability in m^2, ρ is the gas density in kN/m^3, μ is gas dynamic viscosity of gas in Pas, P is the pressure in kPa, β is the compressibility factor of the gas, its variation with temperature and pressure is obtained, and n is effective porosity of the sample.

The equation can be modified by using an exponential variation between the normalized pressure variations with time. The exponential coefficient of the normalized pressure variation with time gives the coefficient of permeability using the simplified solution of Brace and Martin (1968).

$$\frac{\Delta p(t)}{\Delta p(o)} = e^{-\alpha t} \quad (3)$$

$$\alpha = \frac{KA}{\beta \mu L\left(\frac{1}{V_{up}} + \frac{1}{V_{down}}\right)} \quad (4)$$

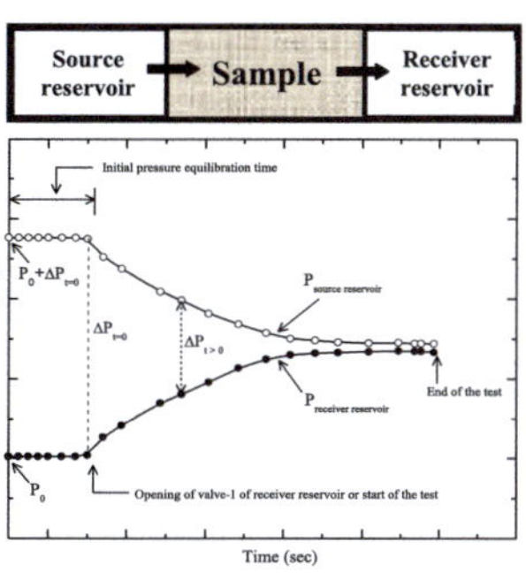

Fig. 1 Principle of pressure decay method

where A is the cross-sectional area of the sample in m^2; L is the length of the sample in m; V_{up} and V_{down} are the volumes of two reservoirs in m^3.

2 Experimental Investigations

2.1 *Material Selection*

In general, clay is well known as the buffer or barrier materials for waste containment because of its low permeability and high sorptive nature. But the use of clay alone is associated with desiccation cracking when subjected to temperature. Therefore, a mixture of sand and bentonite clay is usually used as buffer materials in containment of hazardous wastes. The review of literature revealed that a mixture of 30% bentonite and 70% sand performs efficiently in terms of strength and shrinkage-swelling properties (Dixon et al. 1985; Ada 2007). In view of this, gas permeability experiments are conducted for 30% sand–bentonite mixtures (SB30) along the standard Proctor compaction curve. Fine sand and sodium bentonite are selected for the study. Air permeability test is conducted using nitrogen gas (N_2) as a model gas because it is a relatively inert gas and has very low water solubility (Bouazza and Vangpaisal 2003).

2.2 *Physico-Chemico-Geotechnical Characteristics of Buffer Materials*

Particle size distribution characteristics of the selected materials are obtained by conducting both sieve and hydrometer analyses as per the guide lines presented in ASTM D422 (2007). The results from both sieve and hydrometer analyses are combined to get the grain size distribution characteristics of the chosen materials, and the results are presented in Fig. 2.

The specific gravity (G) of the selected geomaterials are determined with the help of a helium gas pycnometer (Quantachrome, USA), by the guide lines presented in ASTM D 5550 (2006). From the measured solid volume and weight of the

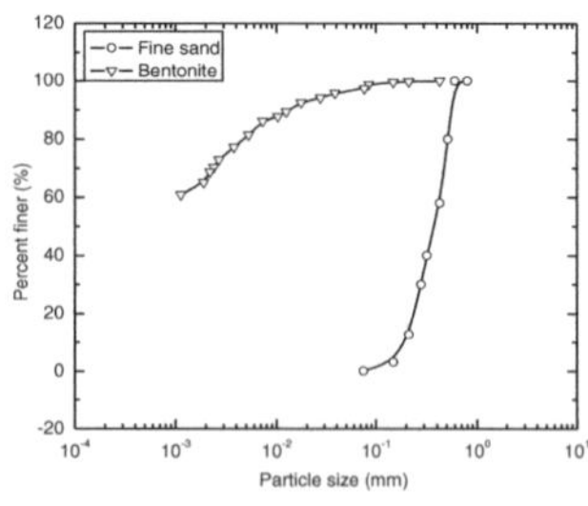

Fig. 2 Particle size distribution characteristics of fine sand and sodium bentonite

sample, the solid density (i.e., specific gravity) is determined using mass–volume relationships and it is obtained as 2.67 for SB30.

Compaction characteristics of the selected geomaterials are obtained by conducting standard Proctor compaction test as per ASTM D698 (2012). Sand–bentonite mixture is prepared at different water contents and the compaction is done. The sand-bentonite mix SB30 is characterized with a maximum dry unit weight of 17.85 kN/m^3 at optimum moisture content of 13.5%.

2.3 Gas Permeability Test Setup

The permeability cell is made up of stainless steel consisting of two cylindrical chambers (hereafter called as source and receiver reservoirs) and a sample holder with an inner diameter of 93 mm. The source and receiver chambers have a length of 190 and 54 mm, respectively, and the length of sample holder was changed depending on the sample thickness. The reservoirs and sample holder are connected together by two intermediate plates and end plates with grooves for proper placing of the chambers. Four plates are connected using four tie rods which are clamped from both sides using washer and wing nut arrangement. O-rings are placed in the grooves of the metallic plates to ensure no leak condition. Both the reservoirs are provided with two inlets: one inlet for inflow and outflow of gas and the other inlet for short circuiting the two chambers to establish equilibrium of pressure initially on both sides of the sample. Both the chambers are connected with pressure sensors with 0–6 bar pressure range, and the pressure in the respective chambers is monitored with the help of readout units and data logging facility. Pressure sensors are calibrated prior to their usage. Figure 3 depicts the photographic view of the experimental setup with pressure sensors and readout units.

Soil sample is compacted in the sample holder using a hand held tamper and is provided with highly pervious porous metal discs on either side of the specimen to avoid soil particles falling into the source and receiving chambers. Gas pressure is directly applied by connecting the valves from the gas supply panel to the inlet and outlet openings of the setup which is controlled by a regulator to maintain constant pressure. The gas pressure supply is controlled by a series of regulators which are used to step down the pressure and allow it to maintain at a specified value. After

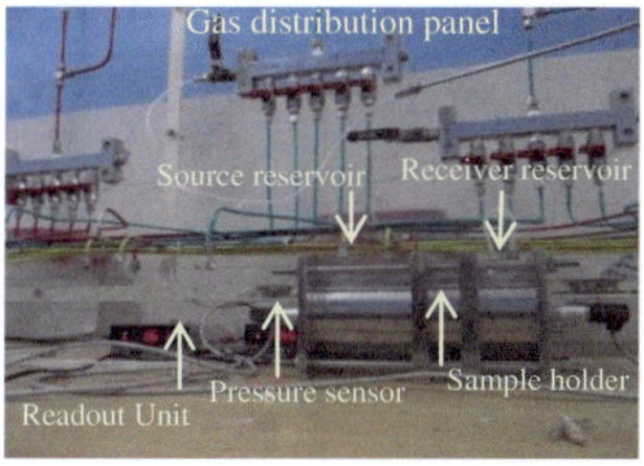

Fig. 3 Gas permeability test setup and gas distribution panel

achieving required pressure, inlet and outlet valves are closed and the pressure in both the reservoirs is monitored. Once the pressure on both the reservoirs becomes constant, it is kept undisturbed for at least three to four hours so that all the pores of the sample filled with gas are at same pressure as the reservoirs. After achieving equilibrium, receiver valve is opened and gas pressure is allowed to drop to the desired lower pressure so that a pressure gradient is applied across the sample. The annular space (5 mm wide and 2 mm deep) between the specimen and sample holder along the circumference is sealed with bentonite paste, to avoid the potential leakage of gas through it. This method helps in preventing gas transport through the interface between inner surface of the sample holder and compacted specimen. The pressure variations in both the reservoirs are monitored with respect to time, and the coefficient of permeability is calculated from the normalized pressure variation with respect to time using the solution proposed by Brace and Martin (1968).

The above procedure is repeated for the samples compacted at different compaction state along the standard Proctor compaction curve. Further, the experiments are conducted for different sample thickness of 30 and 50 mm. The normalized pressure variation with respect to time obtained for sample thickness of 30 and 50 mm is presented in Fig. 4.

It can be noted from Fig. 4 that, even though the trend of variation curve is same, 50-mm-thick sample gives similar exponential coefficients for both the applied pressure gradients. For lesser sample thickness, the exponential value differs and this may be attributed to the slippage that might have occurred in the sample. In view of this, it is better to go for a sample thickness of 50 mm for efficient determination of gas permeability coefficients of buffer materials. Further, the variation of permeability coefficients with volumetric moisture content obtained for sample thickness of 50 and 30 mm is depicted in Fig. 5.

It can be seen from Fig. 5 that the coefficient of permeability decreases with increase in volumetric moisture content. This is due to the fact that as the volumetric moisture content increases the soils particles are packed in a better way with pores filled with moisture. Therefore, the gas finds it difficult to permeate through the sample, and thus, the permeability coefficient decreases with increase in volumetric moisture content.

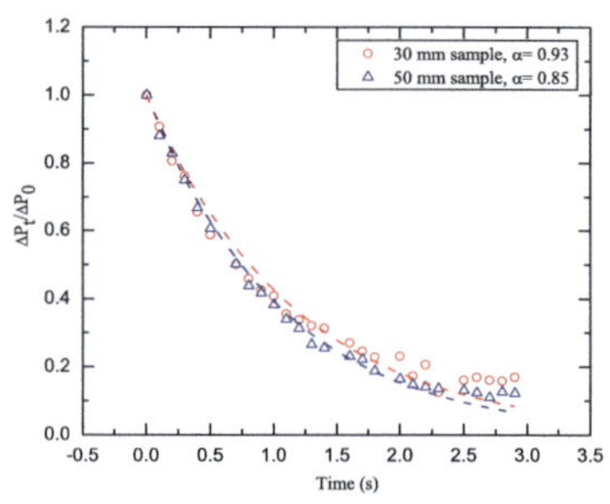

Fig. 4 Normalized pressure variation with respect to time for SB30 for an applied pressure gradient of 0.25 bar

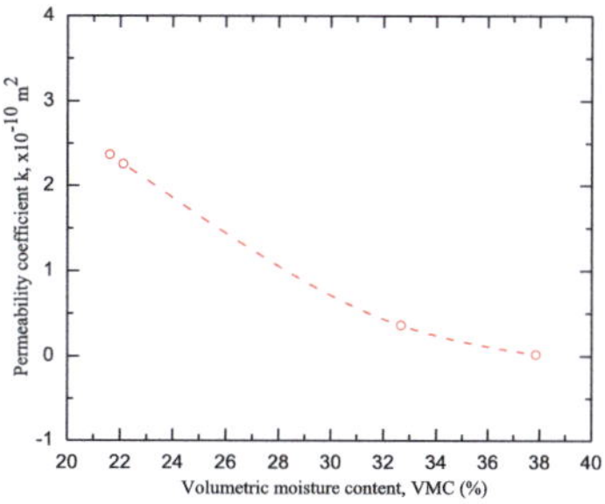

Fig. 5 Coefficient of permeability versus volumetric moisture content for SB30 for sample thickness of 50 mm

3 Summary and Conclusions

The present study evaluated the effect of sample thickness on the laboratory determination of gas permeability of buffer materials. In order to achieve this, the pressure decay method of gas permeability determination has been employed. To evaluate the influence of thickness on the gas permeability characteristics, experiments were conducted at different compaction states of buffer materials using sample thicknesses of 30 and 50 mm. It has been observed that the gas permeability coefficient decreases with increase in volumetric moisture content attributed to the decrease in pores available for the gas to flow. Further, the nature of normalized pressure variation curve with respect to time was observed to be similar for both sample thicknesses of 30 and 50 mm. But the exponential coefficient, α was found to be same for different applied pressure gradients in case of 50-mm-thick samples. This was due to the fact that for less thick samples, the slippage effect is more compared to thicker samples. Therefore, it can be concluded that a minimum of 50-mm-thick samples are required for the determination of gas permeability characteristics of buffer materials.

References

Ada, M. (2007) *Performance assessment of compacted bentonite sand mixtures utilized as isolation material in underground waste disposal repositories*. Master of science thesis, The graduate school of natural and applied sciences, Middle East Technical University, Turkey.

ASTM: D422. (2007). *Standard test method for particle size analysis of soils*. Annual Book of ASTM Standards 04.08, ASTM International, West Conshohocken, PA, USA.

ASTM D5550. (2006). *Standard test method for specific gravity of soil solids by gas pycnometer*. Annual Book of ASTM Standards 04.08, ASTM International, West Conshohocken, PA, USA.

ASTM D698. (2012). *Laboratory compaction characteristics of soil using standard effort*. Annual Book of ASTM Standards 04.08, ASTM International, West Conshohocken, PA, USA.

Billiotte, J., Yang, D., & Su, K. (2008). Experimental study on gas permeability of mudstones. *Physics and Chemistry of the Earth, 33*, S231–S236.

Bouazza, A., & Vangpaisal, T. (2003). An apparatus to measure gas permeability of geosynthetic clay liners. *Geotextiles and Geomembranes, 21,* 85–101.

Brace, W. F., & Martin, R. J. (1968). A test of the law of effective stress for crystalline rocks of low porosity. *International Journal of Rock Mechanics and Mining Sciences, 5,* 415–426.

Calogovic, V. (1995) Gas permeability measurement of porous materials (concrete) by time-variable pressure difference method. *Cement and Concrete research, 25*(5), 1054–1062.

Dixon, D. A., Gray, M. N., & Thomas, A. W. (1985). A study of the compaction properties of potential clay-sand buffer mixtures for use in nuclear fuel waste disposal. *Engineering Geology, 21*, 247–255.

Fedor, F., Hamos, G., Jobbik, A., Mathe, Z., Somodi, G., & Szucs, I. (2008). Laboratory pressure pulse decay permeability measurements of Boda Claystone. *Physics and Chemistry of the Earth, 33*, S45–S53.

Gardner, D. R., Jefferson, A. D., & Lark, R. J. (2008). An experimental, numerical and analytical investigation of gas flow characteristics in concrete. *Cement and Concrete Research, 33*, 360–367.

Moore, T. R., & Roulet, N. T. (1991). A comparison of dynamic and static chambers for methane emission measurements from subarctic fens. *Atmosphere-Ocean, 29,* 102–109.

Biominerlisation as a Remediation Technique: A Critical Review

Surabhi Jain and D. N. Arnepalli

Abstract Rapid industrialization and urbanization cause release of significant quantities of hazardous contaminants, including heavy metals and radionuclides, into the biosphere. Severe accumulation of these contaminants and their exposure deteriorates human health, environment, and biota system. Conventional remediation of heavy metal, radionuclide contaminated soils includes physicochemical extraction, stabilization/solidification/immobilization, soil washing. These techniques demand large quantities of chemical reagents, huge cost apart from the generation of secondary toxic by-products, and hence, the aforementioned techniques become unsuccessful and ineffective. This necessitates an interdisciplinary approach using biomediated processes and/or derived by-products, which enhances remediation process through accelerated biogeochemical phenomenon. Bioremediation is a broad area which involves large matrix of remediation techniques such as bioaccumulation, biosorption, biosparging, bioleaching, biomineralization, phytoremediation. Among all these techniques, biomineralization or microbially induced carbonate mineral precipitation is the most fascinating, promising methods to handle the present-day challenges pertaining to remediation of contaminated soils. In view of this, the current study presents a critical review on mechanisms of microbially induced carbonate precipitation in view of solid-phase sequestration of inorganic contaminants. Further, this study assesses the suitability of various microorganisms along with the associated precipitation processes for transforming soluble inorganic compounds into stable and non-redox sensitive carbonate minerals.

Keywords Bioremediation · Biomineralization · MICP · Contaminated soils

S. Jain (✉) · D. N. Arnepalli
Department of Civil Engineering, Indian Institute of Technology Madras, Chennai 600036, India
e-mail: surabhi.28.09.91@gmail.com

D. N. Arnepalli
e-mail: arnepalli@iitm.ac.in

V. K. Stalin and M. Muttharam (eds.), *Geotechnical Characterisation and Geoenvironmental Engineering*, Lecture Notes in Civil Engineering 16,
https://doi.org/10.1007/978-981-13-0899-4_19

1 Introduction

Large- and small-scale industries are established so hurriedly for gratifying human need and greed across the world. This leads to contribute numerous obligatory products as well as a massive quantity of refuses. The by-products/wastes generated from anthropogenic activities such as industries, excessive fertilizer application, sewage sludge, power plants, and nuclear power reactor contain numerous heavy metals and radionuclides. Safe disposal of these heavy metals and radionuclide is a major concerning factor in view of environmental and ecology system. This necessitates the remediation of hazardous waste by various remediation techniques. The physicochemical approaches such as filtration, precipitation, electrochemical treatment, soil washing and chelating, oxidation/reduction, ion exchange, reverse osmosis, stabilization/solidification are applied for treating heavy metal and radionuclide contaminated soils (Abdel-Sabour 2007). Above-mentioned remediation techniques entail several drawbacks as inefficient, requirement of abundant chemical reagents, energy and cost apart from generation of secondary by-products.

Bioremediation is a cost-effective and environmental-friendly technique which involves chemistry of biological activity to diminish environmental hazards coming from the accumulation of toxic chemicals and other hazardous wastes. Various biological methods such as biosorption, bioaccumulation, biomineralisation, phytoremediation, bioimobilisation are being implemented to remediate contaminated multiphase system as well as to maintain natural environment. Basic principle of bioremediation is decreasing the solubility of pollutant by altering pH, changing the redox reactions, and adsorption of pollutant from contaminated environment. The literature demonstrated that various heavy metals which have been immobilized or biosorbed are being released again into the environment by changing the redox reaction or valance of the pollutant metal by some biomaterials and their chemical activity. In view of this limitation phytoremediation, genetic potential of plants to mitigate toxic and hazardous contaminants from soil or water has been employed throughout the world. Phytoremediation is a lot depended on growing environment such as water availability, climate, temperature, geology, so it cannot be used in dry, harsh regions successfully.

Among all the bioremediation techniques, biomineralisation is most fascinating and promising research area in new generation of geoenvironmental engineering due to its various advantages over daunting problems. The term biomineralisation refers to minerals synthesized by organisms. Biomineralization can be biologically induced mineralization (BIM) or biologically controlled mineralization (BCM). In BIM, microbes amend the geochemical reactions in nearby environment by its uncontrolled metabolic activity, resulting in extra-cellular mineral growth. Whereas in the case of biologically controlled biomineralization (BCM), minerals are formed either intracellular or epicellular and the process is completely regulated.

Microorganisms from different phyla use specific metabolic process and synthesize nearly sixty-four varieties of minerals of phosphorites, carbonates, silicates, iron and manganese oxides, sulfide minerals, and amorphous silica (Knoll 2003).

From all the minerals precipitated by microorganisms, microbially induced carbonate precipitation (MICP) is primary focus till date as carbonate precipitation is abundant in nature with a wide variety of microorganisms and plants involved.

Three types of autotrophic pathways of microbially induced carbonate precipitation include two anaerobic processes and one aerobic process. On the other hand, various heterotrophic pathways such as nitrogen cycle, sulfur cycle and iron reduction are involved in carbonate mineral precipitation. In the nitrogen cycle, three different pathways, namely ammonification of amino acids, reduction of nitrate, and degradation of urea or uric acid, increase pH and induce carbonate ions. Even though different biogeochemical processes are involved in carbonate biomineral precipitation, all are not efficient for bioremediation techniques. Among all the pathways, microbially induced calcite precipitation by urea hydrolysis process or urease-positive microbes is gaining importance to remediate heavy metals and radionuclide (Fujita et al. 2010).

In view of the above-mentioned factors, the present study is emphasized on microbially induced carbonate precipitation (MICP) by urea hydrolysis process. For this purpose, a critical review on calcite precipitation mechanism in urea hydrolysis process and its impact on immobilizing soluble pollutants from soil and aquatic environment have been presented.

2 MICP in Urea Hydrolysis

In urea hydrolysis process, one mole of urea is hydrolyzed to one mole of ammonia and carbamic acid due to the microbial urease enzymes, which further hydrolyzes and generates additional one mole of ammonia and carbonic acid. These products further equilibrate in water and form bicarbonate ion. In urea hydrolysis, mineral formation accelerated due to increase in alkalinity and because of the presence and over saturation of calcium ions, calcium carbonate, or calcite mineral forms.

MICP process can retain heavy metal and radionuclides by adsorbed and/or coprecipitated in calcite lattice, and also, the presence of other metal cation forms carbonate mineral as $PbCo_3$, $SrCo_3$. Concept of mineral precipitation is shown in Fig. 1. MICP technique in urea hydrolysis process is not sensitive to redox potential change, which helps the pollutant not to be released again into the environment make this technique more feasible (Achal et al. 2012a, b, c), as shown below.

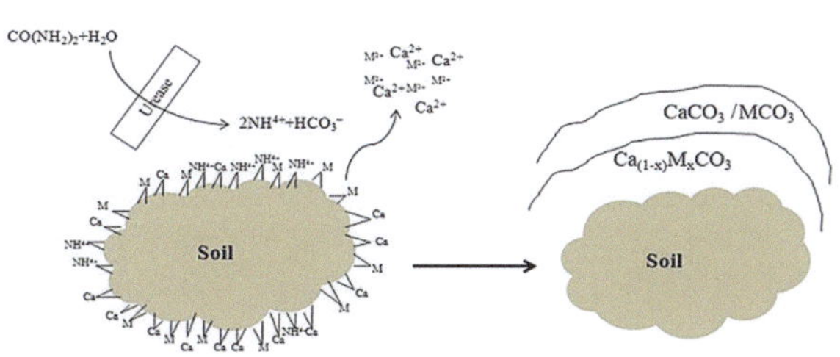

Fig. 1 Schematic view of MICP approach for remediation of contaminated soil

$CO(NH_2)_2 + H_2O \rightarrow NH_2COOH + NH_3$
$NH_2COOH + H_2O^- \rightarrow NH_3 + H_2CO_3$
$H_2CO_3 \rightarrow 2H^+ + 2CO_3^{2-}$
$NH_3 + H_2O \rightarrow NH^{4+} + OH^-$
$Ca^{2+} + 2CO_3^{2-} \rightarrow CaCO_3$ (K_{SP}= 3.8×10^{-9})

K_{SP} is the solubility product.

As toxicity of heavy metals and radionuclide hinder microbial growth, it is necessary to isolate metal tolerant, urease-positive microorganisms for altering soluble pollutants to stable mineral in MICP process. The following section describes different research considered MICP as a remediation technique by using various indigenous and non-indigenous metal tolerant, ureolytic microorganisms.

2.1 Removal of Arsenic

Semi-metalic, toxic arsenic, *As*, is found with different (−3, 0, +3, +5) valencies. The most available form of arsenic is arsenite (III) and arsenate (V).

Achal et al. (2012a, b, c) examined potential of *As*-tolerant isolated bacterium *Sporosarcina ginsengisoli* for calcite production and remediation of *As* contaminated soil. Even though growth of *S. ginsengisoli* was decreased due to the presence of *As*, ample urease production further lead to urea degradation rate and calcite production. Experimental data explained the removal of 96.3% of exchangeable *As* fraction from liquid solution. The five-stage soil sequential extraction showed a radical increment of carbonate bound *As* in bioremediated soil sample.

2.2 Removal of Cadmium

Cadmium, Cd, is in seventh rank in terms of hazardous heavy metals among top-twenty toxic pollutant. Even though heavy metal such as cadmium is needed as a nutrient for the living system, high quantity causes hamper to human and animal health.

In view of this, Kang et al. (2014a, b) isolated *Lysinibacillus sphaericus* and investigated its urease activity and feasibility of Cd bioremediation via carbonate precipitation. Even though growth of *L. sphaericus* hinders in the presence of Cd, urease activity did not get hampered. The study showed 99.95% removal of Cd from sand in 48 h transforming Cd into stable biomineral product. Subsequently, scanning electron microscope and X-ray diffraction analyses confirmed the presence of Cd as carbonate otavite ($CdCO_3$).

Similar way, Kumari et al. (2014a, b) conducted various experiments to investigate the effect of temperature on MICP-based Cd bioremediation by an indigenous ureolytic bacterium, *Exiguobacterium undae*. Urease enzyme activity and growth of

Exiguobacterium undae were not affected by low temperature (10 °C) compared to 25 °C. Tessier sequential extraction result showed soluble Cd fraction got reduced significantly and carbonate fraction got increased by microbially induced calcite precipitation practice compared to controlled samples in both 10 and 25 °C.

2.3 Removal of Copper

Copper, an essential nutrient, can also cause toxicity to plant and human in high concentration, which needs to be remediated. Achal et al. (2011) isolated *Kocuria flava* bacterium from a mining area soil due to its ability of higher urease production and high tolerance against copper. Author explained higher production of urease leads to more urea degradation, high calcite precipitation, and further, facilitates maximum removal of copper. The effect of pH and temperature on growth of bacterium and removable of exchangeable copper is favorable in the acidic medium with a wide range of temperatures, while the optimal temperature was 30 °C. Excess quantity of copper inhibits the bacterial growth due to high toxicity in reverse the data revealed higher percentage of copper removal as the quantity of copper increased in the soil. Similarly, Li et al. (2013) explained removal of copper by using another bacterium known as *Sporosarcina koreensis* and compared the rate of removal of copper is higher than by *Sporosarcina pasteurii.*

2.4 Bioremediation of Chromium

A feasible bioremediation of chromium slag was conducted by MICP technique with urease-positive Bacillus species (Achal et al. 2013). Consolidated bricks were made which shows higher compressive strength and low permeability. The leaching test was done to check releasing of chromium, which showed notable decrease in exchangeable chromium fraction in bio-modified sample. Co-precipitation of chromium in calcite structure forms a stable complex mineral, an effective way to remediate chromium contaminated soil.

Further, Kumari et al. (2014a, b) have evaluated the effect of MICP bioremediation technique for Cr (VI)-contaminated soil by knowing Cr phytoaccumulation. The technique involves participation of ureolytic bacteria *Bacillus cereus*, which arrest the exchangeable Cr (VI) as a stable carbonate biomineral, and consequently, *Pisum sativum* seeds were planted in the treated soil and uptake of chromium was measured. The harvested plant tissue presented a negligible fraction of chromium uptake than the plant in untreated soil zone.

2.5 Bioremdiation of Lead

Achal et al. (2012a, b, c) proposed microbially induced calcite precipitation (MICP) technique as an alternative method for remediating lead-contaminated soil by using urease positive, *Kocuria flava* bacterium. Similar way, Li et al. (2013) evaulated another ureolytic bacteria *Sporosarcina koreensis* for removal of lead and showed 100% removal of exchangable lead fraction from contaminated soil. Very recently, Kang et al. (2015) evaluated the lead remediation by altering the soluble lead to a carbonate biomineral fraction with the help of Pb-resistant bacteria. In this study, ureolytic Pb-resistant bacteria were preferred based on their urease activity, urec gene PCR and minimal inhibitory concentration against lead.

Even though the above-mentioned bioremediation experiments were carried out by pure culture, it is very difficult to maintain and implement a single culture in field. Synergetic effect of mixed bacterial culture enabled to remediate various pollutants in the same region more efficiently as compared to single culture they are more resistant to heavy metals and more robust in harsh environment.

In view of this, Kang et al. (2016) conducted MICP bioremediation of Pb, Cd, and Cu mixture-contaminated soils by various bacterial mixtures. Four bacterial strains with three sets of mixed culture were selected. In this study, experimental result proved higher resistivity and significant amount of calcite production and removal of heavy metals compared to single bacterial strain.

2.6 Bioremediation of Strontium

Microbially induced carbonate mineral precipitation by urea hydrolysis which leads to a solid-phase incorporation of strontium was investigated by *Sporosarcina pasteurii* (Kang et al. 2014a, b). Significant amount of strontium can be captured in calcite lattice make it stable carbonate mineral form.

Fujita et al. (2010) examined the potential of calcite precipitation and co-precipitation of sorbed ^{90}Sr by urease-positive microorganism in the Hanford subsurface, Washington. Similar way, Achal et al. (2012a, b, c) studied the bioremediation of radioactive ^{90}Sr from aquifer sediments/quartz sand with a Sr-resistant Halomonas species.

3 Critical Appraisal

As type of bacterial species used differs with metal tolerance, growth, rate of urea hydrolysis, and calcite precipitation, it has an impact on bioremediation which should be monitored in future. Metal carbonate precipitation and co-precipitation with calcite need much more advance characterization and quantification which enable the understanding of mechanism and other biogeochemical reaction. A lot of study is going on to remediate contaminated environment by MICP, but it is crucial to up-scale and optimize this technique in in situ soil surface. It is essential to understand and overcome the technical hitches for using the biogeochemical processes in situ.

4 Conclusion

This review paper demonstrated microbially calcite formation technique can precipitate various heavy toxic metals and radionuclide by using various indigenous microbial culture. Biomineralisation of calcite provided a unique research need to get more investigation and field implementation.

References

AbdEl-Sabour, M. F. (2007). Remediation and bioremediation of uranium contaminated soils. *Electronic Journal of Environmental, Agriculture and Food Chemistry, 6,* 2009–2023.

Achal, V., Pan, X., Fu, Q., & Zhang, D. (2012b). Biomineralization based remediation of As(III) contaminated soil by *Sporosarcina ginsengisoli. Journal of Hazardous Materials*, 178–184.

Achal, V., Pan, X., Lee, D. J., Kumari, D., & Zhang, D. (2013). Remediation of Cr(VI) from chromium slag by biocementation. *Chemosphere, 93*(7), 1352–1358.

Achal, V., Pan, X., & Zhang, D. (2011). Remediation of copper-contaminated soil by Kocuria flava CR1, based on microbially induced calcite precipitation. *Ecological Engineering, 37*(10), 1601–1605.

Achal, V., Pan, X., & Zhang, D. (2012a). Bioremediation of strontium (Sr) contaminated aquifer quartz sand based on carbonate precipitation induced by Sr resistant Halomonas sp. *Chemosphere, 89*(6), 764–768.

Achal, V., Pan, X., Zhang, D., & Fu, Q. (2012c). Bioremediation of Pb-contaminated soil based on microbially induced calcite precipitation. *Journal of Microbiology and Biotechnology, 22*(2), 244–247.

Fujita, Y., Taylor, J. L., Wendt, L. M., Reed, D. W., & Smith, R. W. (2010). Evaluating the potential of native ureolytic microbes to remediate a ^{90}Sr contaminated environment. *Environmental Science and Technology, 44*(19), 7652–7658.

Kang, C. H., Choi, J. H., Noh, J. G., Kwak, D. Y., Han, S. H., & So, J. S. (2014a). Microbially induced calcite precipitation-based sequestration of strontium by *Sporosarcina pasteurii* WJ-2. *Applied Biochemistry and Biotechnology, 174*(7), 2482–2491.

Kang, C. H., Han, S. H., Shin, Y., Oh, S. J., & So, J. S. (2014b). Bioremediation of Cd by microbially induced calcite precipitation. *Applied Biochemistry and Biotechnology, 172*(6), 2907–2915.

Kang, C. H., Kwon, Y.-J., & So, J.-S. (2016). Bioremediation of heavy metals by using bacterial mixtures. *Ecological Engineering, 89,* 64–69.

Kang, C. H., Oh, S. J., Shin, Y., Han, S. H., Nam, I. H., & So, J. S. (2015). Bioremediation of lead by ureolytic bacteria isolated from soil at abandoned metal mines in South Korea. *Ecological Engineering, 74,* 402–407.

Knoll, A. (2003). Biomineralization and evolutionary history. *Reviews in Mineralogy and Geochemistry, 54,* 329–356.

Kumari, D., Li, M., Pan, X., & Xin-Yi, Q. (2014a). Effect of bacterial treatment on Cr(VI) remediation from soil and subsequent plantation of *Pisum sativum*. *Ecological Engineering, 73,* 404–408.

Kumari, D., Pan, X., Lee, D. J., & Achal, V. (2014b). Immobilization of cadmium in soil by microbially induced carbonate precipitation with *Eexiguobacterium undae* at low temperature. *International Biodeterioration and Biodegradation, 94,* 98–102.

Li, M., Cheng, X., & Guo, H. (2013). Heavy metal removal by biomineralization of urease producing bacteria isolated from soil. *International Biodeteriation and Biodegradation, 76,* 81–85.

Biopolymer-Modified Soil: Prospects of a Promising Green Technology

M. S. Biju and D. N. Arnepalli

Abstract The benefit from using admixtures in soil to improve properties was discovered in ancient times. Various admixtures such as straw, bitumen, lime, salts and pozzolans are conventional additions to soil, while cement, petrochemicals and bacteria are currently being increasingly used in an effort to improve and stabilize soil from both mechanical and chemical aspects. The conventional techniques which utilize cement, lime, petrochemicals, etc., cause significant environmental degradation. With environmental awareness for materials and methods used in ground improvement generally growing, the trend towards using biopolymers as admixtures is expected to increase. This paper gives the concept and theory of ground improvement technique which employs biopolymers and describes the practical application of these techniques. A number of studies have been conducted in the past decades to check the suitability of various biopolymers in improving soil properties. The effectiveness of biopolymers for soil stabilization in agricultural, construction and military applications has been recognized by many researchers. More efficient and scientific usage of these materials for soil improvement requires knowledge about interaction mechanisms involved in the modification of geotechnical properties of soil. Most of the studies in clay–polymer interaction are from the field of medical engineering, where clay particles are suspended in the colloidal form and macromolecules are attached to them in different ways. The fundamental mechanism in biopolymer–soil modification proposed by various researchers is also presented in this paper. The study reveals the prospects of this green technology in the current era of rapid deterioration of natural resources. Furthermore, the need for continuing research on a number of factors which controls the mechanism is suggested.

Keywords Biopolymer · Soil stabilization · Soil–biopolymer interaction

---

M. S. Biju (✉) · D. N. Arnepalli
Department of Civil Engineering, Indian Institute of Technology Madras, Chennai 600036, India
e-mail: bijuvvg@gmail.com

D. N. Arnepalli
e-mail: arnepalli@iitm.ac.in

V. K. Stalin and M. Muttharam (eds.), *Geotechnical Characterisation and Geoenvironmental Engineering*, Lecture Notes in Civil Engineering 16,
https://doi.org/10.1007/978-981-13-0899-4_20

1 Introduction

Enhancing the properties of soil is a long-established technical field. In the studies about cradle of civilization, researchers have noticed the use of a number of soil modification techniques. Antediluvian techniques were there for storing water, for retaining the fertile topsoil, for constructing shelters over loose soil near their agricultural fields and for many other basic needs. Evidences of using admixtures for soil improvement were revealed by archaeologists from different civilizations like Sumerian and Roman. For example, Vitruvius described the application of dried blood for air entrainment and biopolymers for set retardation of gypsum (Rowland et al. 2001). Technologies improved with the advancement of civilization, and materials used for soil improvement also changed to a great extent. In all fields of engineering, the need for better results overruled environmental concerns including protection of nature. Even though many ancient philosophies encouraged growth with the nature, greed of one particular species ignored all those traditional wisdom and used many techniques to conquer biosphere which eventually caused extinction of many other species. The present generation identifies that the development in this style caused the deterioration of environment and it is inevitable to go for corrective measures. The land, water and air have been polluted to the alarming levels, and now it is the time to cleanse them. The need for sustainable technology is a globally accepted fact now.

To improve/stabilize soil, many techniques are there which includes soil modification using physical, mechanical and chemical methods. In many of the engineering applications, it is not viable to use mechanical or physical methods alone as the situation demands the usage of admixtures to enhance the engineering properties of soil under consideration to the desirable value. Using natural admixture like straw was found less effective compared to the cement or petrochemical additives and eventually whole world started using synthetic additives like 'Krilium' and polyvinyl alcohols for soil stabilization (Theng 1982). Use of cement and lime continued along with this.

While considering the environmental impacts of chemical stabilization methods, studies should concentrate on the effects after the application of the chemical as well as the process involved in the production of these chemicals. Studying the impact of soil stabilization using cement will first point to the impact at the time of production of cement. The total quantity of greenhouse gases released per ton of cement production is in the tune of 0.95 tons. With this, the matrix stabilization which prevents the growth of vegetation on the topsoil proves that cement stabilization is not a green technology. The extensive utilization of cement causes quite a lot of related issues in the habitat, such as increased runoff, air pollution and heat island. Moreover, it is not easy to return cement-modified soil to their initial state (Chang et al. 2016).

From the 1950s, geotechnical engineers have started using chemically synthesized macromolecules such as lignosulfonates and polyacrylamides to replace the conventional binders. These comparatively new and effective chemically synthesized polymers raised concerns about contamination and toxicity problems. The need for restrictions to the uncontrolled use of chemically synthesized polymers is globally accepted in the present era (Karol 2003). Electro-osmotic chemical treatment (ECT) has been proved to improve the strength of soft soils, but it requires injection of substantial quantity of chemical chelating agents for an increase in osmotic pressure (Ou et al. 2009). Application of geopolymers for soil stabilization is another growing field, but studies have shown that these novel materials are susceptible to significant strength reduction when saturated, in particular with acidic solutions, due to depolymerization and removal of silicic acid. The conventional techniques which utilize cement, lime, petrochemicals, etc., cause significant environmental impacts. The trend towards employing biopolymers for engineering applications is expected to increase with an increase in environmental awareness.

Polymers or macromolecules are chemical compounds formed by the union of molecules of the same kind (monomers). Biopolymers are environment-friendly polymers that are produced by living organisms. Biopolymers have a wide range of applications: in agriculture, biomedical engineering, food processing, the chemical industry, the energy sector as well as in environmental protection and remediation. Commonly used biopolymers in soil modification include guar gum (derived from the ground endosperm of a legume), xanthan gum (extracellular polysaccharide from bacteria Xanthomonas campestris), chitosan (derived from chitin obtained from crustaceans) and sodium alginate (derived from the cell walls of brown algae). Addition of biopolymer to soil may result in low permeability, higher shear strength characteristics, acceptable swell–shrink behaviour and high erosion resistance. Application of biopolymers to soil as a stabilizing and strengthening agent of aggregates was under consideration in the agricultural engineering from the 1940s (Karimi 1997).

The World Health Organization (1975, 1987) performed toxicity studies for biopolymers guar gum and xanthan gum and found that they do not represent a hazard to health and that there was no need to establish an acceptable daily intake of the substances. Keeping these facts in mind, biopolymer modification of soil can be considered as an environment-friendly technology.

2 Application of Biopolymer in Soil Engineering

Use of biopolymers to reduce soil hydraulic conductivity for groundwater control and containment applications has received attention in recent decades. Different works found that several biopolymers procurable from the market including chitosan, sodium alginate, xanthan gum, poly-hydroxybutyrate, poly-glutamic acid and guar gum were capable of reducing the permeability of soil by up to five orders of magnitude (Khachatoorian et al. 2003; Bouazza et al. 2009; Blauw et al. 2009; Taytak et al. 2012; Wiszniewski et al. 2013).

Erosion and slope stability issues are challenging, especially in the field of cultivation where the use of synthetic polymers will cause significant impacts. Biopolymer-induced stabilization of soil is a very good solution for these problems. Application of xanthan gum, chitosan, starch or polyacrylamide in furrow irrigation can reduce the erosion potential of soil (Sojka et al. 2003; Orts et al. 2007). Application of biopolymers like cellulose, β-glucan and xanthan gum in clayey sand, silty sand, silty clay, loam and silt were found effective in stabilizing slope and controlling erosion (Maghchiche and Immirzi 2010; Larson et al. 2012).

Biopolymers like beta-1,3/1,6-glucan biopolymer, xanthan gum were added to different soils by researchers and found that the unconfined compressive strength enhances more than 150% (Chang and Cho 2012; Liu et al. 2011). Studies on shear strength also show considerable increase. The change is mainly on the cohesion intercept, and difference in angle of internal friction is negligible (Karimi 1997; Liu et al. 2011; Khatami and O'Kelly 2013).

Optimum water content and maximum dry unit weights are important when soil is used as an engineering material. Biopolymer-modified soil shows a higher maximum dry density compared to that of virgin soil, while the change in optimum moisture content is negligible (Karimi 1997; Taytak et al. 2012).

Variation of liquid limit of cohesive soils with biopolymers having different viscosity was studied by many researchers and found that liquid limit increases with the addition of biopolymers. The variation of consistency limits will differ with the properties of biopolymers (Karimi 1997; Nugent et al. 2009; Chen et al. 2013).

3 Stabilization Mechanism in Biopolymer–Soil Geomaterial

From the numerous studies available, it is evident that biopolymers are better alternative for conventional stabilizing agents like cement and synthetic polymers. More efficient and scientific usage of these materials for soil improvement requires the knowledge about the mechanisms involved in biopolymer–soil interaction. Clay–polymer interaction is well studied by many of the biomedical engineers in which clay particles are suspended in the colloidal form and polymers can attach to it in different ways (Theng 2012). In soil improvement, the mechanism is different.

It has been shown by Geoghegan and Brian (1946) that bacterial polysaccharides exert a pronounced binding effect on soil particles. When biopolymers are added to soil, they form a thin coating over the aggregates with secondary bonds. Polar components in the biopolymer may get attached to charged surfaces, enhancing adhesion. For specific biopolymers, ion exchange with soil is also expected (Tingle et al. 2007). In soil–biopolymer mixture, a portion of the stabilizer enters the pore spaces while the other part falls on to the soil particle surface. It is possible for the ionic groups in the macromolecules to react with clay grain. Through ionic, hydrogen or van der Waals bonds formed in the chemical reaction, biopolymers enwrap soil aggregates and interlink to form a viscous and elastic membrane structure. The modified structure causes an improvement in the engineering properties including strength and erosion resistance. The physicochemical reactions among biopolymer and soil particles usually need a few days. The increase in unconfined compressive strength of biopolymer-modified soil, with an increase in curing duration, supports this hypothesis (Liu et al. 2011). The effectiveness of stabilization depends on the interfacial reaction phenomena, the types of forces present at the boundary of soil particle and biopolymer. Chemisorption, hydrogen bonding and physical adsorption are possible at the soil–biopolymer interface.

The five categories of interactions within the soil aggregate, biopolymer and cation system are aggregation of soil particles due to polymer addition; cation-induced interlinking of biopolymers; formation of a interconnected network of polymer and clay via cation bridging; change of thickness in the diffused double layer on clay surfaces and competing adsorption of biopolymer molecules and cations onto clay surfaces (Nugent et al. 2009). The stabilization mechanisms proposed by the researchers are still largely unsubstantiated, but the available data from the literature and laboratory experiments tend to support them.

4 Conclusions

Acceptability of ground improvement techniques increased because of its cost effectiveness and wide-range applicability. The effectiveness of ground improvement techniques has been proven in the past for a wide range of engineering structures such as retaining structures, pavements, highways, runways, ports, bunds, railways, dams, excavations, tunnelling and other infrastructure facilities. The conventional ground improvement techniques which utilize cement, lime, petrochemicals, etc., cause significant environmental impacts. Soil stabilization using biopolymer is an environment-friendly alternative. Studies provide evidence for the capability of biopolymers to strengthen soils, control permeability and for several other applications. Even though biopolymer stabilization got myriads of advantages, still a number of impediments are there to solve. The durability issues and applicability to different soils are yet to be studied.

References

Blauw, M., Lambert, J. W. M., & Latil, M. N. (2009). *Biosealing: A method for in situ sealing of Leakages* (pp. 125–130). GeoSS: Ground Improvement Technologies and Case Histories.

Bouazza, A., Gates, W. P., & Ranjith, P. G. (2009). Hydraulic conductivity of biopolymer-treated silty sand. *Geotechnique, 59,* 71–72.

Chang, I., & Cho, G. C. (2012). Strengthening of Korean residual soil with β-1,3/1,6-glucan biopolymer. *Construction and Building Materials, 30,* 30–35.

Chang, I., Jooyoung, I., & Cho, G. C. (2016). Introduction of microbial biopolymers in soil treatment for future environmentally-friendly and sustainable geotechnical engineering. *Sustainability, 8*(3), 1–23.

Chen, R., Zhang, L., & Budhu, M. (2013). Biopolymer stabilization of mine tailings. *Journal of geotechnical and geoenvironmental engineering, 139,* 1802–1807.

Geoghegan, M. J., & Brian, R. C. (1946). Influence of some bacterial polysaccharides on the binding of soil particles. *Biochemical Journal, 43*(1), 5–13.

Karimi, S. (1997). A study of geotechnical applications of biopolymer treated soil with emphasis on silt' (Ph.D. Thesis). Civil Engineering Department, University of Southern California, Los Angeles.

Karol, R. H. (2003). *Chemical grouting and soil stabilization* (3rd ed.). New York, USA: CRC Press.

Khachatoorian, R., Ioana, G. P., Chang-Chin, K., & Yen, T. F. (2003). Biopolymer plugging effect: Laboratory-pressurized pumping flow studies. *Journal of Petroleum Science and Engineering, 38*(1–2), 13–21.

Khatami, H. R., & O'Kelly, B. C. (2013). Improving mechanical properties of sand using biopolymers. *Journal of Geotechnical and Geoenvironmental Engineering, ASCE, 139*(8), 1402–1406.

Larson, L. S., John, K. N., Christopher, S. G., Milton, B., & Catherine, C. N. (2012). Biopolymers as an alternative to petroleum-based polymers for soil modification. *US Army Corps of Engineers*, ERDC TR-12-8.

Liu, J., Bin, S., Hongtao, J., Huang, H., Wang, G., & Kamai, H. (2011). Research on the stabilization treatment of clay slope topsoil by organic polymer soil stabilizer. *Engineering Geology, 117,* 114–120.

Maghchiche, A., Haouam, & Immirzi, B. (2010). Use of polymers and biopolymers for water retaining and soil stabilization in arid and semiarid regions. *Journal of Taibah University for Science, 4,* 9–16.

Nugent, R. A., Guoping, Z., & Robert, P. G. (2009). Effect of exopolymers on the liquid limit of clays and its engineering implications. *Journal of the Transportation Research Board, 2101,* 34–43. https://doi.org/10.3141/2101-05.

Orts, W. J., Aicardo, R., Sojka, R. E., Gregory, M. G., Syed, H. I., Kurt, E., et al. (2007). Use of synthetic polymers and biopolymers for soil stabilization in agricultural, construction, and military application. *Journal of Materials in Civil Engineering ASCE, 19,* 58–66.

Ou, C. Y., Chien, S. C., & Chang, H. H. (2009). Soil improvement using electroosmosis with the injection of chemical solutions: Field tests. *Canadian Geotechnical Journal, 46,* 727–733.

Rowland, I. D., & Thomas, N. H. (Eds.) (2001). *Ten books on architecture by vitruvius.* New York, USA: Cambridge University Press.

Sojka, R. E., Entry, J. A., Orts W. J., Morishita, D. W., Ross, C. W., & Horne, D. J. (2003). Synthetic- and bio-polymer use for runoff water quality Management in irrigated agriculture. In *Diffuse Pollution Conference*, 17–21 Aug 2003, Dublin, Ireland.

Taytak, B., Pulat, H. F., & Yukselen, A. (2012). Improvement of engineering properties of soils by biopolymer additives. In *3rd International Conference on New Developments in Soil Mechanics and Geotechnical Engineering,* 28–30 June 2012, Near East University, Nicosia, North Cyprus.

Theng, B. K. G. (1982). Clay-polymer interactions: Summary and perspectives. *Clays and Clay Minerals, 30,* 1–10.

Theng, B. K. G. (2012). *Formation and properties of clay-polymer complexes*. Amsterdam, The Netherlands: Elsevier.

Tingle, J. S., Newman, J. K., & Larson, S. L. (2007). Stabilization mechanisms of non-traditional additives. *Journal of the Transportation Research Board, 1989,* 59–67.

Wiszniewski, M., skutnik, Z., & Cabalar, A. F. (2013). Laboratory assessment of permeability of sand and biopolymer mixtures. *The Journal of Warsaw University of Life Sciences, 45*(2), 217–226.

Factors Influencing Zeta Potential of Clayey Soils

K. Nikhil John and D. N. Arnepalli

Abstract Electro-kinetic properties of colloidal substance can be studied in terms of its zeta potential, which indicates the stability of the colloidal system. Numerous investigations have been made in the past several decades in areas of electro-kinetic remediation and stabilization of fine-grained soils. A proper understanding of the underlying mechanism of the above processes demands a thorough knowledge of the zeta potential of the system. Further, the electro-kinetic process can significantly alter the physio-chemical and electrical properties of the clay-water-electrolyte system which is also manifested as a change in the zeta potential value. Various environmental factors that affect the zeta potential include temperature, electrolytic concentration, cation valency and pH of the medium. The investigations made in view of understanding the role of zeta potential in determining electro-kinetic efficiency of various soils are widely scattered and no attempts have been made so far to interpret the available data, making it difficult to arrive at any conclusive inference. In this context, the present study attempts to evaluate the investigations carried out, by the previous researchers, to identify the factors that are influencing zeta potential and its role on electro-kinetic properties of clay minerals. In addition, zeta potential measurements are conducted on kaolinitic type and Na-bentonite soils over a wide range of pH and the results are compared with the data available in the literature.

Keywords Zeta potential · Soil electro-kinetics · Electro-osmosis

K. Nikhil John (✉) · D. N. Arnepalli
Department of Civil Engineering, Indian Institute of Technology Madras, Chennai 600036, India
e-mail: njkollannur@gmail.com

D. N. Arnepalli
e-mail: arnepalli@iitm.ac.in

V. K. Stalin and M. Muttharam (eds.), *Geotechnical Characterisation and Geoenvironmental Engineering*, Lecture Notes in Civil Engineering 16,
https://doi.org/10.1007/978-981-13-0899-4_21

1 Introduction

When two different phases come in contact in a way that there exists excess surface charges, ions or ionic groups or phase polarity in either or both the phases, then a non-uniform charge distribution develops at the phase interface. Clay minerals, possessing uniform negative electrostatic potential at their surface, when mixed with an aqueous solution, the dissolved ions present and/or the polarized water itself will accumulate at the interface boundary. The counter ions present in the aqueous phase decays the mineral surface potential in a systematic manner until it becomes constant in the bulk of the solution. The potential decay, which normally occurs over a distance of 5–200 nm, is influenced mainly by the electrolyte chemistry (Hunter 1981).

The negative charge possessed by clay mineral face is mostly due to isomorphic substitutions and bond breakages within the clay lattice. But the mineral edges are amphoteric, and the nature of charge is determined by the electrolyte pH (Tombacz and Szekeres 2004). The distribution of the ions at the interface boundary can be explained in terms Gouy-Chapman double layer theory (Fig. 1). As per this model, a layer of counter charges and polarized water adsorbed strongly to the mineral surface is present, called the Stern layer. Following the Stern layer, there is a diffused layer where the ions are less tightly held. The decay of surface potential varies linearly in the Stern region, thereafter following an exponential trend (Fig. 1). When the free water within the pore space flows under gravity, there will be a notional boundary, termed as the plane of shear, up to which the water acts as a single entity, abstaining from the bulk flow. The electrical potential along this plane is known as zeta potential, ζ.

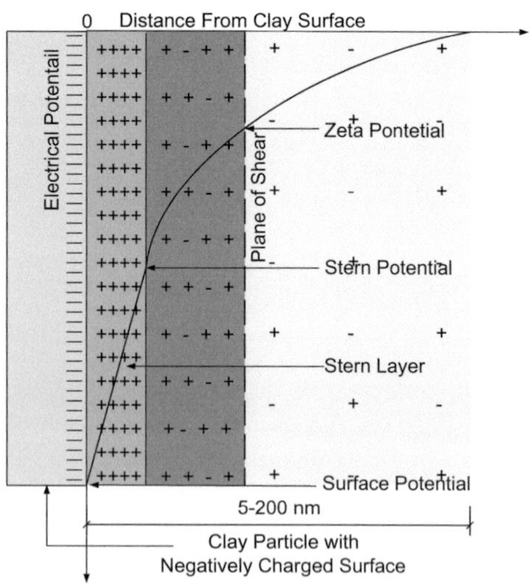

Fig. 1 Schematic representation of diffused double layer model (modified from Hunter 1981)

Electro-kinetic effects characterized by the relative tangential motion of one phase with respect to the other can be studied in terms of the zeta potential of the associated colloids. The aforesaid effects, including electro-osmosis, electro-phoresis, streaming potential and sedimentation potential, are having special significance in the stabilization and remediation of problematic and contaminated soils. In addition, the zeta potential is extremely important in analysing stability of any colloidal system. Though colloidal stability demands more attention in fields like pharmaceuticals, emulsions, ceramics, bio-medicine, there are areas in geotechnical engineering such as the performance of drilling fluid, stability and dewatering of mine tailings where the assessment of colloidal stability is inevitable (Au and Leong 2013; Kumar et al. 2014). Taking into account the paramount role played by zeta potential in a wide arena of geotechnical engineering, a detailed review of available literature is carried out in the present manuscript to identify the factors that influence zeta potential of clay minerals.

2 Factors Influencing Zeta Potential

Since zeta potential represents the electrical potential at the plane of shear, any element that can change the surface potential on the clay mineral or change the potential decay pattern can intern influence the zeta potential value. Otherwise stated, zeta potential connotes the diffused double layer (DDL) thickness of the colloidal particles. With this in view, the zeta potential response on various influencing parameters is tried to apprehend by considering their effect on the DDL thickness, given by the equation

$$\frac{1}{k} = \left(\frac{DkT}{8\pi n_0 \varepsilon^2 v^2}\right)^{\frac{1}{2}} \tag{1}$$

where $\frac{1}{k}$ is the DDL thickness, D is the dielectric constant of the medium, T is temperature, ϵ is unit electric charge, n_0 is the electrolyte concentration, and υ is the ionic valency.

2.1 *Influence of pH on Zeta Potential*

During electro-kinetic treatment of soil, the pore water undergoes electrolysis due to current flow, producing H+ and OH− ions at the cathode and anode, respectively, and thereby changes the soil pH. This necessitates the study of variation of zeta potential with pH, since the change in pH can affect the suitability of the soil for the intended treatment. Also, there are instances where the soil pH changes such as treatment with lime or cement, where the zeta potential measurement gives valuable insight towards the change in fabric arrangement.

Zeta potential variation of clay minerals over wide range of pH can be explained in terms of the DDL theory. At low pH, excess of H+ ions compress the double layer, and a subsequent reduction of zeta potential is observed. Reverse trend is visible at high pH where the DDL expands due to OH− adsorption into the interface (Yukselen and Kaya 2003).

In the present study, variation of zeta potential with pH is assessed for two soils, a commercially available white clay (with kaolinite as the major mineral) and Na-bentonite. The details of the basic characterization carried out are given in Table 1. Soil samples are prepared at different pH as per the standard practice (Min et al. 2013), and the zeta potential measurement is made with Horiba Nano-Particle Analyzer SZ-100.

Zeta potential exhibited a higher pH dependency in the case of Na-bentonite, with an increase in negativity upto pH around nine and thereafter reversing the trend (Fig. 2). The response is in contrary with the literature data, where montmorillonitic soils are reported to showcase only nominal change with pH, resulting out of the OH− affinity of the mineral edges at high pH (Delgado et al. 1986; Avena et al. 1990; Kaya and Yukselen 2005). The behaviour is rationalized with the fact that in the case of montmorillonite mineral, the total surface area contribution from the amphoteric edges, which are pH dependent, is approximately 1% (Delgado et al. 1986; Sondi et al. 1996; Yalçin et al. 2002). The same argument fits well to explain the absence of an iso-electric point (IEP) for montmorillonitic soils. But in the present case, the conflicting response can be partly attributed to the variation in the mineral composition.

For kaolinite, the literature reports significant reliance of zeta potential value with pH and an IEP was evident in many of the reported studies (Ma and Pierre 1999). In the present work, though the zeta potential variation of kaolinitic soil with pH was less prominent compared to bentonite, the obtained response was comparable with the literature (Fig. 2).

A trend of reduction in zeta potential negativity in the alkaline pH is found to be common for both the soils considered. This is assumed to be because of the dissolution of clay minerals at high pH. Since mineral dissolution is a time-dependent process and unlike in the literature, for the present study, an equilibration time of 24 h was provided before the measurement, and the high valency minerals might have dissolved into the solution, affecting the DDL thickness and thereby reducing the zeta potential.

Table 1 Properties of selected geomaterials

	LL	PL	SSA	CEC	Minerals present
Kaolinitic soil	52	29	11	65	Quartz, kaolinite, dickite
Na-Bentonite	350	27	282	130	Montmorillonite, rutile, quartz

LL liquid limit (%), *PL* plastic limit (%), *SSA* specific surface area (m^2/g of soil), *CEC* cation exchange capacity (meq/100 g of soil)

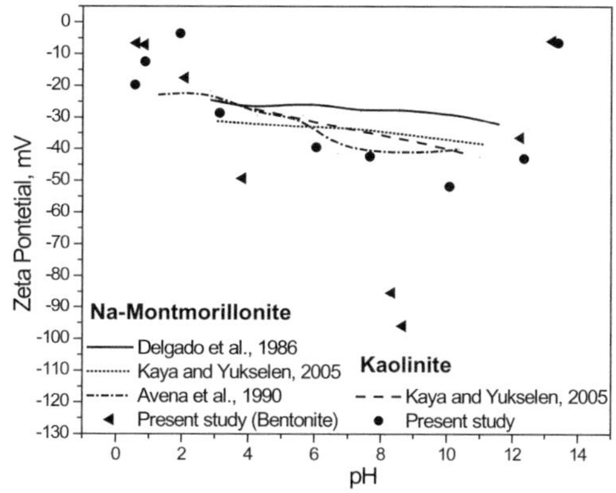

Fig. 2 Variation of zeta potential with pH

2.2 *Influence of Electrolyte Type and Concentration*

Studies reported a reduction in the zeta potential value with increase in electrolyte concentration and cation valency (Fig. 3) (Delgado et al. 1986; Yukselen and Kaya 2003). The response is obvious due to the reduction in DDL thickness in both the cases. The observations made by Kaya and Fang (1996) can also be accounted in similar lines where the zeta potential negativity increased with dielectric constant of the medium.

In addition, studies show that hydrated ionic radius also affects significantly the zeta potential value (Yukselen and Kaya 2011). Larger the ionic radius, thicker will be the DDL and the corresponding zeta potential will be more negative. Along similar lines, the effect of various additives on zeta potential can be studied. Akbulut and Arasan (2010) investigated the zeta potential variation of high plasticity clay with varying percentages of lime, cement and fly ash. Zeta potential became increasingly negative with lime content. The behaviour is expected as

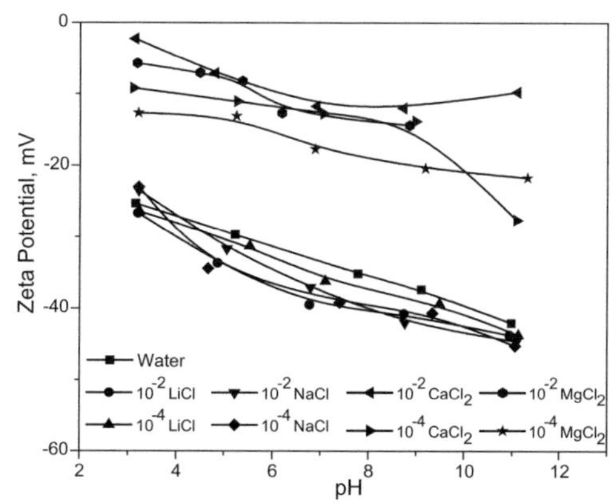

Fig. 3 Zeta potential variation with electrolyte type and concentrations (modified from Yukselen and Kaya 2003)

higher lime content increases the soil pH, causing subsequent OH^- adsorption. In the case of cement and fly ash, a reverse trend is observed.

2.3 *Effect of Surfactants and Chelating Agent*

Chelating agents and surfactants are excessively being employed in recent times for enhancing electro-kinetic remediation of contaminated soils. Though a thorough understanding of their effects on zeta potential is inevitable to assess the treatment effectiveness, the aspect is very less addressed in the literature.

Chelating agents are chemical compounds that undergo complexation with metal contaminants, thereby preventing them from precipitation. Popov et al. (2004) studied the effect of the chelating agent, hydroxyethane diphosphonic acid (HEDPA) on the zeta potential of clay loam. The addition of the ligand compound increased the zeta potential negativity. Similar results were obtained in the study carried out by Gu et al. (2009) on natural clay from Shanghai. The possible reasons for the behaviour include adsorption of the chelating compounds onto the mineral surface, dissolution of the hydrous oxide films on the clay mineral surface and adsorption of monovalent ions instead of multivalent ions since the later is being masked because of the chelation (Popov et al. 2004; Gu et al. 2009).

Surfactants are amphiphilic compounds used to enhance the dissolution of contaminants in the soil. Yalçin et al. (2002) reported that the zeta potential becomes less negative with anionic surfactant concentration in bentonites. In contrary to this, the investigations carried out by Kaya and Yukselen (2005) on kaolinitic and montmorillonitic soils show that anionic surfactant addition increases the zeta potential negativity. Whereas zeta potential positivity increased with cationic surfactants, while non-ionic surfactants did not bring any significant change.

2.4 *Effect of Temperature*

Studies on the effect of non-ambient temperature on zeta potential value demand attention in the case of reservoir minerals where fabric structure is of prime importance. Ramachandran and Somasundaran (1986) studied the effect of temperature on zeta potential value of quartz and kaolinite. The authors observed an increase in the magnitude of the zeta potential with temperature for quartz mineral. Similar observation was made by Rodríguez and Araujo (2006) and Winsniewska (2011). For kaolinite, the effect was mostly pH dependent where the positivity and negativity of zeta potential increased in acidic and basic pH, respectively, (Fig. 4).

The temperature dependence of quartz owes to the dissolution of the amorphous silanol group and the subsequent adsorption of $H_3SiO_4^-$, since the reaction kinetics are influenced by temperature. For kaolinite, the temperature dependency attributes

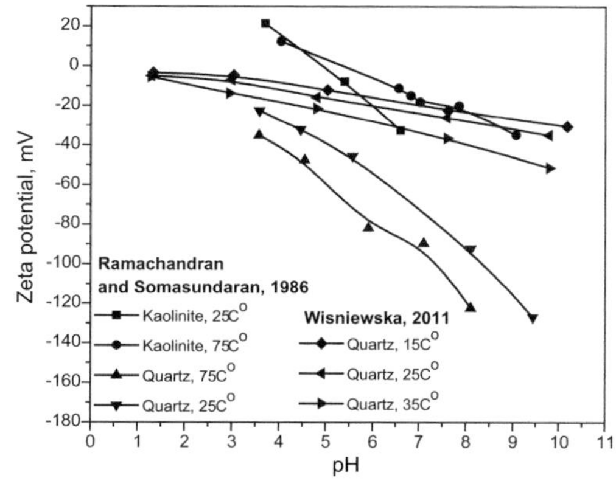

Fig. 4 Temperature effect on zeta potential

to species distribution in the mineral at different pH, Al_3^+ and $Al(OH)_2^+$ in acidic range, H_4SiO_4 and $Al(OH)_3$ at neutral and $H_3SiO_4^-$ in the basic range (Ramachandran and Somasundaran 1986).

3 Conclusion

A detailed review of factors affecting zeta potential of clay minerals has been presented. The major influencing factors comprise solution pH, electrolyte type and concentration, surfactants and chelating agents and temperature. In addition, the variation of zeta potential with pH is measured for kaolinitic soil and Na-bentonite and the results are compared vis-á-vis with the literature data.

References

Akbulut, S., & Arasan, S. (2010). The variations of cation exchange capacity, pH, and zeta potential in expansive soils treated by additives. *International Journal of Structural and Civil Engineering Research, 1*(2), 139–154.

Au, P. I., & Leong, Y. K. (2013). Rheological and zeta potential behaviour of kaolin and bentonite composite slurries. *Colloids and Surfaces A: Physicochemical and Engineering Aspects, 436*, 530–541.

Avena, M. J., Cabrol, R., & De Pauli, C. P. (1990). Study of some physicochemical properties of pillared montmorillonites: Acid-base potentiometric titrations and electrophoretic measurements. *Clays and Clay Minerals, 38*(4), 356–362.

Delgado, A., Caballero, F. G., & Bruque, J. M. (1986). On the zeta potential and surface charge density of montmorillonite in aqueous electrolyte solutions. *Journal of Colloid and Interface Science, 113*(1), 203–211.

Gu, Y. Y., Yeung, A. T., Koenig, A., & Jiang Li, H. (2009). Effects of chelating agents on zeta potential of cadmium-contaminated natural clay. *Separation Science and Technology, 44*(10), 2203–2222.

Hunter, R. J. (1981). *Zeta potential in colloidal science principles and applications*. London: Academic Press.
Kaya, A., & Fang, H. Y. (1996). Characterization of dielectric constant on fine-grained soil behavior. *Sampling Environment Media, STP, 1282,* 303–314.
Kaya, A., & Yukselen, Y. (2005). Zeta potential of clay minerals and quartz contaminated by heavy metals. *Canadian Geotechnical Journal, 42*(5), 1280–1289.
Kumar, S., Bhattacharya, S., & Mandre, N. R. (2014). Characterization and flocculation studies of fine coal tailings. *Journal of the Southern African Institute of Mining and Metallurgy, 114*(11), 945–949.
Ma, K., & Pierre, A. C. (1999). Clay sediment-structure formation in aqueous kaolinite suspensions. *Clays and Clay Minerals, 47*(4), 522–526.
Min, F., Zhao, Q., & Liu, L. (2013). Experimental study on electrokinetic of kaolinite particles in aqueous suspensions. *Physicochemical Problems of Mineral Processing, 49*(2), 659–672.
Popov, K., Kolosov, A., Ermakov, Y., Yachmenev, V., Yusipovich, A., Shabanova, N., et al. (2004). Enhancement of clay zeta-potential by chelating agents. *Colloids and Surfaces A: Physico-chemical and Engineering Aspects, 244*(1–3), 25–29.
Ramachandran, R., & Somasundaran, P. (1986). Effect of temperature on the interfacial properties of silicates. *Colloids and Surfaces, 21,* 355–369.
Rodríguez, K., & Araujo, M. (2006). Temperature and pressure effects on zeta potential values of reservoir minerals. *Journal of Colloid and Interface Science, 300*(2), 788–794.
Sondi, I., Biscan, J., & Pravdic, V. (1996). Electro-kinetics of pure clay minerals revisited. *Journal of Colloid and Interface Science, 178,* 514–522.
Tombacz, E., & Szekeres, M. (2004). Colloidal behavior of aqueous montmorillonite suspensions: The specific role of pH in the presence of indifferent electrolytes. *Applied Clay Science, 27,* 75–94.
Wiśniewska, M. (2011). The temperature effect on electrokinetic properties of the silica-polyvinyl alcohol (PVA) system. *Colloid and Polymer Science, 289*(3), 341–344.
Yalçin, T., Alemdar, A., Ece, O. I., & Gungor, N. (2002). The viscosity and zeta potential of bentonite dispersions in presence of anionic surfactants. *Materials Letters, 57*(2), 420–424.
Yukselen, Y., & Kaya, A. (2003). Zeta potential of kaolinite in the presence of alkali, alkaline earth and hydrolyzable metal ions. *Water, Air, and Soil pollution, 145,* 155–168.
Yukselen, Y., & Kaya, A. (2011). A study of factors affecting on the zeta potential of kaolinite and quartz powder. *Environmental Earth Sciences, 62*(4), 697–705.

Experimental Analysis of Salt Diffusion in Compacted Clays by Through-Diffusion and Half Cell Technique

Partha Das, S. R. Man Parvesh and T. V. Bharat

Abstract The estimation of the model parameters namely effective diffusion coefficient and retardation factor of a potential landfill liner material was presented in this paper using the experimentally measured salt concentration data. Experimental data of concentration variation of time and spatial distance in compacted bentonite was obtained using two diffusion measurement techniques viz. through-diffusion and half cell technique, respectively. The bentonite was subjected to the same concentration gradient and compacted density in both the experimental methods to compare the results and understand underlying mechanism in the diffusion tests. The measured data from the laboratory diffusion techniques was analysed using a Graphical User Interface (GUI)-based Dot-net application CONTRADIS. The CONTRADIS was used to estimate the model parameters by the inverse analysis. The application uses the solution of the forward analysis and stochastic algorithm for the inverse analysis. The retardation factor obtained theoretically was validated using laboratory batch sorption tests.

Keywords Through-diffusion · Half cell · Clay · CONTRADIS

1 Introduction

The barrier materials used in the landfill liners are mostly clayey soil having very high plasticity and hydraulic conductivity as low as 10^{-9} cm/s. In such a case, the solute transport will be mostly governed by diffusion and the flow due to advection will be negligible (Barone et al. 1992; Rowe and Booker 1985; Shackelford et al.

P. Das (✉) · S. R. Man Parvesh · T. V. Bharat
Department of Civil Engineering, IIT Guwahati, Guwahati 781039, Assam, India
e-mail: partha.das@iitg.ernet.in

S. R. Man Parvesh
e-mail: manparveshsr@gmail.com

T. V. Bharat
e-mail: tvb@iitg.ernet.in

V. K. Stalin and M. Muttharam (eds.), *Geotechnical Characterisation and Geoenvironmental Engineering*, Lecture Notes in Civil Engineering 16,
https://doi.org/10.1007/978-981-13-0899-4_22

1989). In most of the municipal solid waste landfills or highly toxic and hazardous waste landfills, flow due to diffusion is considered to be the significant transport mechanism (Kau et al. 1999). As such for the effective design of landfill liners, knowledge of the parameter governing the diffusive mechanism which is described by diffusion coefficient becomes essential. Also, flow through the plastic compacted clays is influenced by the sorption characteristics of the soils. The sorption potential is best described by a parameter known as retardation factor (Shackelford and Daniel 1991). Hence, proper estimation of both the model parameters which are the effective diffusion coefficient and retardation factors becomes important. Diffusion coefficients are obtained experimentally from laboratory diffusion tests, and the retardation factors are obtained from the equilibrium batch tests (Shackelford et al. 1989, 1991). However, estimation and comparison of the model parameters by different available methods are scarcely conducted.

In this paper, two different laboratory techniques namely through-diffusion and half cell technique were used to estimate the model parameters. The tests were performed on duplicate specimens at a particular density and under same concentration gradient to understand the migration rate of a particular type of ion, to have a comparative analysis of the model parameters, in which not much literature is available.

The paper also describes the development of a dot-net application-based software package CONTRADIS that was built for the purpose of analyzing experimental results obtained from through-diffusion and half cell technique. This software is capable of using the contaminant transport data from experiments and predicting the diffusion and linear sorption parameters of the soils.

2 Governing Mechanism

The governing equation describing the transport of solute through saturated soil is formulated utilizing Fick's second law which is given as (Shackelford and Daniel 1991; Bharat 2014; Bharat et al. 2012):

$$\frac{\partial c}{\partial t} = \frac{D^*}{R_d}\frac{\partial^2 c}{\partial x^2}$$

where R_d is the retardation factor, and D^* is the effective diffusion coefficient.

3 Materials and Methods

Bentonite soil rich in montmorillonite mineral having liquid limit of 393% was used in the present study. The specific surface area and the cation exchange capacity of the soil are 495 m^2/g and 68 meq/100 g, respectively.

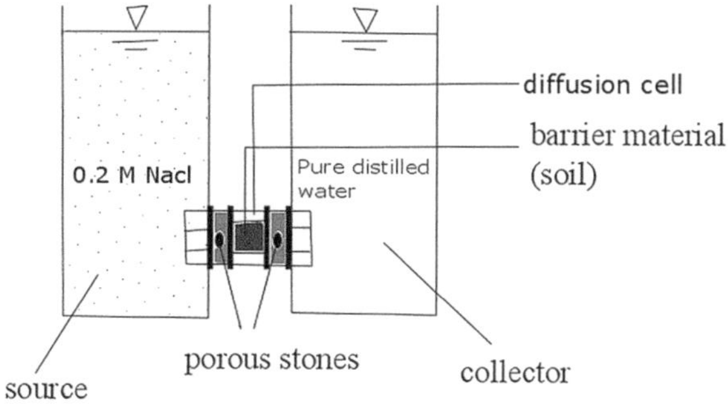

Fig. 1 Through-diffusion set-up

The diffusion cell, made of poly(methyl methacrylate) glass tubes having diameter of 2.5 cm and thickness of 1 cm, was used for the diffusion testing.

3.1 Through-Diffusion Technique

Bentonite soil compacted a particular density of 1.5 g/cc was placed in the diffusion cell and saturated for a period of 21 days. The clay plug was placed in contact with two reservoirs namely source and collector reservoir as depicted in Fig. 1. At the end of the saturation period, diffusion testing was started by placing NaCl solution of 0.2 M in the source reservoir and pure distilled water in collector reservoir. Due to this concentration gradient, one-dimensional diffusive mass flux would occur from source to collector reservoir. Both the reservoir concentrations were analysed with time to obtain the plot of relative concentration (c/c_0) of sodium ions in both the reservoirs with time. c_0 is the initial concentration which is 0.2 M or 12,000 ppm.

3.2 Half Cell Technique

Two half cells of diameter 3.9 and 4 cm length were used in this technique. In both the half cells, soil is compacted at 1.5 g/cc where one half cell was saturated with 0.2 M NaCl and the other half cell is saturated with pure distilled water. After saturation, both the cells were brought in contact with each other and diffusion process took place from the tagged half cell with NaCl to the untagged half cell (Fig. 2). At the end of the diffusion testing period, both the half cells were dismantled and the soil in the cells was sliced at 5 mm thickness and analysed for concentration variation along the length of the half cells.

Based on various initial and boundary conditions, analytical solutions are developed for both the methods (Bharat 2014; Robin et al. 1987).

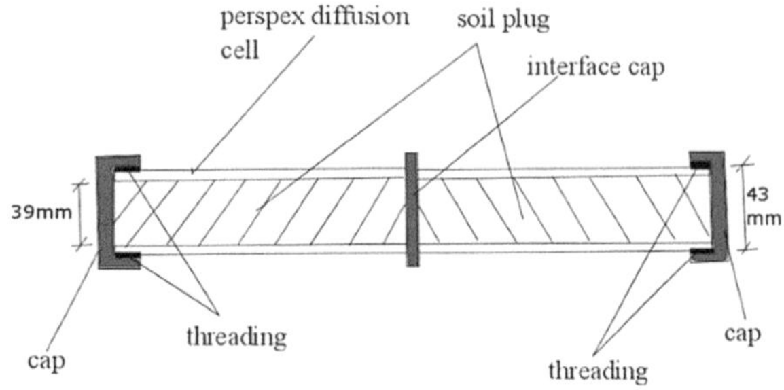

Fig. 2 Half cell set-up

3.3 Batch Test

Batch study was carried out in the present study to determine one of the model parameters experimentally which is the retardation factor and compare with the theoretical result. The soil-to-solution ratio maintained in the test is 1:20. Batch equilibrium test was performed on air-dried soil to determine the sorption characteristics of the clay soil with respect to sodium ions. Ten gram of the soil is mixed with 200 mL of NaCl of different initial concentrations. Six different initial concentrations are maintained in six different soil samples having weight of 10 g. The six different initial concentrations that were analysed are 2, 4, 6, 8, 10 and 12 g/L. The soil solution is mixed thoroughly in an orbital shaker for 48 h as equilibrium time is assumed to be 48 h. After thorough mixing, the slurries are taken in a centrifuge tube and centrifuged for 2000 rpm for 10 min. The supernatant from each tube is analysed for the equilibrium concentration of the sodium ion by flame photometer. The results were plotted to determine the adsorption isotherm. The sorbed concentration of sodium ions in the soil was plotted against the equilibrium concentration of the ion in the solution to get the sorption isotherm which is given as:

$$q = \frac{(C_0 - C_e) \times V}{M_s} \tag{1}$$

where V is the volume of the solution, and M_s is the mass of the soil sample.

4 CONTRADIS

The name CONTRADIS stands for "CONtaminant TRAnsport due to DIffusion in Soils." CONTRADIS is a software package that was built for the purpose of analyzing the experimental readings obtained from various methods that have been used by the previous studies in the field of contaminant transport in soils.

This application-based software package was developed to overcome the shortcomings of the existing software POLLUTE which is a commercially available software and is based on semi-analytical solutions. However, the present application CONTRADIS can perform inverse analysis and estimate the model parameters using stochastic optimization algorithm (Bharat et al. 2012).

The CONTRADIS utilizes the analytical solutions (Bharat 2014; Robin et al. 1987) of both the above-described experimental methods and can generate theoretical concentration profile by performing inverse analysis. This software package was verified on the synthetic data obtained from the calculated concentration value as an input data. With the input parameter of D^* and $R_{d,}$ a theoretical concentration profile is obtained and the actual estimate of the model parameters are taken to be the one for which the experimental profile fits well the theoretical profile with minimum RMSE.

5 Results and Discussions

5.1 Results from Through-Diffusion Technique

The concentration profile was obtained for a period of 60 days for both the reservoirs. It was observed that the concentration in the source reservoir decreased with time and concentration in the collector reservoir increased with time. The experimental data was used in CONTRADIS for the inverse analysis which compares it with the theoretical obtained data by minimizing the RMSE error. The fitting of the theoretical and the experimental concentration profiles as obtained from the software by inverse analysis is given in Fig. 3. The effective diffusion coefficient so obtained is 7.7×10^{-6} cm^2/s and retardation factor is 10.

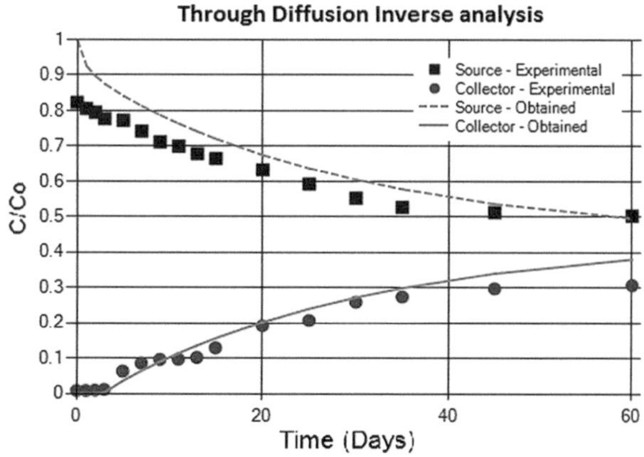

Fig. 3 CONTRADIS conc. profile for through-diffusion test

5.2 Results from Half Cell Technique

After dismantling the half cells at the end of the diffusion testing period, the soil plugs were sliced and analysed for the sodium ion concentration after mixing the soil with 1 M ammonium acetate and mixing it for a period of 16 h. Ammonium acetate replaces all the adsorbed and the free exchangeable ions present in the soil layer. The mixture was then centrifuged and the supernatant so obtained was analysed for total sodium ion concentration for all the slices. The mobile ion or the free ion concentration was obtained by subtracting the total ion concentration from the adsorbed concentration for the same dried mass in the particular slice. A total of 16 soil slices were analysed. The concentration profile along the length of the two half cell obtained from the experimental half cell technique shows that the profile reaches to the ends of the half cell which is finite porous medium case. Hence, the theory of finite porous medium case is used in CONTRADIS, and the experimental data was analysed in the software package for getting the model parameters by inverse analysis. The theoretical data was fitted with the experimental data with minimum root mean square error. The best-fit curve showing the theoretical concentration data and experimental data is shown in Fig. 4. The effective diffusion coefficient so obtained is 1.51×10^{-5} cm^2/s and retardation factor is 1.5.

5.3 Results from Batch Test

Batch test results were utilised to obtain both linear and nonlinear sorption isotherms. Figures 5 and 6 show that the linear sorption isotherm does not fit the experimental data as accurately as nonlinear Freundlich model does. This is because of the fact that at higher concentration, the sorption isotherms are highly nonlinear. However for simplicity of solving the governing differential equation (1), the sorption isotherm is considered linear. Because of this assumption, we can see that

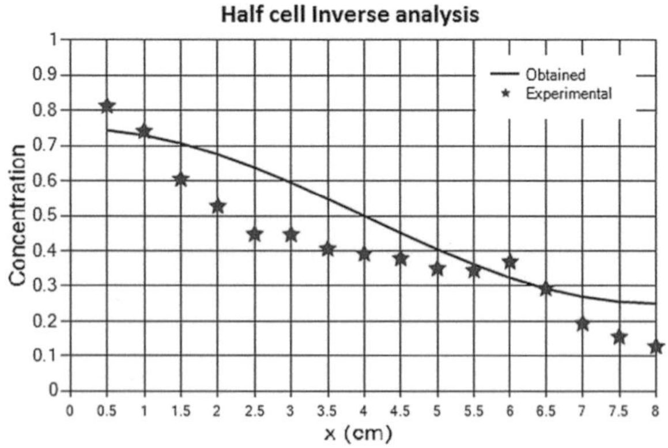

Fig. 4 CONTRADIS conc. profile for half cell test

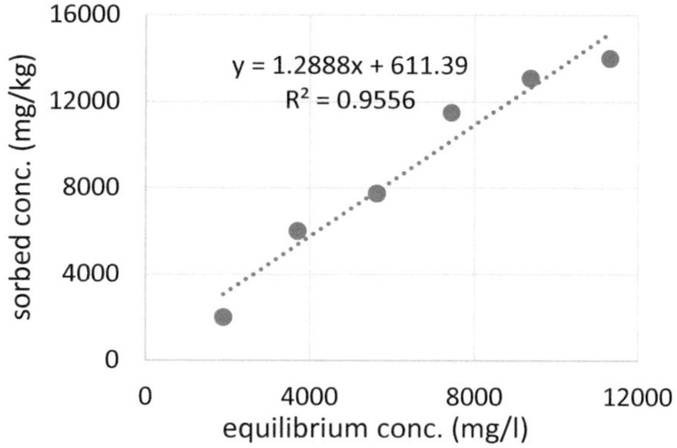

Fig. 5 Linear isotherm model

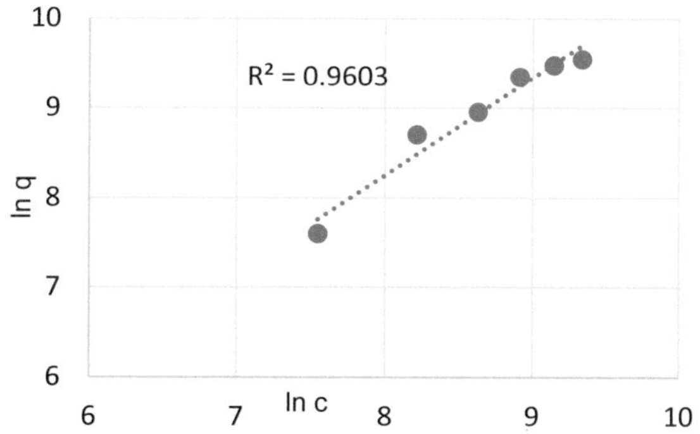

Fig. 6 Freudlich isotherm model

in Figs. 3 and 4 experimental data and the theoretical data could not be as well-fitted as is expected. The equations describing the process to determine the retardation factor from linear and Freundlich model can be found elsewhere (Shackelford et al. 1989, 1991). The retardation factor from the linear model obtained experimentally is 4.8, and the retardation factor obtained from the linearized Freundlich model is 6.1.

6 Summary and Conclusions

Measured salt concentration profiles with time and length of the soil plugs were obtained for two different laboratory techniques viz. through-diffusion and half cell technique, respectively. A new software package named CONTRADIS was developed which utilizes the analytical solutions of the governing differential equation, for both these techniques to determine the model parameters by inverse analysis using optimization algorithm. In order to validate one of the model parameters namely retardation factor, laboratory batch equilibrium test was conducted. The test results show that the model parameters obtained by through-diffusion technique were more reliable as the retardation factor obtained experimentally nearly matched the theoretically obtained value. The results of the half cell technique might not be considered as realistic owing to experimental error like improper contact between the half cells. However, more tests need to be conducted on various contaminating species using both these techniques to understand the reliability of the model parameters which would be useful in the design of landfill liners.

References

Barone, F. S., Rowe, R. K., & Quigley, R. M. (1992). A laboratory estimation of diffusion and adsorption coefficients for several volatile organics in a natural clayey soil. *Journal of Contaminant Hydrology, 10,* 225–250.

Bharat, T. V. (2014). Analytical model for 1D contaminant transport through clay barriers. *Environmental Geotechnics, 1*(4), 210–221. (2013).

Bharat, T. V., Sivapullaiah, P. V., & Allam, M. M. (2012). Robust solver based on modified particle swarm optimization for improved solution of diffusion transport through containment facilities. *Expert Systems with Applications, 39*(12), 10812–10820.

Kau, P. M. H., Binning, P. J., Hitchcock, P. W., & Smith, D. W. (1999). Experimental analysis of fluoride diffusion and sorption in clays. *Journal of Contaminant Hydrology, 36,* 131–151.

Robin, M. J. L., Gillham, R. W., & Oscarson, D. W. (1987). Diffusion of strontium and chloride in compacted clay-based materials. *Soil Science Society of America Journal, 51,* 1102–1108.

Rowe, R. K., & Booker, J. R. (1985). 1-D pollutant migration in soils of finite depth. *Journal of Geotechnical Engineering, 111*(4), 479–499.

Shackelford, C. D., & Daniel, D. E. (1991). Diffusion in saturated soil. II. Results for compacted clay. *Journal of Geotechnical Engineering, 117*(3), 485–506.

Shackelford, C. D., Daniel, D. E., & Liljestrand, H. M. (1989). Diffusion of inorganic chemical species in compacted clay soil. *Contaminant Hydrology, Amsterdam, the Netherlands, 4*(3), 241–273.

A Study on Influence of pH and Organic Chemical on the Retention Capacity of Red Earth

A. Subhashini, Y. Sudheer Kumar and P. Hari Krishna

Abstract Improper dumping practices may consume more land and contaminate the surrounding environment. There is a necessity to design a proper sanitary landfill in order to control the environmental effects. If a locally available soil meets the criteria of landfill material, the locally available material can be used to construct the compacted clay liner (CCL). This paper presents the feasibility study on the utilization of the locally available red earth as a landfill liner based on its contaminant sorption capacity. The potential for retention of lead (Pb^{2+}) by red earth from the Warangal city in India is examined. Lead solution of varying initial concentrations and pH values were used for the analysis. In addition, an attempt is made to observe the influence of organic chemical (EDTA) on the retention capacity of red earth. Batch sorption tests were conducted for a single salt solution with and without organic chemical, and results were presented. From this study, it is revealed that with an increase in the initial concentration of the lead solution from 10 to 30 mg/L, the percentage removal was decreasing. Whereas with an increase in pH from 2 to 7, the maximum increase in percentage removal was observed as 23% at maximum initial concentration (30 mg/L). With the presence of organic chemical, at neutral pH value, the decrease in the sorption capacity of soil was at least 36% at maximum initial concentration (30 mg/L).

Keywords Red earth · Adsorption · Heavy metal · EDTA · Lead
Batch sorption

A. Subhashini (✉) · Y. Sudheer Kumar · P. Hari Krishna
Department of Civil Engineering, NIT Warangal, Warangal, India
e-mail: subhashinianathan@gmail.com

Y. Sudheer Kumar
e-mail: dr.sudheernitw@gmail.com

P. Hari Krishna
e-mail: phknitw@gmail.com

V. K. Stalin and M. Muttharam (eds.), *Geotechnical Characterisation and Geoenvironmental Engineering*, Lecture Notes in Civil Engineering 16,
https://doi.org/10.1007/978-981-13-0899-4_23

1 Introduction

Landfills are designed and constructed to contain discarded waste, and they play a significant role in the waste management practice. But soil and groundwater systems are still getting spoiled due to the ingression of toxic heavy metals from the leachate generated within the landfill system. To reduce the pollution caused by leachate or industrial effluents, there is a need for an effective liner material to contain the toxic heavy metals in terms of their less permeability as well as good sorption capacity. Also, physicochemical properties of the liner material such as redox potential, pH value, and moisture content will affect the rate of adsorption of heavy metals. Sorption studies done by Adamcova (1999) suggested that sorption capacity of geomaterials in real time was better predicted by composite heavy metal solution rather than a single heavy metal solution. Later, desorption studies by Gomes et al. (2001) indicated that under certain conditions such as metal complex formation, changing the acidity, ionic strength and redox potential, desorption of heavy metals occurs. Research studies by Pickering (1986) revealed that addition of complexing ligands to the standard salt solutions of heavy metals can cause the mobility of metal ions through soil.

In the present study, efforts were being made to understand the sorption capacity of the red earth for the heavy metal "lead" combined with a complexing ligand, EDTA with varying concentration and pH values.

2 Soil and Contaminant Solutions

Red earth available in the Warangal city of Telangana was used for the present study, and properties are shown in Table 1.

Table 1 Physical and chemical properties of red earth

Physical properties	Chemical composition (%)
Gravel (%) = 6	SiO_2 = 61.9
Sand (%) = 27	MgO = 0.34
Silt (%) = 41	Al_2O_3 = 8.40
Clay (%) = 26	$CaCO_3$ = 24.98
Liquid limit (%) = 38 Plastic limit (%) = 21 Plasticity index (%) = 17 Soil classification-CI	Fe_2O_3 = 4.38
MDD (g/cc) = 1.80 OMC (%) = 19.7	
FSI (%) = 5	
$K^* = 0.141 \times 10^{-7}$ cm/s	

K permeability

The properties of red earth are conforming with the specifications required for liner material as detailed in studies of Daniel (1993) and Rowe et al. (1995). Research data on contaminated site by Yarlagadda et al. (1995) reported heavy metal concentration values ranging from 10 mg/L. In this study, standard salt solutions of known concentrations ranging from 10 to 30 mg/L are prepared from lead nitrate salt of analytical grade. EDTA is chosen for this study owing to its properties of complexing ligand as suggested by Pickering (1986) and its considerable presence in leachate (nearly 30%) as reported by European Amino-Carboxylates Producers Committee.

3 Experimental Program

3.1 Batch Sorption Studies

A series of batch sorption tests were conducted on red earth according to ASTM D 4646 (2008) to determine its removal efficiency for the lead with and without EDTA solution. Sample preparation involves mixing of selected quantity of red earth (say 5 g) with 100 mL of lead solution of varying initial concentrations (10, 15, 20, 25, and 30 mg/L). Later, pH of the soil solution matrix (2, 4, 6, and 7) was stabilized to required value. The filtered clear solution obtained was analyzed using inductively coupled plasma-optical emission spectrophotometer (ICP-OES) for determining the concentration of heavy metal retained by the red earth. In the present study, the volume of EDTA added was 5% of that of the volume of lead solution to prepare Pb-EDTA mixture.

4 Results and Discussions

4.1 Soil Contaminated with Lead Solution

Figure 1 shows the variation of the percentage removal of lead with respect to pH for concentrations ranging from 10 to 30 mg/L.

From this figure, it is observed that the removal efficiency of the soil for lead is increasing with an increase in the pH of the soil solution matrix. This is in accordance with research works of McBride 1994 and Sparks 1995 where studies showed that increasing the soil pH increases the cationic heavy metal retention through adsorption and inner sphere surface complexation. Removal efficiency of soil decreases with an increase in the initial concentration of lead solution due to decrease in the number of sorption sites available on the soil with an increase in the number of metal ions in the heavy metal solution. The increase in the value of the

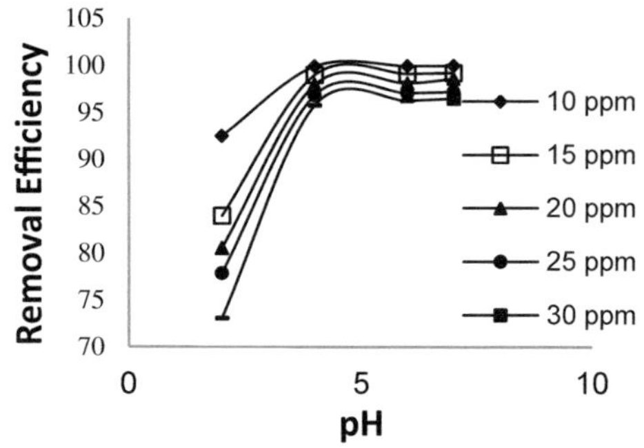

Fig. 1 Removal efficiency variation with pH

removal efficiency, with an increase in pH, was found to be in the range of 7–23% for all the initial concentrations.

4.2 Soil Contaminated with EDTA Mixed Lead Solution

Figure 2 shows the variation of removal efficiency of soil upon addition of 5% of EDTA to lead solution (10–30 mg/L).

In contrast to the pattern observed in Fig. 1 (without EDTA), there is a constant decrease in the removal efficiency with an increase in the pH upon addition of EDTA. This is because most of the lead ions will remain complexed to the EDTA species at higher pH values as indicated by studies of Vohra (2010). The soluble metal complexes thus formed can migrate through soil easily, thus leading to lower values of percentage removal.

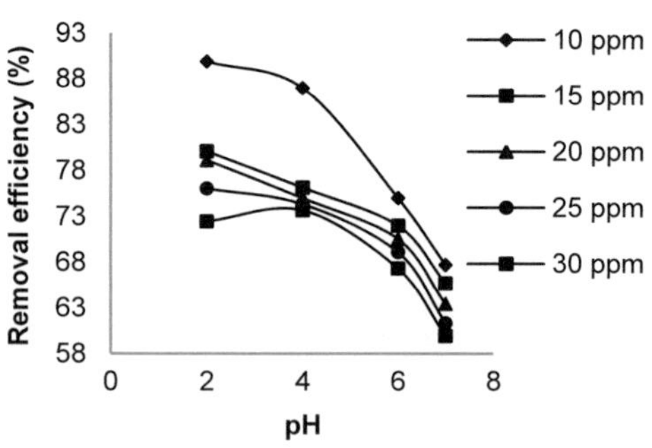

Fig. 2 Effect of EDTA on variation of removal efficiency with pH

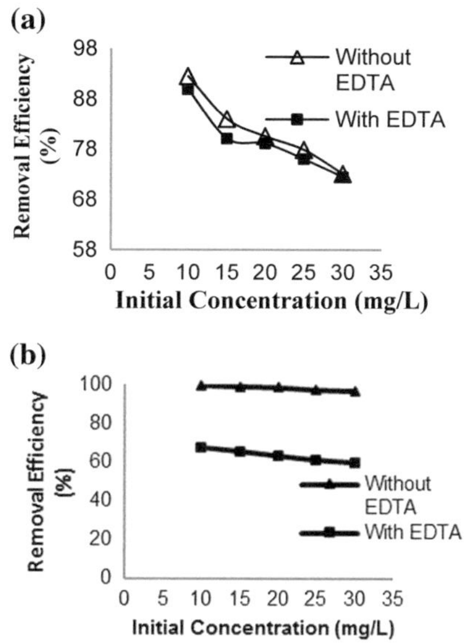

Fig. 3 **a** Effect of EDTA on removal efficiency at pH 2. **b** Effect of EDTA on removal efficiency at pH 7

4.3 Comparative Studies with and Without EDTA

Figure 3a, b shows the influence of EDTA on removal efficiency of soil by comparing with and without EDTA conditions at pH values of 2 and 7. As observed from the figures, the decrease in removal efficiency upon addition of EDTA is insignificant at a low pH of 2 as compared to pH 7. Metal complexation process is highly affected by hydrogen ion concentration. Hence at lower pH, increase in H^+ ions promoted co-precipitation of hydrous oxides of iron which can adsorb Pb and prevent its complexation with EDTA. However at pH 7, due to chelation effect of EDTA soluble metal complexes were formed which could not be retained by soil. The phenomenon of chelation was best given by the studies of Farrah and Pickering (1977) and Peters and Shem (1992) as the nature of EDTA to bond with the heavy metal ions by means of its six donor atoms resulting in the formation of a metal complex.

5 Conclusions

The following conclusions are made based on the laboratory studies conducted on red earth to retain lead. The removal efficiency of lead by the red earth is increasing with an increase in the pH value of the solution. The presence of EDTA resulted in the formation of stable metal–ligand complexes in the solution, thus hindering the heavy metal from being adsorbed onto the soil. Even under the organic chemical presence (EDTA), the removal efficiency of the soil toward lead is about 60%. These findings are indicating that red earth can be used as a liner material even in the zones where organic chemical ingression is present.

References

Adamcova, A. (1999). Heavy metal retention capacity of natural clay liners of landfills. In *Geoenvironmental engineering* (pp. 247–254). London: Thomas Telford.

ASTM D 4646-03. (2008). 24 H Batch type measurement of contaminant sorption by soils and sediments.

Daniel, D. E. (1993). *Geotechnical practices for waste disposal*. London: Chapman and hall.

Farrah, H., & Pickering, W. F. (1977). The sorption of lead and cadmium species by clay minerals. *Australian Journal of Chemistry, 30,* 1417–1422.

Gomes, P. C., Fontes, M. P. F., Silvan, A. G., Mendonca, E. S., & Netto, A. R. (2001). Selectivity sequence and competitive adsorption of heavy metals by Brazilian soils. *Soil Science Society of American Journal, 65,* 1115–1121.

Peters, R. W., & Shem, L. (1992). Adsorption/desorption characteristics of lead on various types of soil. *Environmental Progress, 11,* 234–240.

Pickering, W. F. (1986). Metal ion speciation-soils and sediments (a review). *Ore Geology Reviews, I,* 83.

Rowe, et al. (1995). *Clay barrier systems of waste disposal facilities*. London: E & SPON.

Vohra, M. S. (2010). Adsorption of lead, ethylenediaminetetraacetic acid and lead-ethylenediaminetetraacetic acid complex onto granular activated carbon. *International Journal of Environmental Science and Technology, 7*(4), 687–696. (Autumn 2010).

Yarlagadda, P. S., et al. (1995). Characteristics of heavy metals in contaminated soils. *ASCE Journal of Environmental Engineering,* 276–286.

Effect of Induced Osmotic Suction on Swell and Hydraulic Conductivity of an Expansive Soil

M. Julina and T. Thyagaraj

Abstract Compacted expansive soils are widely used as engineered barriers in waste contaminant applications like landfills, brine ponds, and nuclear waste disposal sites. These liners are designed for very low hydraulic conductivity ($<1 \times 10^{-7}$ cm/s). Percolation of chemical waste or leachate results in physico-chemical changes in compacted expansive soils which increases the hydraulic conductivity. This paper brings out the changes in swelling behavior and hydraulic conductivity of compacted expansive soil induced with osmotic gradients using NaCl and $CaCl_2$ solutions. Multiple identical soil specimens placed in oedometer assemblies were inundated with distilled water, 0.4 and 4 M NaCl (monovalent cations), and 0.4 and 4 M $CaCl_2$ (divalent cations) salt solutions and allowed to swell under a surcharge pressure of 12.5 kPa. Void ratio–water content plots were also traced during swelling process. Falling head permeability tests were conducted on swollen soil specimens in rigid wall oedometer permeameters under a hydraulic gradient (i) of 20. The experimental results showed that the swell potentials reduced and hydraulic conductivity increased with the increase in induced osmotic suction.

Keywords Swell potential · Void ratio · Hydraulic conductivity

1 Introduction

Compacted expansive soils are used as base and cover liners in waste contaminant applications. A well-constructed liner protects groundwater from leachate contamination. In order to achieve this, the clay liners are designed for low hydraulic conductivity ($<1 \times 10^{-7}$ cm/s). Leachate percolating through solid waste in

M. Julina · T. Thyagaraj (✉)
Department of Civil Engineering, Indian Institute of Technology Madras, Chennai 600036, India
e-mail: ttraj@iitm.ac.in

M. Julina
e-mail: julina.lina15@gmail.com

V. K. Stalin and M. Muttharam (eds.), *Geotechnical Characterisation and Geoenvironmental Engineering*, Lecture Notes in Civil Engineering 16,
https://doi.org/10.1007/978-981-13-0899-4_24

landfills contains dissolved or suspended chemicals. The design of clay liners based on distilled water as the inundating fluid does not replicate the conditions that exist during the life span of landfills. Hence, the behavior of compacted expansive soils exposed to chemicals needs to be studied. The properties of compacted expansive soils are altered upon exposure to chemicals due to physicochemical changes (Rao et al. 2006; Thyagaraj and Rao 2010; Thyagaraj et al. 2016).

Studies related to the effect of pore fluid on the swelling behavior have showed that the increase in concentration of pore fluid resulted in reduction of swell potentials and void ratios (Jo et al. 2001; Rao et al. 2006; Thyagaraj and Rao 2010; Thyagaraj et al. 2016). Osmotic gradient arises due to the concentration difference between inundating fluid and pore fluid (usually distilled water) and results in diffusion of salts. Increase in salt concentration in pore fluid results in suppression of diffused double layers, and the suppression further increases with the increase in cation valency (Yong and Wankentin 1966).

In addition to the swelling behavior, the concentration of inundating salt solution has greater influence on the hydraulic conductivity as well (Jo et al. 2001; Petrov and Rowe 1997; Lin and Benson 2000). Compacted expansive soils are characterized by double structure, with macropores (inter-aggregate) and micropores (pores within the aggregates) (Gens and Alonso 1992; Musso et al. 2013). Increase in the pore fluid concentration reduces the diffused double layer thickness and the size of micropores thereby increasing the macropore volume and hence results in higher hydraulic conductivity (Mesri and Olson 1971; Musso et al. 2013).

This paper presents the effect of induced osmotic suction on swell potential, time–swell plots, void ratio–water content plots, and hydraulic conductivity of compacted expansive soil. To bring out the effect of induced osmotic suction on swell potential, identical soil specimens, prepared with distilled water, were inundated with distilled water (DW), sodium chloride (NaCl), and calcium chloride ($CaCl_2$) solutions in separate oedometer assemblies. The void ratio–water content plots were traced during swelling process. Falling head permeability tests were also carried out to determine the hydraulic conductivity of compacted soil specimens subjected to induced osmotic suction.

2 Experimental Program

2.1 Material Properties

The expansive soil collected from Kishkinta, near West Tambaram, Chennai, Tamil Nadu, India, was air dried, pulverized, and passed through 2-mm sieve for testing. Table 1 summarizes the properties of expansive soil used in the present study. NaCl and $CaCl_2$ laboratory grade salts were used for the preparation of inundating solutions.

Table 1 Properties of expansive soil

Property	Value
pH	7.52
Specific gravity, G_s	2.71
Atterberg limits	
Liquid limit (%)	78
Plastic limit (%)	27
Shrinkage limit (%)	9
Grain size distribution (%)	
Sand	3
Silt	15
Clay	82
Unified soil classification	CH
Free swell index (%)	105
Compaction characteristics	
Maximum dry density (mg/m^3)	1.34
Optimum moisture content (%)	27

2.2 *One-Dimensional Oedometer Swell Tests*

The effect of induced osmotic suction on the swell potential of compacted soil specimens was brought out by conducting one-dimensional oedometer swell tests. Initially, the pulverized soil was pre-wetted with required amount of distilled water and allowed to equilibrate for 48 h. After attaining the target water content, the pre-wetted soil was statically compacted into the oedometer rings (30-mm height and 75.4-mm diameter) with initial height of 20 mm. Identical soil specimens prepared with standard Proctor maximum dry density (MDD) and optimum moisture content (OMC) were used for testing. Soil specimens were allowed to swell under a surcharge pressure of 12.5 kPa in the oedometer assembly. The swell potential was calculated as:

$$\text{Swell}(\%) = \frac{H}{H_i} \times 100$$

where ΔH is the increase in height of soil specimen during swelling process and H_i is the initial height of soil specimen.

The effect of induced osmotic suction on swelling behavior was brought out by inundating the identical compacted specimens prepared using distilled water with distilled water, NaCl solutions (0.4 and 4 M), and $CaCl_2$ solutions (0.4 and 4 M), in separate oedometer assemblies. Table 2 presents the details of inundating fluid. For the same concentration, osmotic suction value is more for divalent salt solution than monovalent salt solution. The dimensions of identical soil specimens were measured during intermediate stages of swelling. The calculated void ratios and

Table 2 Details of compacted specimens

Fluid used for specimen preparation	Inundating fluid	Induced osmotic suction, $\Delta\pi$ (kPa)
DW	DW	100
DW	0.4 M NaCl	1,951
DW	4 M NaCl	19,512
DW	0.4 M $CaCl_2$	2,927
DW	4 M $CaCl_2$	29,268

corrected water contents were used to trace the void ratio–water content plots during the swelling.

Induced osmotic suction is calculated using van't Hoff equation as

$$\Delta\pi = (M_2 - M_1)iRT$$

where $(M_2 - M_1)$ is the difference in concentration between the pore water (DW) and the corresponding inundating fluid in mole/liter, R is universal gas constant (8.32 L kPa/mole K), T is absolute temperature (K), and i is van't Hoff factor (i = 2 and 3 for NaCl and $CaCl_2$ solutions, respectively).

After complete saturation in the rigid wall oedometer permeameters, the hydraulic conductivity was determined under a hydraulic gradient, i, of 20 as per ASTM D 5856-15.

3 Results and Discussion

3.1 Swell Potential

Figures 1 and 2 compare the time–swell plots of distilled water inundated specimens with NaCl and $CaCl_2$ solutions of different concentrations, respectively. The initial (<0.5 min), primary (0.5–250 min), and secondary (after 250 min) swelling stages were well identified in specimens inundated with distilled water, NaCl, and $CaCl_2$ solutions. However, the specimen inundated with 4 M $CaCl_2$ solution showed a delayed initial swell. In general, the initial and primary swell stages are relatively rapid and are independent of induced osmotic suction, whereas the diffusion of salts at the end of primary and secondary swelling stages resulted in reduction of swell potential in salt solution inundated specimens in comparison with the specimen inundated with distilled water (Rao et al. 2006). The reduction of swell potential was more with divalent Ca^{2+} ions than monovalent Na^{+} ions at both concentrations. The soil specimen inundated with salt solutions with higher initial induced osmotic suction had lower swell potential. The reduction in swell potential of salt solution inundated specimens is due to the suppression of diffuse double layer happening at the microstructural level (Musso et al. 2013).

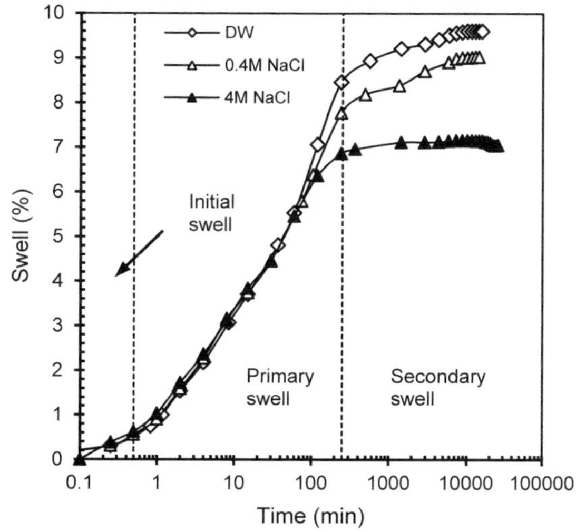

Fig. 1 Time–swell plots of compacted soil specimens inundated with distilled water and NaCl solutions

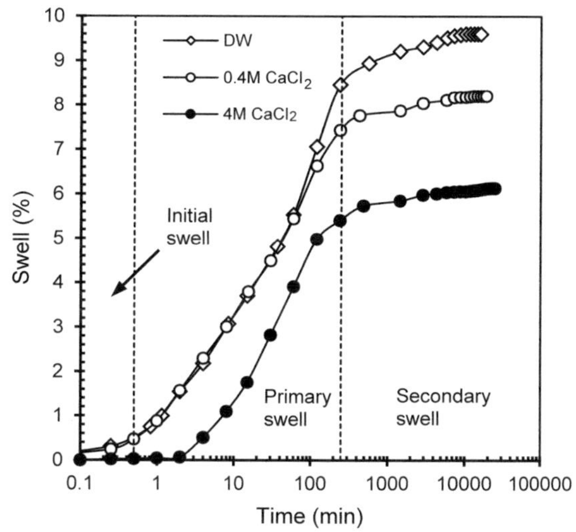

Fig. 2 Time–swell plots of compacted soil specimens inundated with distilled water and $CaCl_2$ solutions

3.2 *Void Ratio–Water Content Plots*

Figure 3 presents the void ratio–water content plots during swelling for the soil specimens inundated with distilled water and 0.4 M salt solutions. The swollen void ratios and swollen water contents of compacted specimens reduced with the increase in initial induced osmotic suction. During the initial swelling process, the distilled water inundated soil specimens showed a void ratio variation parallel to zero air voids line. With the increase in initial induced osmotic suction, the void ratio–water content curves became flatter. Similar results were reported by

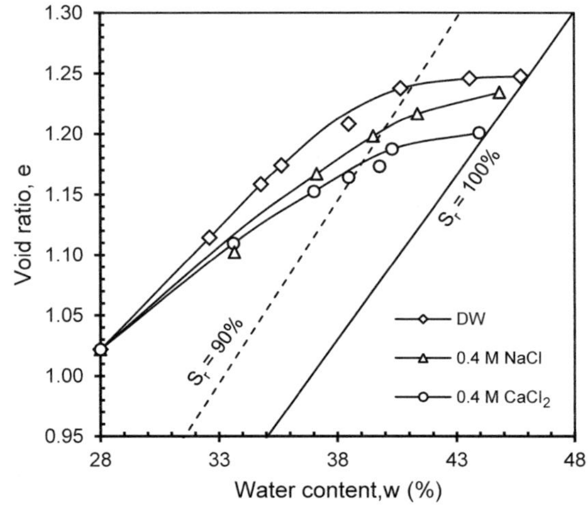

Fig. 3 Void ratio–water content plots during swelling for compacted soil specimens inundated with distilled water and 0.4 M salt solutions

Thyagaraj and Rao (2010) for specimens inundated with NaCl solutions. At any given water content, the void ratio decreased with the increase in induced osmotic suction which is attributed to the suppression of diffused double layer thickness.

3.3 *Hydraulic Conductivity*

In order to bring out the effect of induced osmotic suction on the hydraulic conductivity of compacted expansive soil, the rigid wall permeability tests were conducted on compacted specimens at swollen state. The inundating fluid was used as permeating fluid during permeability tests. Figure 4 presents the variation of hydraulic conductivity with the concentration of permeating solution. It can be seen that the hydraulic conductivity increases with increase in initial induced osmotic

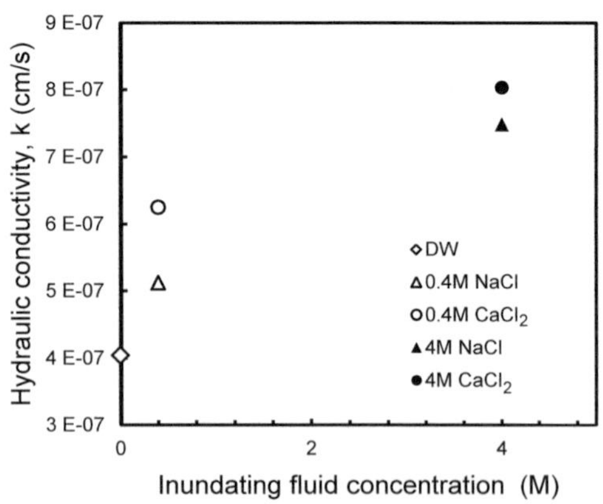

Fig. 4 Variation of hydraulic conductivity with inundating fluid concentration for compacted soil specimens

suction when the soil specimens were inundated and permeated with salt solutions. At a given concentration, the soil specimens inundated and permeated with $CaCl_2$ solution with higher induced osmotic suction value had higher hydraulic conductivity than NaCl solution inundated and permeated soil specimen. Mishra et al. (2005) also reported that the divalent cations were more effective in increasing the hydraulic conductivity than the monovalent cations. The difference in hydraulic conductivity between higher concentration (4 M) monovalent and divalent salt solutions was less when compared to the difference at lower concentration (0.4 M).

The increase in hydraulic conductivity value was significant for the specimen inundated with higher concentration solution in comparison with the specimens inundated with lower concentration salt solutions and distilled water. Higher the concentration of pore fluid, higher is the reduction of diffused double layer thickness and the micropores and hence results in higher permeability owing to the increase in the size of macropores (Musso et al. 2013).

4 Summary and Conclusions

Interaction of leachate alters the property of expansive soil and has significant impact on swelling and hydraulic properties of expansive soil. Based on the experimental results, the following observations were drawn:

Swell potential of soil specimens inundated with calcium chloride solutions was lower than the sodium chloride solutions and distilled water inundated soil specimens. Also, the swell potential of compacted expansive soil reduced with increase in initial induced osmotic suction.

Void ratios and water contents of compacted specimens reduced with the increase in the initial induced osmotic suction during the swelling process. The suppression of diffused double layers, i.e., micropores, resulted in reduction of swell potentials and void ratios.

Hydraulic conductivity of compacted expansive soil increased with the increase in induced osmotic suction and cation valence.

Additional investigations are in progress to evaluate the effect induced osmotic suction on hydraulic conductivity during alternate wetting and drying cycles. Also, the variation of swell potential and void ratio–water contents plots during alternate wetting and drying cycles needs to be studied.

References

ASTM D5856-15. (2015). *Standard test method for measurement of hydraulic conductivity of porous material using a rigid-wall, compaction-mold permeameter*. West Conshohocken, PA: ASTM International.

Gens, A., & Alonso, E. E. (1992). A framework for the behaviour of unsaturated expansive clays. *Canadian Geotechnical Journal, 29*(6), 1013–1032.

Jo, H. Y., Katsumi, T., Benson, C. H., & Edil, T. B. (2001). Hydraulic conductivity and swelling of nonprehydrated GCLs permeated with single-species salt solutions. *Journal of Geotechnical and Geoenvironmental Engineering, ASCE, 127*(7), 557–567.

Lin, L., & Benson, C. H. (2000). Effect of wet-dry cycling on swelling and hydraulic conductivity of GCLs. *Journal of Geotechnical and Geoenvironmental Engineering, ASCE, 126*(1), 40–49.

Mesri, G., & Olson, R. E. (1971). Mechanisms controlling the permeability of clays. *Clays and Clay Minerals, 19,* 151–158.

Mishra, A. K., Ohtsubo, M., Li, L., & Higashi, T. (2005). Effect of Salt Concentrations on the permeability and compressibility of soil-bentonite Mixtures. *Journal of the Faculty of Agriculture, Kyushu University, 50*(2), 837–849.

Musso, G., Romero, E., & Vecchia, G. D. (2013). Double-structure effects on the chemo-hydro-mechanical behaviour of a compacted active clay. *Geotechnique, 63*(3), 206–220.

Petrov, R. J., & Rowe, R. K. (1997). Geosynthetic clay liner (GCL)-chemical compatibility by hydraulic conductivity testing and factors impacting its performance. *Canadian Geotechnical Journal, 34,* 863–885.

Rao, S. M., Thyagaraj, T., & Thomas, H. R. (2006). Swelling of compacted clay under osmotic gradients. *Geotechnique, 56*(10), 707–713.

Thyagaraj, T., & Rao, S. M. (2010). Influence of osmotic suction on the soil-water characteristic curves of compacted expansive clay. *Journal of Geotechnical and Geoenvironmental Engineering, ASCE, 136*(12), 1695–1702.

Thyagaraj, T., Thomas, S. R., & Das, A. P. (2016) Physico-chemical effects on shrinkage behavior of compacted expansive clay. *International Journal of Geomechanics, ASCE,* 06016013.

Yong, R. N., & Wankentin, B. P. (1966). *Introduction to soil behaviour*. New York: The Macmillan Company.

Attenuation Characteristics of Laterite-Fly Ash-Bentonite Mix as Liner

T. Muhsina and S. Chandrakaran

Abstract The liners play an important role in controlling migration of contaminant present in the leachate in waste containment system such as landfills. Compacted clay liners are widely used because of their low permeability and large attenuation capacity. Nowadays the lack of availability of natural clay with satisfactory engineering properties resulted to look for an alternative material for liner. Commercial clay such as bentonite is commonly used for liner material. Fly ash is an industrial waste produced from thermal power plant, and disposal of this material is a serious problem. The present study is carried out to evaluate the capability of laterite-fly ash-bentonite liner to attenuate contaminant in municipal solid waste leachate. Laterite soil for the present study is collected from NIT Calicut campus. A series of laboratory tests were conducted to investigate the combined effect of fly ash and bentonite added at different percentage of lateritic soil as landfill liner. Suitable laterite-fly ash-bentonite mixes satisfying liner criteria were identified, and batch adsorption tests are performed on this mixes with hexavalent chromium as model contaminant. 78% removal efficiency was achieved for laterite soil mixed with 30% fly ash and 5% bentonite.

Keywords Laterite · Fly ash · Bentonite · Land fill liner · Adsorption

1 Introduction

With increasing economic growth and rapid urbanization of developing country like India, the amount of municipal solid waste generated increases rapidly and their safe disposal is an environmental challenge. Different treatment processes can be used

---

T. Muhsina · S. Chandrakaran (✉)
Department of Civil Engineering, National Institute of Technology,
Calicut 673601, India
e-mail: chandra@nitc.ac.in

T. Muhsina
e-mail: muhsinatm5@gmail.com

V. K. Stalin and M. Muttharam (eds.), *Geotechnical Characterisation and Geoenvironmental Engineering*, Lecture Notes in Civil Engineering 16,
https://doi.org/10.1007/978-981-13-0899-4_25

such as incineration, disinfection, and landfilling. Landfill is the more attractive due to its less investment, if a suitable disposal area can be found. An important component of a landfill is a layer of compacted, low permeability soil that is intended to act as a hydraulic barrier and prevents the leachate from contaminating the ground water. Many experimental and analytical studies have been reported regarding utilization and application of laterite, fly ash, and bentonite as liner material and adsorption studies on soil with various admixtures (Hettiaratchi et al. 1999; Bradl 2004; Bellir et al. 2005; Chalermyanont et al. 2008; Mohammed and Naik 2009; Amadi 2010; Sree et al. 2010; Amadi 2012; Amadi and Eberemu 2013; Varma et al. 2013; Satyanarayana et al. 2013; Shankara et al. 2014). Most of these studies mainly focused on the suitability of laterite, fly ash, bentonite, and their combinations as a low permeability barrier in engineering landfill. But not much study has been conducted on their capabilities for attenuation and retention of contaminant in the leachate. This study is to verify the efficacy of fly ash stabilization in enhancing compatibility between lateritic soil and solid waste leachate. It will give the feasibility of using fly ash-bentonite-soil mix for attenuation and retention of contaminants in the leachate using hexavalent chromium as the model contaminant.

2 Materials Used for the Study

2.1 Laterite Soil

The laterite soil used in the study was locally available soil. The basic properties of laterite are given in Table 1.

2.2 Fly Ash

The fly ash used in the investigation was class F fly ash collected from Mettur thermal power plant, Tamil Nadu. The basic properties of fly ash are given in Table 2.

Table 1 Basic properties of laterite

S. No.	Properties	Value
1	Specific gravity	2.6
2	% fines	64
3	% sand	26
4	% gravel	10
5	Liquid limit (%)	40
6	Plastic limit (%)	20
7	Maximum dry unit weight (kN/m^3)	16.4
8	Optimum moisture content (%)	18
9	Permeability (cm/s)	5.12E-05

Table 2 Basic properties of fly ash

S. No.	Properties	Value
1	Specific gravity	2.07
2	Maximum dry unit weight (kN/m^3)	12.8
3	Optimum moisture content (%)	24

Table 3 Basic properties of bentonite

S. No.	Properties	Value
1	Specific gravity	2.5
2	% clay	65
3	% silt	34
4	% sand	1
5	Liquid limit (%)	400
7	Plastic limit (%)	56

2.3 Bentonite

The bentonite used in the investigation was sodium bentonite. The basic properties of bentonite are given in Table 3.

3 Experimental Investigations

3.1 Lateritic Soil-Fly Ash Mixtures

Lateritic soil and fly ash of different proportions (10, 20, 30, and 40%) were combined to check the design criteria of liner materials. Various laboratory tests like compaction, falling head permeability, UCC strength were conducted, and results are given in Table 4.

Table 4 Compaction, permeability, and UCC with fly ash content

% laterite	% fly ash	$_{dmax}$ (kN/m^3)	OMC (%)	*k* (cm/s)	UCC (kPa)
100	0	16.4	18	5.12E-05	245
90	10	16.3	18.6	7.63E-06	264
80	20	15.8	19	5.81E-06	335
70	30	15.5	20	3.40E-06	235
60	40	15	20.5	2.34E-06	220

3.2 Lateritic Soil-Fly Ash-Bentonite Mix

Bentonite content of 3 and 5% was incorporated with different proportions of lateritic soil-fly ash mixtures to check liner material design criteria.

Figure 1 shows variation of plasticity index of soil with fly ash and bentonite. It was observed that liquid limit and plasticity index of the soil increased with increase in the bentonite content.

Figure 2 shows variation of permeability of soil with fly ash and bentonite. The permeability of soil gets decreased by the addition of bentonite because it forms a gel and fills most of voids.

Figure 3 shows variation of UCC of soil with fly ash and bentonite. UCC of samples decreased while adding bentonite, and the reduction is partly attributed to the increase in clay content that filled void between soil particles, which lowered the frictional resistance between the soil particles at their contact points. However, the recorded values of the various soil mixtures exceeded the general specification for the performance of liner.

Fig. 1 Plasticity index of soil with fly ash and bentonite

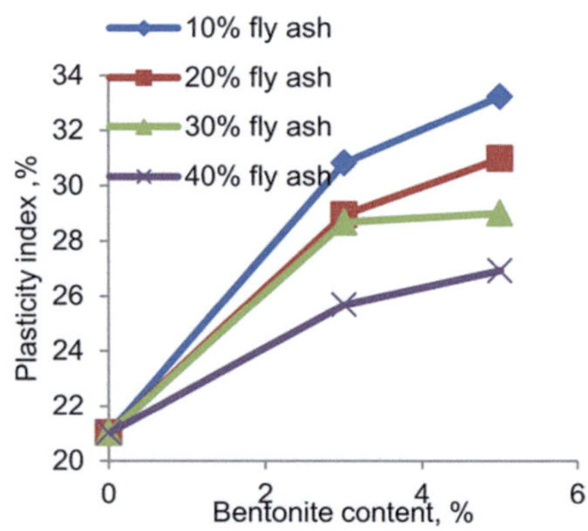

Fig. 2 Permeability of soil with fly ash and bentonite

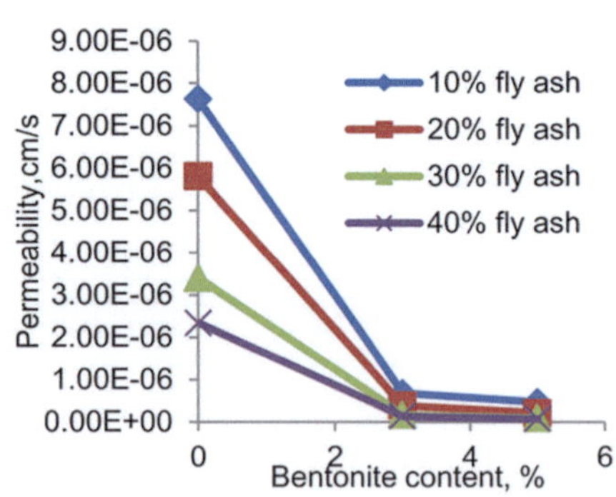

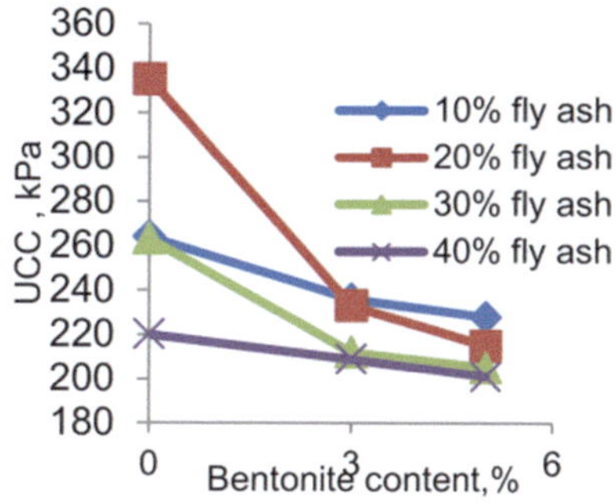

Fig. 3 UCC of soil with fly ash and bentonite

Figure 4 shows variation of volumetric shrinkage of soil with fly ash and bentonite. Volumetric shrinkage increased by the addition of bentonite. But it is less than acceptable limits (<4%).

3.3 *Determination of Best Mix*

From the studies best mixes satisfying geotechnical criteria for liner is found out. The best mixes in Table 5 can be used for finding out the attenuation capacity for major heavy metals in solid waste leachate.

3.4 *Adsorption Characteristics*

For the present study metal solutions were prepared synthetically for investigating adsorption characteristics of blended laterite soil. For hexavalent chromium the adsorbate solution was prepared as follows: About 2.828 g potassium dichromate ($K_2Cr_2O_7$) was dissolved in 1 liter distilled water to get chromium (Cr (VI))

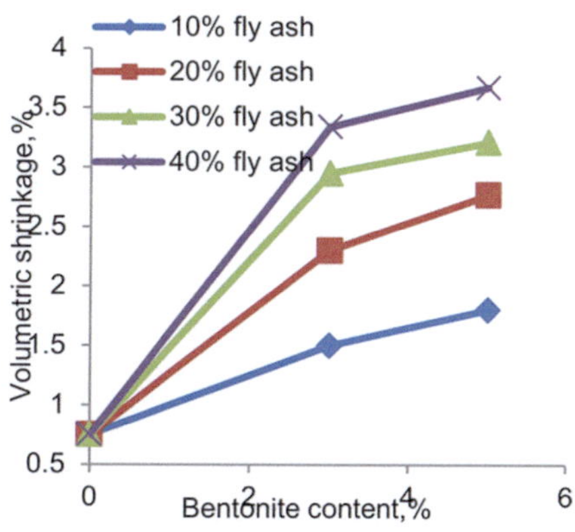

Fig. 4 Volumetric shrinkage of soil with fly ash and bentonite

Table 5 Best mixes satisfying all the geotechnical criteria for liner materials

Laterite (%)	Fly ash (%)	Bentonite (%)	*k* (cm/s)	UCC (KPa)	PI (%)	VS (%)
67	30	3	1.70×10^{-7}	212	28	2.9
65	30	5	9.90×10^{-7}	209	29	3.2
57	40	3	1.40×10^{-7}	205	26	3.3
55	40	5	8.00×10^{-7}	201	27	3.7

concentration of 1000 mg/L. Both low and high concentrations of Cr (VI) solutions were prepared by diluting with deionized water. Solutions of 5–35 ppm (Cr (VI)) were prepared for the analysis.

3.4.1 Batch Adsorption Test

The study was carried out for determination of sorption capacity of soil mixes using Cr (VI) as contaminant, optimum dosage of adsorbent, and isotherm parameters for the optimum dosage. Figure 5 shows percentage adsorption of Cr (VI) as a function of optimum dosage.

3.4.2 Adsorption Isotherms

For developing adsorption isotherms, the equilibrium adsorption data obtained through batch studies were used. After determining concentration of Cr (VI) in aliquot using spectrophotometer, the Cr (VI) uptake by blended soil was calculated using the following equation:

$$S = \frac{(C_0 - C_e)V}{m} \tag{1}$$

Fig. 5 Percentage adsorption of Cr (VI) as a function of optimum dosage

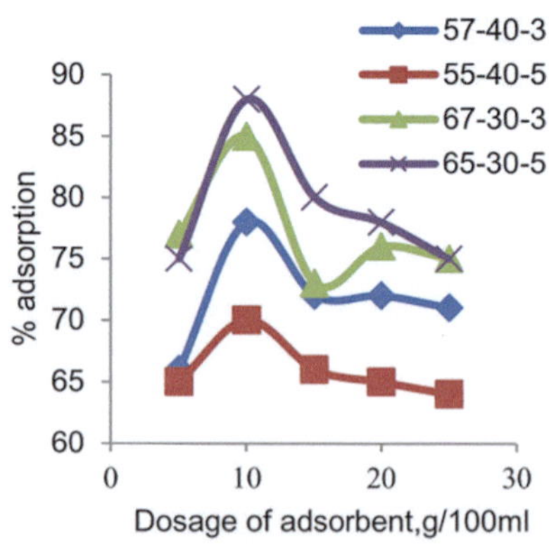

where S is the amount of adsorbate adsorbed (mg adsorbate/kg of adsorbent), C_0 and C_e are the initial and final concentration of adsorbate (mg/L), V is the volume of solution taken (L), and m is the mass of adsorbent taken (g).

There are several adsorption isotherm models. Among them the most commonly used ones are linear isotherm, Langmuir isotherm, and Freundlich isotherm.

The linear adsorption isotherm can be established by plotting a graph of q versus C_e. The relationship can be expressed as

$$q = K_d C_e \tag{2}$$

where K_d is the distribution coefficient which is a direct measure of separation of contaminant between solid and liquid phases.

The Langmuir equation can be expressed as

$$q_e = \frac{abC_e}{(1 + bC_e)} \tag{3}$$

where a, number of moles of solute adsorbed per unit weight of adsorbent; b, a constant related to energy of adsorption.

Freundlich isotherm is the most commonly used isotherm to describe the adsorption characteristics of the adsorbent used in water and wastewater treatment. It is expressed as:

$$q = \frac{x}{m} = K_f C_e^{1/n} \tag{4}$$

where K_f, Freundlich capacity factor; $\frac{1}{n}$, Freundlich intensity factor; $\frac{x}{m}$, amount of solute adsorbed per unit mass of adsorbent (mg/g).

Various adsorption isotherm parameters for chromium adsorption are listed in Table 6. By comparing the regression values of various isotherms, it can be seen that adsorption is suitable for both Langmuir and Freundlich isotherms. However, Langmuir model is best fitted one due to greater R^2 value; i.e., monolayer

Table 6 Isotherm parameters for Cr^{6+}

Type of isotherm	Parameters	Soil proportions			
		67-30-3	65-30-5	57-40-3	55-40-5
Linear isotherms	K_d (L/kg)	12.4	14.58	10.2	8.39
	R^2	0.5	0.793	0.81	0.82
Freundlich isotherm	K_f (mg/kg)	43	24	22	17.95
	n	2	2.5	1.48	1.5
	R^2	0.95	0.96	0.995	0.966
Langmuir isotherm	a (mg/kg)	676	1000	733	500
	b (L/kg)	0.17	0.05	0.06	0.07
	R^2	0.99	0.98	0.998	0.978

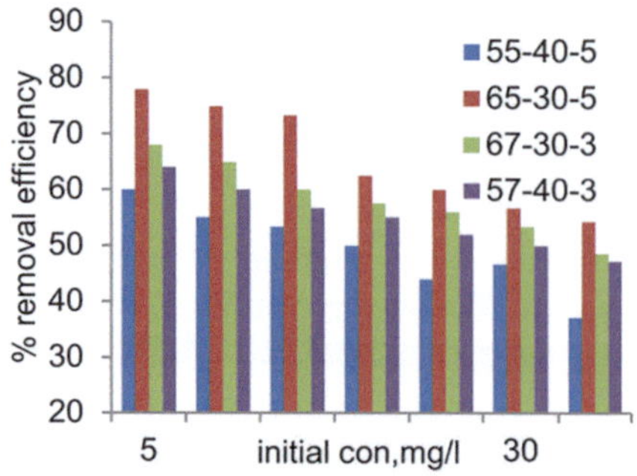

Fig. 6 Variation of initial concentration versus % adsorption

adsorption is taking place in this case. The value of n is found to be greater than 1, which indicates favorable adsorption. Figure 6 shows variation of initial concentration with % adsorption.

4 Conclusions

Addition of fly ash from 0 to 40% to the soil resulted in the reduction of permeability from 10^{-5} to 10^{-6} cm/s.

Fly ash content up to 20% increases the unconfined compressive strength of soil from 245 to 335 kPa. The maximum dry unit weight decreased and optimum moisture content increased with increasing fly ash content.

Addition of 3–5% bentonite to soil-fly ash mix causes significant reduction in the permeability. Unconfined compressive strength of laterite-fly ash mix decreased with the addition of bentonite. Volumetric shrinkage increased more than 300% when fly ash and bentonite is added to laterite soil.

Batch adsorption test on mixes satisfying geotechnical criterion with hexavalent chromium as model contaminant showed that the soil mixes possess good adsorption capacities.

78% removal efficiency was achieved for laterite soil mixed with 30% fly ash and 5% bentonite.

By comparing the regression values of various isotherms, it was noticed that adsorption is suitable for both Langmuir and Freundlich isotherms.

Based on the test results, it can be concluded that the lateritic soil, fly ash, and bentonite mixes can be effectively used as a liner material to control the mitigation of contaminant present in the leachate.

References

Amadi, A. A. (2010). Hydraulic conductivity tests for evaluating compatibility of lateritic soil—Fly ash mixtures with municipal waste leachate. *Geotechnical and Geological Engineering*, 259–265.

Amadi, A. A. (2012). Utilisation of fly ash to improve the engineering properties of lateritic soil. *International Journal of Materials Engineering Innovation, 3*(1), 78–88.

Amadi, A. A., & Eberemu, A. O. (2013). Characterization of geotechnical properties of lateritic soil-bentonite mixtures relevant to their use as barrier in engineered waste landfills. *Nigerian Journal of Technology, 32*(1), 93–100.

Bellir, K., Bencheikh-Lehocine, M., Meniai, A. H., & Gherbi, N. (2005). Study of the retention of heavy metals by natural material used as liners in landfills. *Desalination*, 111–119.

Chalermyanont, T., Arrykul, S., & Charoenthaisong, N. (2008). Potential use of laterite and marine clay soils as landfill liners to retain heavy metals. *Waste Management, 29,* 117–127.

Varma, G., Singh, R. K., & Sahu, V. (2013). A comparative study on the removal of heavy metals by adsorption using fly ash and sludge: A review. *International Journal of Application or Innovation in Engineering & Management, 2*(7), 45–56.

Bradl, H. B. (2004). Adsorption of heavy metal ions on soils and soils constituents. *Journal of Colloid and Interface Science*, 1–18

Hettiaratchi, J. P. A., Achari, G., Joshi, R. C., & Okoli, R. E. (1999). Feasibility of using fly ash admixtures in landfill bottom Liners or vertical barriers at contaminated sites. *Journal of Environmental Science & Health Part A*, 1897–1917

Satyanarayana, P. V. V., Harshitha, A., & Sowmya Priyanka, D. (2013). Utilization of red soil bentonite mixes as clay liner materials. *International Journal of Scientific & Engineering Research, 4*(5), 876–882.

Mohammed, S., & Naik, M. (2009). Adsorption characteristics of metals in aqueous solution by local materials with additives as liners for waste containment facilities. *Journal of Water and Environment Technology, 8*(1), 29–50.

Shankara, Maya, N., Sivapullaiah, P. V., & Mohammed, S. A. S. (2014). Sorption of iron and copper on sand bentonite fly ash mixtures. *International Journal of Research in Chemistry and Environment, 4*(2), 1–8.

Sree, D., Ajitha, A. R., & Evangeline, Y. S. (2010). Study on amended soil liner using lateritic soil. In *Indian Geotechnical Conference, GEO trend, Bombay* (pp. 381–384).

An Experimental Investigation on the Geoengineering Properties of Pond Ash-Bentonite Mixes

Suryaleen Rout and Suresh Prasad Singh

Abstract An engineered landfill necessitates an impervious barrier to control the intrusion of groundwater into the landfill facilities and migration of leachate to the surrounding ground. The liner material should have sufficient shear strength along with highly impermeable. This paper aims at investigating certain features of a novel material that may serve as a landfill liner material as a substitute to sand-bentonite mixture. In this investigation, bentonite was blended with coarse pond ash ranging from 0 to 30% by weight of the total material in 5% intervals. The compaction characteristics, strength behavior, and permeability characteristics of these mixes are evaluated. It is observed that with bentonite content varied from 20 to 30%, the average hydraulic conductivity reduces from 3×10^{-7} to 9×10^{-9} cm/s for samples compacted to MDD at OMC corresponding to a compactive effort of 595 kJ/m^3. However, when the samples are compacted with 2674 kJ/m^3 energy; the hydraulic conductivity as mentioned above is achieved at a bentonite content of 15%. With the gradual addition of bentonite the cohesion of the mixture increases whereas the angle of internal friction decreases. Hence, it is concluded that sand-bentonite mixes can effectively be replaced by pond ash-bentonite mixture in the landfill liners.

Keywords Bentonite · Pond ash · Landfill · Permeability and strength parameters

---

S. Rout (✉) · S. P. Singh
Department of Civil Engineering, National Institute of Technology Rourkela, Rourkela 769008, Odisha, India
e-mail: suryaleen1512@gmail.com

S. P. Singh
e-mail: spsingh@nitrkl.ac.in

V. K. Stalin and M. Muttharam (eds.), *Geotechnical Characterisation and Geoenvironmental Engineering*, Lecture Notes in Civil Engineering 16,
https://doi.org/10.1007/978-981-13-0899-4_26

1 Introduction

Wastes are generated more and more with increasing industrialization and population growth. Based on the safety level, these wastes can be controlled by different options such as waste reduction, separation and recycling, resources recovery through waste processing, waste transformation, and environmentally sustainable disposal on land. The most frequently used disposal option for solid waste is the landfill because of its low cost and efficiency. The core component of a waste disposal facility is its liner system. The design of liner is made so as to isolate the waste from the environment minimizing the passage of leachate into the groundwater. To ensure this the important characteristics for compacted landfill liners are selection of materials, hydraulic conductivity, and strength. Usually, soil rich in clay minerals is used as compacted liner materials for their low hydraulic conductivity which is required to be less than 1.0×10^{-7} cm/s. Instead of clay, mixture of expansive soil such as bentonite with fly ash and pond ash can be used as compacted barriers. The current work focuses on finding an accurate mixture of pond ash and bentonite that can be used as compacted clay liner. Even though pond ash and bentonite have extremely opposite properties, when mixed together they show complimentary behavior. Pond ash is highly permeable, but with the addition of bentonite the hydraulic conductivity can be reduced. Bentonite being an excellent sealant helps check the passage of leachate to the groundwater (Mishra et al. 2010). The bulk availability of pond ash helps to reduce the cost of raw materials required for liner as well as providing their safe disposal on a large scale. A number of researchers have used the pond ash and its engineering properties. This paper investigates the geoengineering properties of pond ash-bentonite mixes and its suitability as a landfill liner material. The non-availability of sand at all landfill sites triggers a search for an alternate material. Pond ash is an industrial waste and has the distinct advantage of a geotechnical material as it resembles the natural sand. Moreover, it has the similar mineralogical composition to conventional earth materials and needs special attention for its effective utilization rather than disposing it by consuming large land area leading to the loss of agricultural production.

2 Materials and Methodology

2.1 Bentonite

Bentonite is primarily composed of the montmorillonite group of minerals. Commercial sodium-based bentonite procured from local market is used in this experimental investigation.

2.2 Pond Ash

Pond ash used in the present investigation was collected from ash pond of National Thermal Power Plant (NTPC), Talcher, Odisha. The samples were mixed thoroughly to bring homogeneity and dried at an oven temperature of 105–110 °C. These were stored in airtight containers for subsequent use. The physical properties of the bentonite and pond ash are presented in Table 1.

2.3 Methodology

In this present study, synthetic mixes are prepared by adding bentonite of 5 to 30% by weight at 5% intervals with the pond ash. In this manner, six different mixes were synthesized. The OMC and MDD of these mixes corresponding to both standard and modified Proctor energy are determined as per IS: 2720 (Part 7) 1980 and IS: 2720 (Part 8) 1983, respectively. Prior to the compaction tests, these synthetic materials were mixed with initial water and were kept in an airtight container for 48 h for proper distribution of water in the mix. To find out the shear strength parameters synthetic mixes were compacted in standard Proctor mold to their respective MDD at OMC using both the standard and modified Proctor energy. Undisturbed specimens of size 60 mm × 60 mm × 25 mm were recovered from the mold and were sheared at a rate of 0.2 mm/min. The shear parameters (i.e., c and Φ values) were found out from the plots between normal stress and shear stress. These tests were performed as per IS: 2720 (Part 13) 1986. Unconfined compressive strength tests on specimens compacted to their corresponding MDD at OMC was performed according to IS: 2720 (Part 10) 1991. The cylindrical test specimens were of size 50 mm in diameter and 100 mm in height, and these were sheared at an axial strain rate of 1.25 mm/min. Hydraulic conductivity of different pond ash-bentonite mixes, compacted to their respective MDD at OMC, are evaluated as per IS: 2720(Part 17) 1987.

Table 1 Physical property of pond ash and bentonite

Physical parameters	Pond ash	Bentonite
Fines content, <75 μm (%)	1.99	96.93
Fine sand (%)	64.28	3.07
Medium sand (%)	33.73	0
Uniformity coefficient, C_u	1.15	–
Coefficient of curvature, C_c	2.91	–
Specific gravity	2.23	2.7
Liquid limit	–	301
Plastic limit	–	65
Plasticity index	NP	236
Soil type	SP	CH

3 Results and Discussion

3.1 Compaction Characteristics

Figure 1 shows that with the addition of bentonite, the OMC value decreases up to an optimum bentonite content after that it remains almost unchanged. Figure 2 shows the variation of maximum dry unit weight with bentonite content. It is found that with the addition of bentonite to pond ash, maximum dry unit weight value increases up to an optimum bentonite content after that it remains almost constant. This trend is observed for samples compacted with either standard or modified Proctor energy. The optimum content of bentonite for both the cases is found to be 20%. This may be due to the fact that when bentonite percentage is small, the finer particles of bentonite mostly go into the intra-void spaces present in the conglomerated ash particles. As the percentage of bentonite is further increased, these particles filled up the inter-particular void spaces available in the pond ash sample. This increases the maximum dry unit weight value. Further increase of bentonite

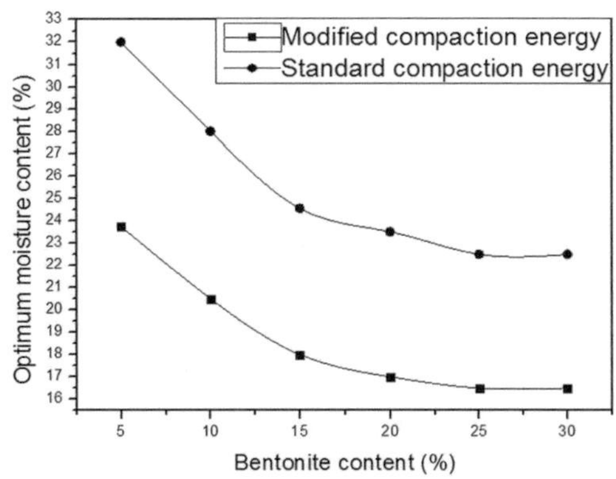

Fig. 1 Variations of optimum moisture content with bentonite content

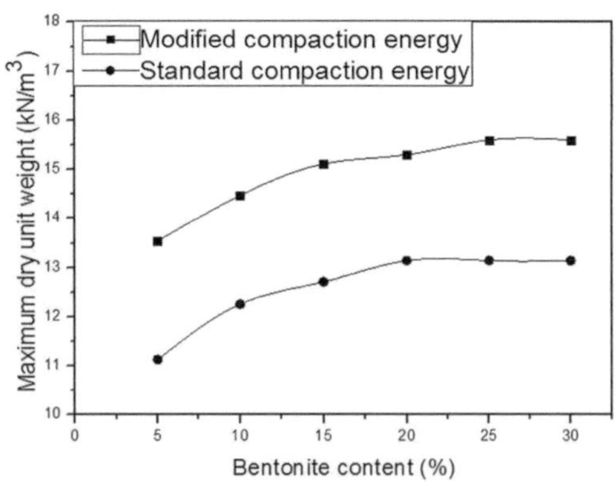

Fig. 2 Variations of maximum dry unit weight with bentonite content

creates a space for itself forcing the ash particles to move apart, thus stabilizing the maximum dry unit weight value.

3.2 *Hydraulic Conductivity*

The variation of hydraulic conductivity with bentonite content is shown in Fig. 3. It is obvious that an increase of bentonite content in the compacted mixtures lowers the hydraulic conductivity values. This has been verified by many studies. Since bentonite alone has characteristics of impervious material and has extremely low hydraulic conductivity values in the order of 10^{-10}–10^{-12} cm/s, it is expected to have lower hydraulic conductivities as its percentage in the mixture increases. The experimental data shows that the reduction in hydraulic conductivity with bentonite content is not linear. Initially, the rate of reduction is mild followed by a sharp decrease in the hydraulic conductivity value. At low bentonite content, the finer bentonite particles get adjusted in the intra-particle void space of pond ash; thus, there is an appreciable reduction in capillary void space. However, as its content increases it reduces the capillary void and thus has a sharp reduction in the hydraulic conductivity value.

3.3 *Unconfined Compressive Strength*

Unconfined compressive strength test was performed for sample prepared with both standard and modified compaction energy. Compressive strength of pond ash-bentonite mixture compacted for both the energy were found to be in the range of 52–197 kPa and 99–691 kPa, respectively, as shown in Fig. 4. As per the general requirement of isolation material such as liner, the compressive strength should have a value more than 200 kPa. The above strength is accomplished for the samples containing 15% or above bentonite and prepared with modified compaction energy.

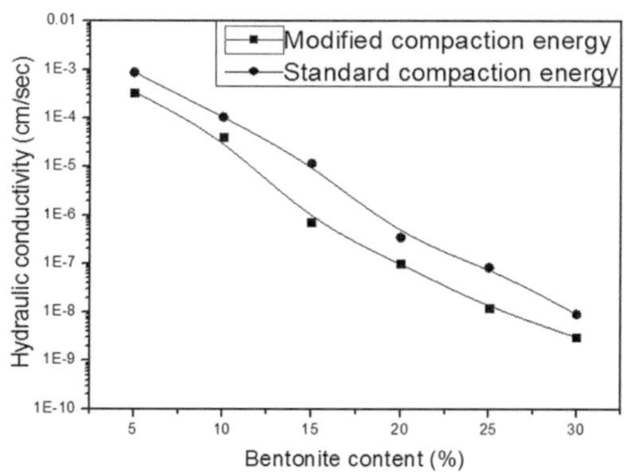

Fig. 3 Variations of hydraulic conductivity with bentonite content

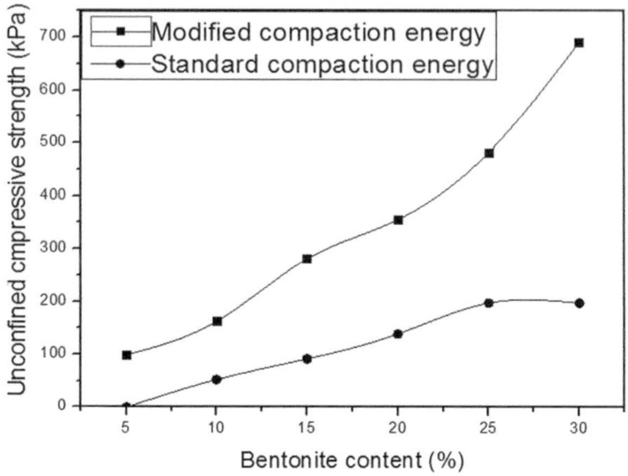

Fig. 4 Variations of unconfined compressive strength with bentonite content

3.4 Shear Strength Parameters

Shear strength parameters such as unit cohesion (c) and internal angle of friction (ϕ) were found to be in the range of 16.27–63.81 kPa and 42.5°–28.4° for mixtures prepared with standard compaction energy, 12.67–89.59 kPa and 44.6°–30.9° for the mixtures prepared with modified compaction energy when the bentonite content of the mixture is changed from 5 to 30%. Figures 5 and 6 present the variation of unit cohesion and angle of internal friction with bentonite content. It is seen that as the bentonite content increases, the cohesion increases and internal friction angle decreases. This may be due to the fact that as pond ash particles are irregular in shape, there exists a good interlocking between the particles and, hence, the angle of internal friction is high. When a smaller amount of bentonite is added to the mixture, the finer particle of bentonite tries to fill the intra-void spaces present in the conglomerated ash particles. But as the bentonite content increases, the excess bentonite coats the surfaces of the ash particle. As bentonite particles are lubricative in nature, the bentonite-coated pond ash particles losses their contact and slips over

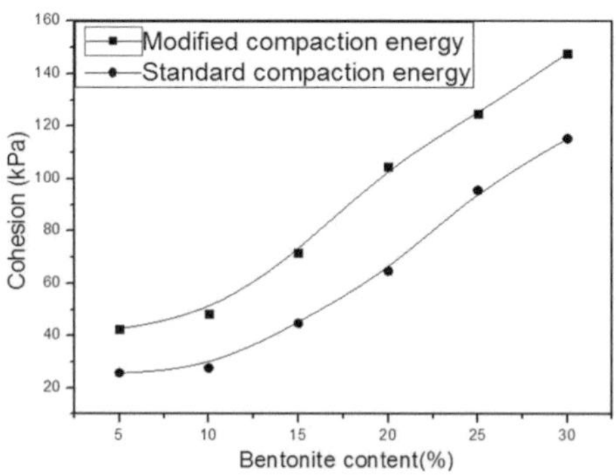

Fig. 5 Variations of cohesion with bentonite content

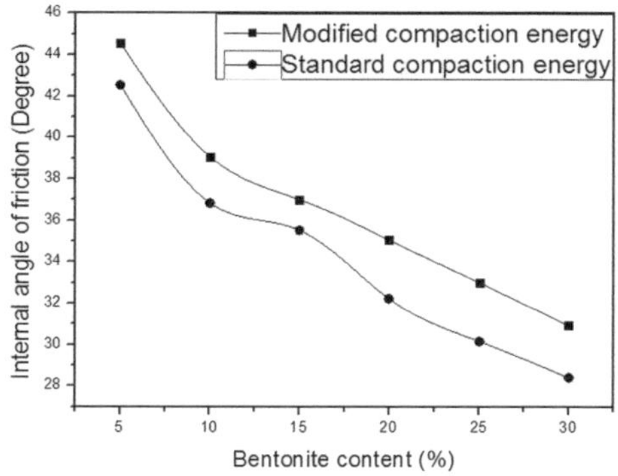

Fig. 6 Variations of internal angle of friction with bentonite content

each other and hence frictional angle decreases. However, due to the presence of strong inter-particular attractive forces between the bentonite particles the mixture gradually develops cohesion with an increase in bentonite content.

4 Conclusion

The suitability of pond ash-bentonite mixes as landfill liner material was investigated in this work. Based on the experimental investigation, the following conclusions are drawn:

- Samples compacted with standard compaction energy the maximum dry unit weight increases from 11.13 to 13.14 kN/m^3 and the corresponding OMC of the mixtures decreases from 32 to 22.5% as the bentonite content varies from 5 to 30%. However, with modified compaction energy, the maximum dry unit weight increases from 13.54 to 15.6 kN/m^3 and the corresponding OMC decreases from 23.75 to 16.5%
- Hydraulic conductivity values of bentonite-pond ash mixes compacted with both standard and modified compaction energy were found to be in the range of 8.83×10^{-4}–9.23×10^{-9} cm/s and 3.42×10^{-4}–5.15×10^{-9} cm/s with a change in bentonite content from 5 to 30%, respectively.
- Compressive strength of pond ash-bentonite mixture compacted for both the energy was found to vary from 52 to 197 kPa and 99 to 691 kPa, respectively.
- Shear strength parameters such as cohesion (c) and internal friction angle (ϕ) were found to be in the range of 16.27–63.81 kPa and 42.54°–28.4° for mixtures prepared with standard compaction energy. These values are 12.672–89.59 kPa and 44.55°–30.90° for the mixtures prepared with modified Proctor energy.

From this study, it can be concluded that sand-bentonite mixture can effectively be replaced by pond ash-bentonite mixture when sand is not easily available and wastes are to be utilized.

References

Akcanca, F., & Aytekin, M. (2012). Effect of wetting–drying cycles on swelling behavior of lime stabilized sand–bentonite mixtures. *Environmental Earth Sciences, 66*(1), 67–74.

Alam, J., Khan, M. A., Alam, M. M., & Ahmad, A. (2012). Seepage characteristics and geotechnical properties of flyash mixed with bentonite. *International Journal of Scientific Engineering Research, 3*(8), 1–11.

Mishra, A. K., Ohtsubo, M., Li, L. Y., & Higashi, T. (2010). Influence of the bentonite on the consolidation behaviour of soil–bentonite mixtures. *Carbonates and Evaporites, 25*(1), 43–49.

Influence of Engineering Behaviour of Coal Ash on Design of Ash Dykes

Aali Pant, Manoj Datta and Venkata Ramana Gunturi

Abstract Currently, nearly 170 million tonnes of coal ash is generated annually in India from 85 thermal power plants, and because of the limited availability of land, vertical expansions of ash dykes are unavoidable. The present paper compares the variability in engineering behaviour of coal ash from two different thermal power plants (TPP) A and B and highlights the influence of the variability in coarse particle content on design of ash dykes. Using the geotechnical properties of compacted ash as well as hydraulically deposited ash, ash dyke sections have been designed and a series of stability runs are carried out to map the factor of safety at various stages of ash dyke raising. The results indicate that slopes of ash dykes have to be made flatter as bottom ash content decreases in pond ash due to increased utilization or separate storage of bottom ash.

Keywords Ash dyke · Slope stability · Upstream method

1 Introduction

Indian coal is of low grade having high ash content of the order of 30–45% generating large quantity of coal ash at coal-based thermal power stations in the country. At present, nearly 184.14 million tons of coal ash is being generated annually in India and more than 70,000 acres of land is presently occupied by ash ponds. Similar trends can also be observed in many other developing countries. Although the scope for the use of coal ash in cement, concrete, brick making,

A. Pant (✉) · M. Datta · V. R. Gunturi
Department of Civil Engineering, Indian Institute of Technology Delhi, New Delhi 110016, India
e-mail: aali.pant@gmail.com

M. Datta
e-mail: mdatta@civil.iitd.ac.in

V. R. Gunturi
e-mail: ramana@civil.iitd.ac.in

V. K. Stalin and M. Muttharam (eds.), *Geotechnical Characterisation and Geoenvironmental Engineering*, Lecture Notes in Civil Engineering 16,
https://doi.org/10.1007/978-981-13-0899-4_27

soil-stabilization treatment and other applications has been well recognized (Ahmaruzzaman 2010; Naganathan et al. 2015), and the utilization of coal ash has increased from 6.64 million ton in 1996–97 to a level of 102.54 million ton in 2014–15 (CEA 2015), but till date huge quantities of coal ash have to be disposed off suitably on land by creating engineered ash pond to take care of environmental concerns.

This disposal problem can be minimized by utilizing ash in large earthworks for geotechnical applications like construction of highways, embankments and filling low-lying areas, and by reclaiming old abandoned ash ponds for constructing parks, parking areas and low weight structures. However, increasing use of ash in geotechnical engineering applications requires detailed understanding of its engineering behaviour.

Due to limited availability of land, the ash pond dykes are raised in stages once it is filled with ash to take care of further quantities of generated ash. Apart from the drainage facilities, the shear strength characteristics and compacted density of the coal ash used as construction material for raising of dyke plays an equally important role in stability of the raised dyke.

2 Objective and Scope

The present paper compares the variability in engineering behaviour of coal ash from two different thermal power stations (TPP) A and B and highlights the influence of the variability on design of ash dykes.

Experimental investigations have been carried out on representative ash samples from TPP-A and TPP-B. Influence of grain size variation in the shear strength parameters and the compaction characteristics of the samples are studied from laboratory results.

Using the geotechnical properties of compacted ash as well as hydraulically deposited ash, a four-stage ash dyke section has been analysed. Slope stability in steady-state seepage condition with water inside the pond up to full height of the hydraulically deposited ash has been carried out. A series of stability runs are carried out for static and seismic condition to check for the impact of shear strength on the downstream slope of incrementally raised dykes.

3 Variability in Coal Ash

Coal ash produced in a thermal power plant is basically of two types: bottom ash and fly ash. Bottom ash is the coarser ash which is collected from the bottom of a furnace, after being ground in a clinker grinding unit. Fly ash is the finer ash which is collected from hoppers beneath electrostatic precipitators (ESPs). The ratio of bottom ash to fly ash production in thermal plants is approximately 20:80 (ACAA 2001).

Conventionally, the unutilized volumes of ash (both bottom ash and fly ash) are mixed with large amounts of water and then disposed onto the ash ponds, which have already occupied 65,000 acres of valuable land in India (Parswal et al. 2003) and many million acres of land all over the world. This ash is referred to as pond ash. It is the ash stored in ash ponds deposited by hydraulic fill method. In the ash ponds, spatial variation of the pond ash characteristics can be observed. Near the slurry inflow point, the ash is predominantly sand sized while near the decant pond zone the ash is predominantly silt sized (Skarzynska et al. 1989; Datta et al. 1996).

4 Design Issue

The coal ash from the filled ash ponds is generally used as construction material in the raising of dyke by upstream method in order to save the borrow material and its transportation cost from the borrow site. The raising of dyke at one time may vary from 3 to 5 m and the final total height may reach up to 20–25 m.

Bottom ash is the coarser ash that is being widely used by the construction industry. The amount of bottom ash discharged into the ash pond is thereby reducing. At many thermal plants in India, separate lagoons have been constructed for storage of bottom ash and fly ash.

At present, in the ash ponds at some sites 20% bottom ash is present in pond ash, while at other places its amount is less than 10% and even nil at other locations. Thus, most of the ash deposited in the lagoons is of fine sized (silt size). In this paper, the issue of stability of slopes with reference to the variability of ash deposited in the pond has been addressed.

5 Experimental Investigation

Ash samples collected from the ash ponds of two different thermal power plants (TPP) in India: TPP-A and TPP-B were considered for the investigations to study their variability in engineering behaviour. In addition, one more set of samples, without bottom ash were obtained from TPP-B which will be referred to as TPP-B(F).

Representative ash samples were tested for complete geotechnical characterization. All the ash samples are non-plastic in nature. Specific gravity values of ash from TPP-A, TPP-B and TPP-B(F) were 2.29, 2.24 and 2.22, respectively. It can be observed that the specific gravity values of all the three ash samples are considerably lower than that of the natural soils (e.g., Yamuna sand). Nature of particles and their grain size distribution were studied for all the materials. Both sieve analysis and hydrometer analysis were performed.

The pond ash samples from TPP-A have higher sand content (50–55%) and can be classified as silty sand. On the other hand, pond ash samples from TPP-B have predominantly silt content (10–15% sand content) and can be classified as sandy silt. Since TPP-B(F) does not have bottom ash mixed in it and hence has only 5–10% sand content in it.

Compaction characteristics of the ash samples have been studied using standard proctor and vibratory table tests as per the Indian standard specifications. The results of grain size distribution and compaction test are reported in Table 1.

Effective-stress strength parameters c' and ϕ' of the ash samples were evaluated by performing direct shear tests on saturated samples in the loose and dense state. Loose samples were prepared by slow pouring and then saturating while dense samples were prepared by rodding and tamping and then saturating. The local soil is saturated silty clay, and the soil parameters have been obtained from the site investigation report. Table 2 shows the material properties used for the stability analysis.

Table 1 Engineering properties of ash samples

Property	TPP-A	TPP-B	TPP-B (F)
Sand size (4.75–0.075 mm), %	52.5	10.27	5.87
Silt size (0.075–0.002 mm), %	45.3	84.86	80.43
Clay size (<0.002 mm), %	0.9	4.86	13.7
MDD (std. proctor), kN/m^3	12.7	12.6	11.9
OMC, %	27.5	26.9	30.2
MDD (vibratory), kN/m^3	12.6	11.9	10.7
Minimum dry density, kN/m^3	8.4	8.1	8.0

Table 2 Material properties used in stability analysis

Materials		Total unit weight (kN/m^3)	c'(kPa)	ϕ' (°)
Foundation soil		20.00	10	31
Starter dyke (local soil)		19.00	9	28
TPP-A	C[a]	16.20	0	39
	HD[b]	13.7	0	29
TPP-B	C[a]	16.00	0	34
	HD[b]	13.50	0	28
TPP-B (F)	C[a]	15.50	0	30
	HD[b]	13.3	0	27

[a]Compacted ash (at OMC)
[b]Hydraulically deposited ash; fully saturated

6 Upstream Raising of Ash Dykes

Dykes of ash pond are raised by upstream method to take the volumes of further ash in stages. Since the pond ash is abundantly available on the construction site itself, it is the economical choice to use the ash as construction material for raising the ash dyke embankment. Top cover of natural earth is provided to take care of erosion and gully formation For analysis purpose, a four-stage dyke raising was considered by upstream method on the starter dyke. Following are the salient features of the ash dykes:

The downstream slope of the starter dyke has an inclination of 2.5H:1V.

The downstream slope of all the four raising has equal inclination. For each of the ash type, stability has been analysed for three cases characterized by different slopes of the ash dyke raisings:

- Case 1: Inclination of the four ash dyke raising is 3H:1V.
- Case 2: Inclination of the four ash dyke raising is 2.5H:1V.
- Case 3: Inclination of the four ash dyke raising is 2H:1V.

The starter dyke and each raising are 5 m high making the total height of ash dyke 25 m above ground level. The crest width of each raising is of 6 m. One more set of stability analysis has been done with the crest width of 3 m.

7 Results of Stability Analysis

Slope stability analysis was performed for both static and seismic conditions (Zone III) using GEOSLOPE version Slope/W 2007. The design values of horizontal seismic coefficient and vertical seismic coefficient have been calculated using the seismic coefficient method and taken as 0.12 and 0.06, respectively. The minimum acceptable factor of safety (FOS) for steady-state seepage condition in static and with seismic loading condition is kept 1.5 and 1.0, respectively.

The FOS obtained for the different ash samples are listed in Tables 3, 4, 5 and 6. As observed, FOS increases with increase in the shear strength parameter of the ash raising for both static and seismic condition. As the percentage of coarser particles

Table 3 FOS in static condition (crest width = 6 m)

Case	Average slope	Ash type		
		TPP-A	TPP-B	TPP-B (F)
Case 1 (3H:1V)	4.7H:1V	1.871	1.768	1.651
Case 2 (2.5H:1V)	4H:1V	1.628	1.549	1.455
Case 3 (2H:1V)	3.7H:1V	1.484	1.391	1.191

Table 4 FOS in seismic condition (crest width = 6 m)

Case	Average slope	Ash type		
		TPP-A	TPP-B	TPP-B (F)
Case 1 (3H:1V)	4.7H:1V	1.144	1.082	1.013
Case 2 (2.5H:1V)	4H:1V	1.064	1.009	0.946
Case 3 (2H:1V)	3.7H:1V	1.006	0.958	0.902

Table 5 FOS in static condition (crest width = 3 m)

Case	Average slope	Ash type		
		TPP-A	TPP-B	TPP-B (F)
Case 1 (3H:1V)	4H:1V	1.605	1.512	1.409
Case 2 (2.5H:1V)	3.5H:1V	1.352	1.276	1.189
Case 3 (2H:1V)	3H:1V	1.229	1.155	1.075

Table 6 FOS in seismic condition (crest width = 3 m)

Case	Average slope	Ash type		
		TPP-A	TPP-B	TPP-B (F)
Case 1 (3H:1V)	4H:1V	1.041	0.989	0.923
Case 2 (2.5H:1V)	3.5H:1V	0.933	0.882	0.823
Case 3 (2H:1V)	3H:1V	0.883	0.835	0.776

reduce, the raisings get restricted to have a flatter slope. As the slopes become steeper, slopes made of coal ash with higher shear strength remain stable. ($FOS_{static} > 1.5$ and $FOS_{seismic} > 1.0$).

With crest width of 6 m (Tables 3 and 4), dykes constructed by either of the three ash samples remain stable when the downstream slope of the ash dyke is 3H:1V (average slope of 4.7H:1V). As the downstream slope becomes steep, 2.5H:1V, ash dyke constructed using fly ash, TPP-B(F), become susceptible to failure with FOS of less than 1.5. Thus, it is not advisable to construct ash dyke raising with fly ash or pond ash of equivalent shear strength parameters with a slope of 2.5H:1V. For much steeper slope, i.e., 2H:1V, ash dyke constructed using any of the three ash samples does not remain stable in static steady-state seepage condition (FOS of 1.484, 1.39, 1.19).

With crest width of 3 m (Tables 5 and 6), the average slope of the ash dyke becomes steep. For 3 m of crest width, dykes constructed by fine ash (TPP-B(F)) fails for all the cases analysed. The ash dyke constructed with ash from TPP-A to TPP-B remains stable when the downstream slope of the ash dyke is 3H:1V (average slope of 4H:1V). As the downstream slope becomes steep, 2.5H:1V, ash dyke

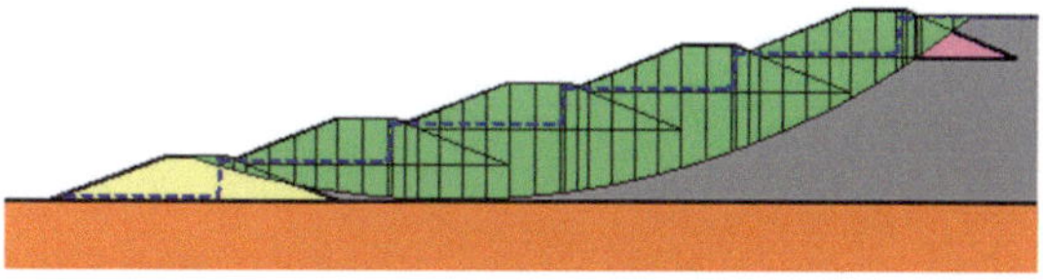

Fig. 1 Critical failure surface of TPP-A ash dyke for berm width of 6 m; FOS 1.628 (Case 2)

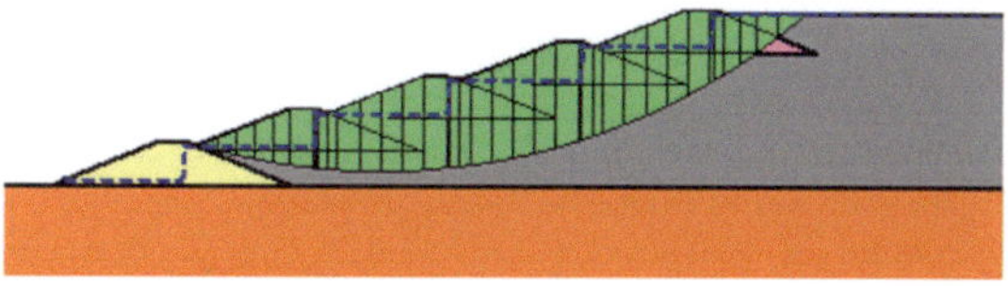

Fig. 2 Critical failure surface of TPP-A ash dyke for berm width of 3 m; FOS 1.352 (Case 2)

constructed using either of the ash samples become susceptible to failure with FOS of less than 1.5.

Increasing the crest width reduces the average slope of a dyke. As seen in Tables 5 and 6, with crest width of 3 m, the average slope of the dyke increases. Hence, only those raising constructed of ash TPP-A and TPP-B remain stable for static steady-state seepage condition when the downstream slope of the ash dyke is 3H:1V (average slope of 4H:1V). In all the other cases, the raisings are unstable in static as well as seismic condition.

Typical failure surfaces can be seen for static condition for the TPP-A coal ash in Figs. 1 and 2. It is noted that critical failure surfaces for all results of stability analysis look similar with almost the entire failure surface passing through the underlying hydraulically deposited ash which has lower strength than the compacted ash in the incrementally raised dykes.

8 Conclusions

At a time when the riverbed sand is getting scarce and expensive, use of bottom ash as a substitute for sand encourages sustainable development. As the utilization of bottom ash in the construction industry is gaining acceptance, the coarse fraction in the pond ash is decreasing.

Strength parameters are largely influenced by the percentage of coarse ash particles. From the experimental investigation, it is evident that as the percentage of bottom ash decreases in pond ash, the strength parameters of pond ash reduce. Thus, with reduced coarse fraction, slopes of the dyke raisings cannot be as steep as with pond ash containing 20% bottom ash particles. Also, it can be concluded that larger berm widths are required in order to achieve stability of dykes constructed on ash having lower percentage of coarse fraction.

References

Ahmaruzzaman, M. (2010). A review on the utilization of fly ash. *Progress in Energy and Combustion Science, 36*(3), 327–363.

American Coal Ash Association (ACAA). (2001). *CCP Survey*. Retrieved from https://www.acaa-usa.org/PDF/ACAA2001CCPS Survey.pdf.

Datta, M., Singh, A., & Kaniraj, R. (1996). Spatial variation of ash characteristics in an ash pond. In V. S. Raju, et al. (Eds.), *Ash pond and ash disposal systems* (pp. 111–119). Delhi: Narosa Publishers.

Naganathan, S., Almamon, Y., Omer, M., & Kamal, N. (2015). Performance of bricks made using fly ash and bottom ash. *Construction and Building Materials, 96,* 576–580.

Parswal, I. S., Makan, O. P., Atrea, A. K. (2003). Eco friendly ash management in the form of ash mounds. In: Proceedings CBIP Third International Conference-Fly Ash Utilisation and Disposal, New Delhi, India, vol. III, pp. 10–16.

Report on Fly Ash Generation at Coal/Lignite Based Thermal Power Stations and its Utilization in the Country for the year 2014–15. Central Electricity Authority, New Delhi October, 2015.

Skarzynska, K. M., Rainbow, A. K. M., Zawisza, E. (1989). Characteristics of ash in storage ponds. In: Proceedings of the 12th International Conference of Soil Mechanics and Foundation Engineering, Brazil, pp. 1915–1918.

Virus Transport Through Heterogeneous Unsaturated Zone in Guwahati City in Assam Under Transient State Condition

Mamata Das and Triptimoni Borah

Abstract Virus present in the groundwater is considered to be an important agent for waterborne diseases in India. In order to predict how far viruses can be transported and how long they can remain infective in soil and groundwater is desirable for proper management of the placement of sources of contamination so that they will not have an impact on drinking-water wells. With respect to that, a one-dimensional virus fate and transport model is developed for transient heterogeneous unsaturated flow to identify the transport parameters in the unsaturated zone. Simulation of virus transport in groundwater aquifer is necessary for predicting the vertical movement of virus in an aquifer and to implement remedial measures to inactive the virus present in the groundwater. The model involves solution of the advection–dispersion equation, which additionally considers virus inactivation rate in the solution. In case of unsaturated porous media, the transport of virus is responsible for some of the parameters such as linear distribution coefficient, hydrodynamic dispersion coefficient and inactivation coefficient for both aqueous and sorbed virus. As there are often changes in the state and content of soil water during flow, it is considered to be a highly nonlinear problem and for such it becomes necessary to solve the flow equation before solving the virus transport equation. In this study, finite element scheme computer-coded software, HYDRUS-1D is used to simulate the one-dimensional flow equation and virus transport equation. This study is mainly carried out for a particular location of Guwahati City, Assam, India. The viruses that is been employed in this study were the male-specific RNA coliphage MS2, and the Salmonella typhimurium phage, PRD1. For simulating the partial differential equation of virus transport equilibrium, solute transport model is selected with Crank–Nicholson as time weight scheme and Galerkin finite elements as space weight scheme. The purpose of this research is to determine the role that unsaturated flow conditions play in virus sorption and inactivation during transport through different

M. Das (✉) · T. Borah
Department of Civil Engineering, Assam Engineering College, Guwahati, Assam 781013, India
e-mail: mamtadas4@gmail.com

T. Borah
e-mail: btriptimoni@gmail.com

V. K. Stalin and M. Muttharam (eds.), *Geotechnical Characterisation and Geoenvironmental Engineering*, Lecture Notes in Civil Engineering 16,
https://doi.org/10.1007/978-981-13-0899-4_28

soil type. The effects of the moisture content variation on virus transport in unsaturated porous media were also investigated. The results obtained after several simulations indicate that the concentration of virus is affected by the moisture content and the heterogeneity of the soil profile during its flow through unsaturated zone. The model developed in this study can successfully simulate the virus transport through heterogeneous unsaturated columns.

Keywords Virus · Heterogeneous · Transient · HYDRUS · Unsaturated Crank–Nicholson · Galerkin · Advection · Dispersion

1 Introduction

Hazardous wastes, sewage sludge, fertilizers and pesticides from the ground surface are removed by filtering action of the unsaturated zone. This may result in high contents of organic matters and clay, which undergoes biological degradation, transformation of contaminants and sorption. Therefore, the unsaturated zone can be considered as a buffer zone for protecting the groundwater. Virus that is present in wastewater and sewage sludge is mainly the pathogenic microbial organisms which may lead to waterborne diseases. In order to understand spatial and temporal movement of these microorganisms, one has to simulate the fate and transport processes. These two processes are mainly governed by four processes: advection, hydrodynamic dispersion, inactivation process and adsorption onto the soil matrix. In this zone, the virus sorption and inactivation are influenced by the soil moisture content and subsurface temperature fluctuations (Vilker and Burge 1980; Vilker 1981; Thompson and Yates 1999), so the flow equation is solved before solving the transport equation.

Several mathematical models are developed for predicting the movement of viruses in porous media. (Tim and Mostaghimi 1991) developed a numerical model, VIROTRANS for simulating vertical movement of water and virus through soil treated with wastewater effluents and sewage sludge. (Yates and Ouyang 1992) developed a predictive model of virus fate and transport, VIRTUS which studies the effect of temperature-dependent inactivation rate, inactivation rates for sorbed versus free virus and the effect of soil type. (Sim and Chrysikopoulos 2000) developed a one-dimensional numerical model to investigate the effect of moisture content variation on virus transport on a homogeneous unsaturated porous media. They found that at low moisture content, the transport of virus is highly affected by the irreversible sorption of virus onto air–liquid interfaces. Anders and Chyrysikopoulos (2009) conducted laboratory-scale virus transport experiments in columns packed with sand under saturated and unsaturated condition. They found that the liquid to liquid–solid and liquid to air–liquid interface mass transfer rates increase as the saturation level was reduced. A mass conservative fully implicit finite difference model simulating moisture flow in the unsaturated zone with the hybrid finite volume model for virus transport is developed (Ojha et al. 2012). The accuracy of the numerical scheme is tested for both advection and dispersion-dominated transport.

This work provides a numerical model of virus fate and transport process through heterogeneous unsaturated porous media. The viruses employed in this study are MS2, the male-specific RNA coliphage and the Salmonella typhimurium phage, PRD1. In order to solve the moisture flow and virus transport equation, a one-dimensional computer code software HYDRUS-1D is used that is based on finite element scheme. This study is carried out in a 'Parbatkha' location of Guwahati City, Assam, India.

2 Model Development

2.1 *Governing Equation*

The mass conservation equation for the transport of water and virus through variably saturated media under transient flow condition can be written as:

Richard's flow equation (Tim and Mostaghimi 1991):

$$\frac{\partial \theta}{\partial t} = \frac{\partial}{\partial z}\left[K(\theta)\left(\frac{\partial \psi}{\partial Z} + 1\right)\right] \tag{1}$$

Virus transport Equation (Sim and Chrysikopoulos 2000):

$$\begin{aligned}\frac{\partial}{\partial t}(\theta C) + \frac{\partial}{\partial t}(\rho C^*) + \frac{\partial}{\partial t}(a_v C^0) = \frac{\partial}{\partial Z}\left(D^W \frac{\partial C}{\partial Z}\right) + \frac{\partial C}{\partial Z}\left(D^g \frac{\partial C}{\partial Z}\right) \\ - q\left(\frac{\partial C}{\partial Z}\right) - \lambda \theta C - \lambda^* \rho C^* - \lambda^0 a_v C^0\end{aligned} \tag{2}$$

From Eqs. (1) and (2), ψ is the pressure head; θ is the volumetric moisture content; K is the hydraulic conductivity; Z is the vertical coordinate; t is the time in days. C is the virus concentration in liquid phase, C^* is the virus concentration at liquid–solid interface, C^0 is the virus concentration at air–liquid interface, ρ is the bulk density of the soil, a_v is the air content, D^w and D^g are the dispersion coefficient of liquid and gaseous state, q is the water flux, λ is the first-order inactivation rate coefficient in aqueous virus, λ^* is the first-order inactivation rate coefficient of sorbed viruses at liquid–solid interfaces, λ^0 at air–liquid interfaces.

From Eq. (1), it is clear that the equation is nonlinear as both the hydraulic conductivity and moisture content depend on the pressure head. So the constitutive relation given by Van Genuchten (1980) is considered in this study.

Constitutive Relationships

The relationship proposed by van Genuchten's (1980) gives the relationship between $\theta - \psi$ and $K - \theta$ as follows

$$\theta(\psi) = \theta_r + \frac{\theta_s - \theta_r}{(1 + |\alpha\psi|^n)^m} \tag{3}$$

$$K(\psi) = K_s S_e^l \left[1 - \left(1 - S_e^{1/m}\right)^m\right]^2 \tag{4}$$

$$S_e = \frac{\theta - \theta_r}{\theta_s - \theta_r}$$

where θ_r is the residual water; θ_s is the saturated water content; n and m (=1 − 1/n) are unsaturated soil parameters, l is the pore connectivity parameter; K_s is the saturated hydraulic conductivity; and S_e is effective saturation.

2.2 *Initial and Boundary Condition for Moisture Flow*

$$t = t_0, \quad \psi = \psi_0 = -1000 \text{ cm} \quad 0 \leq z \leq L \tag{6}$$

$$t > 0, \quad \left(\frac{\partial\psi}{\partial z} = 0\right) \quad z = L \tag{7}$$

$$\left|-k\frac{\partial\psi}{\partial z} - k\right| \leq E \tag{8}$$

where ψ_0 is the initial pressure head potential and E is the maximum potential rate of infiltration or evapotranspiration under the current atmospheric conditions. Equation (6) indicates that the initial pressure head potential is uniformly distributed within the medium, whereas Eq. (7) gives an atmospheric boundary condition with surface runoff and Eq. (8) implies a free draining bottom boundary.

2.3 *Initial and Boundary Condition for Virus Transport*

$$C(0, z) = C^*(0, z) = C^0(0, z) = 0 \tag{9}$$

$$\frac{\partial C(t, \infty)}{\partial Z} = 0 \tag{10}$$

$$C(0, t) = C_0 \tag{11}$$

$$-D_Z \frac{\partial C}{\partial Z} + qC = qC_0 \tag{12}$$

where the initial condition given by Eq. (9) establishes that there is no initial liquid phase and adsorbed virus concentrations within the porous medium. The lower boundary condition given by Eq. (10) preserves concentration continuity for a vertical soil profile. For the upper boundary condition, Eq. (11) represents the case of constant concentration at inlet, while Eq. (12) represents the case of constant flux at inlet.

2.4 *Input Data*

See Tables 1 and 2.

Table 1 Different soil parameters that are considered in this study

Type of soil	Loam	Silt clay
θ_r	0.078	0.07
θ_s	0.43	0.36
K_s (cm/day)	24.96	0.48
l	0.5	0.5
n	1.56	1.09
α (cm^{-1})	0.036	0.005
ρ (g/cm^3)	1.43	1.24

Table 2 Solute transport parameters used in this study

Parameter	Value	Units	Reference
α_z	15	cm	HYDRUS-1D tutorial
D_z^w	0.825	cm^2/day	-do-
D_z^w	7128	cm^2/day	-do-
H	0.066		-do-
PRD1			
K_d	0.041	ppm	Anders and Chrysikopoulos (2009)
λ	0.0021	day^{-1}	-do-
λ^*	0.00,204	day^{-1}	-do-
λ^0	0.12	day^{-1}	-do-
MS2			
K_d	0.137	ppm	Anders and Chrysikopoulos (2009)
λ	0.06	day^{-1}	-do-
λ^*	0.066	day^{-1}	-do-
λ^0	0.0708	day^{-1}	-do-

3 Model Simulation

The flow and transport model for a heterogeneous soil profile up to a depth of 160 cm (up to 100 cm, it is loamy soil type, and after that, it consists of silt clay soil till 160 cm) from the ground surface is developed. This model is simulated for 300 days of the year 2013. The soil profile is divided into ten equal zones such that $\Delta z = 16$ cm and at the centre of each zone a node (*Ni*) is considered. The node *N*1 is at 8 cm, *N*2 at 24 cm, *N*3 at 40 cm, respectively, from the ground surface. The result obtain will give the variation of pressure head, water content, and concentration of both the virus with time. In this model, the source of virus MS2 and PRD1 is assumed to be active with concentration unit ($C_0 = 1$) for 200 days and after that, it is been considered as inactive ($C_0 = 0$) for the remaining 100 days.

Figures 1 and 2 provide the variation of pressure head and water content for different observation node with respect to its depth. The important observation from the figure is that there is an increase in water content and pressure head in loamy soil type as compared to the silt clay soil type. The reason may be as the rate of infiltration in silt clay is less than the loam soil so more amount of water remains onto the loamy soil profile.

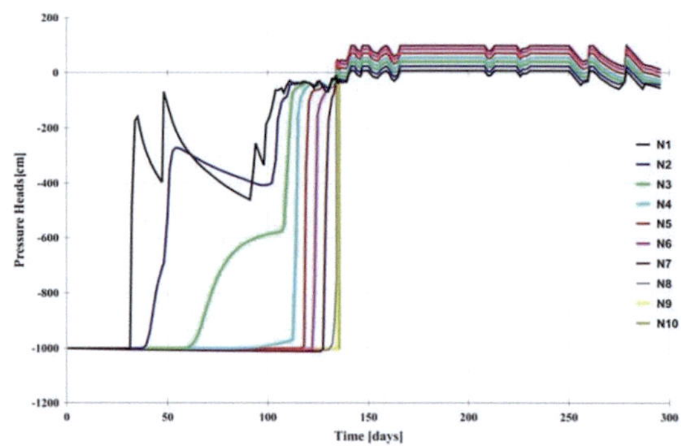

Fig. 1 Variation of nodal pressure head (ψ) for different time period (days)

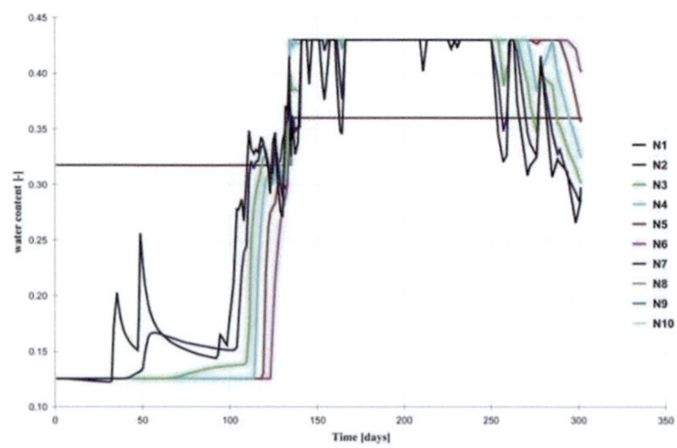

Fig. 2 Variation of nodal water content (θ) for different time period (days)

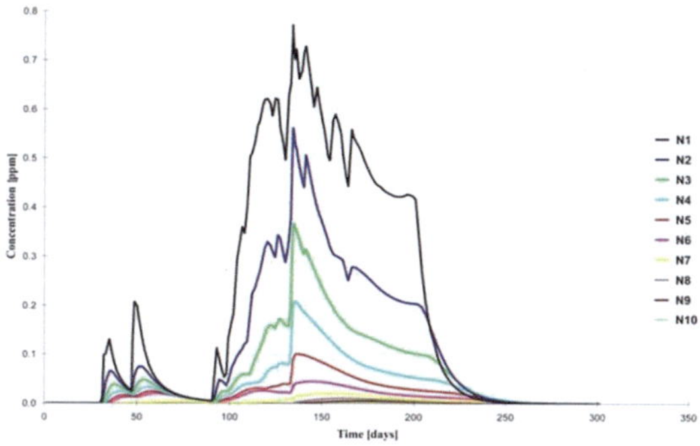

Fig. 3 Variation of concentration MS2 virus at different observation node and time period

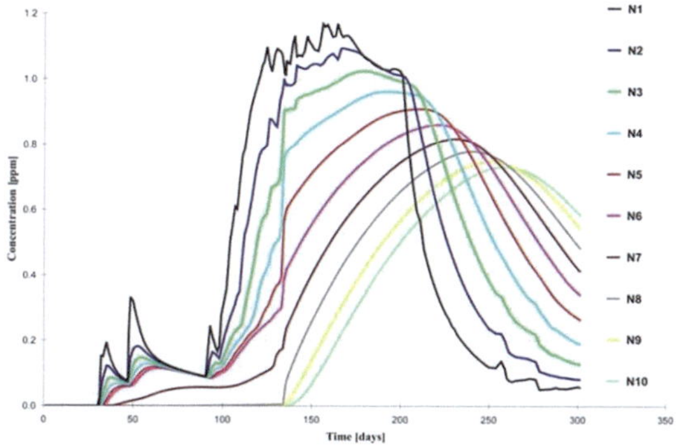

Fig. 4 Variation of concentration PRD1 virus at different observation node and time period

Figure 3 and 4 show the concentration variation of virus MS2 and PRD1 for different observation nodes. The figure indicates that decrease in the aquifer moisture content leads to a substantial reduction in the liquid-phase virus concentration. Thus, moisture content is one of the relevant parameters for both aqueous and sorbed virus.

4 Summary

In case of heterogeneous soil, a change in all the flow parameters and concentration is observed when the soil layer changes from loamy soil to silt clay soil type. It is observed that the viruses were transported more rapidly and in higher concentrations in loamy soil than the silty clay soil. The reason is that the hydraulic conductivity is higher in loamy soil than the silt clay soil so water and contaminants get transported faster in loamy soil. Powelson and Gerba (1994) suggested that size of the viruses may be considered as another important parameter if the porous media is highly heterogeneous, where virus may move faster than conservative tracers. From the result, it is observed that the virus MS2 is found to be transported faster than the virus PRD1. Since the size of MS2 virus is small (≈25 nm) compared to the PRD1 virus (≈62 nm) so the virus MS2 will get transported faster than the PRD1 virus.

References

Anders, R., Chrysikopoulos, C. V. (2009). Transport of viruses through saturated and unsaturated columns packed with sand. *Transport in Porous Media, 76*, 121–138.

Ojha, C., Hari Prasad, K., Ratha, D., & Surampalli, R. (2012). Virus transport through unsaturated zone: Analysis and parameter identification. *Journal of Hazardous, Toxic, and Radioactive Waste*, 96–105. https://doi.org/10.1061/(asce)hz.2153-5515.0000102

Powelson, D. K., & Gerba, C. P. (1994). Virus removal from sewage effluents during saturated and unsaturated flow through soil columns. *Water Resource, 28*, 2175–218.

Sim, Y., & Chrysikopoulos, C. V. (2000). Virus transport in unsaturated porous media. *Water Resource Research*, *36*(1), 173–179.

Thompson, S.S., Yates, M. V. (1999). Bacteriophage inactivation at the air–water–solid interface in dynamic batch systems. *Applied Environmental Microbiology, 65*(3), 1186–1190.

Tim, S. U., & Mostaghimi, S. (1991). Model for predicting virus movement through soils. *Ground Water, 29*(2), 251–259.

Van Genuchten, M. T. (1980). A closed-form equation for predicting the hydraulic conductivity of unsaturated soils. *Soil Science Society of America Journal, 44*(5), 892–898.

Vilker, V. L. (1981). *Simulating virus movement in soils, modeling waste renovation: Land treatment* (pp. 223–253). New York: Wiley.

Vilker, V. L., & Burge, W. D. (1980). Adsorption mass transfer model for virus transfer in soils. *Water Resource, 14*(7), 783–790.

Yates, M. V., & Ouyang, Y. (1992). Virtus, a model of virus transport in unsaturated soils. *Applied Environmental Microbiology, 58*(5), 1609–1616.

Behaviour Thresholds of Quarry Dust–Bentonite Mixes

Pradeep Raghu and Rakesh J. Pillai

Abstract Bentonite was added to a locally available rock dust to study the behaviour thresholds of the binary mix. Thresholds may be defined as the critical fines content above which the system behaviour changes drastically. By understanding the threshold values of such binary mixtures will help us to have a better control over its behaviour. In this study, the hydraulic conductivity and the swelling behaviour of quarry dust mixed with increasing fractions of bentonite were studied. Index properties of quarry dust were determined by standard sieve analysis, mini-compaction and by cone penetration method. The specific gravity of the quarry dust was about 2.65. The addition of bentonite to quarry dust has found to increase the maximum dry density and reduce the hydraulic conductivity by several orders. The swell potential of the various mix proportions was studied in an oedometer. Samples were of 14 mm thickness and were given a hydration period of 24 h and were compacted at their OMC and MDD. Mixes with low bentonite contents such as a 5% and 10% showed no swelling. The substantial swelling was only observed with 15% bentonite and more.

Keywords Bentonite · Quarry dust · Thresholds

1 Introduction

Almost all soils in the world are a combination of one or more types of soils. The success of any mixture design depends upon the mix ratio. A soil with too less fine fraction will be highly porous. This is because of the inability of the coarser fractions to form a compact system. But such soils might have a relatively better

P. Raghu (✉)
Vardhaman College of Engineering, Hyderabad 500018, India
e-mail: pradeepraghuofficial@gmail.com

R. J. Pillai
National Institute of Technology, Warangal 506004, India
e-mail: rakeshpilla@gmail.com

V. K. Stalin and M. Muttharam (eds.), *Geotechnical Characterisation and Geoenvironmental Engineering*, Lecture Notes in Civil Engineering 16,
https://doi.org/10.1007/978-981-13-0899-4_29

rigidity and less compressibility as compared to those with high fines content. It could be seen that a dominance of any of the fraction will result in a compromise of some or any favourable soil properties. Therefore, we can understand there exists a suitable ratio of fines to the coarse fraction at which we can successfully attain any design requirements. Binary mixtures are those that comprises of two soil types. Binary mixtures find applications in various fields like a filter for dams, low permeability applications like waste retention plants and landfills, and also in flexible pavements. The fines to the coarse fraction of these binary mixtures have found to control its behaviour under mechanical and thermal loadings. For better decision-making and to have a control over the mix behaviour, we need to have a thorough knowledge of its fundamental behaviour. In this study, a binary mixture of quarry dust and bentonite is adopted in order to identify the behavioural thresholds, as in the transitional mix ratio at which a shift from the sand-like to clay-like property occurs. A behavioural threshold may be defined as a critical fines content at which a sudden change in the fines content will have a significant effect on the mixture properties.

1.1 Behaviour Threshold

As explained earlier, a behaviour threshold is that transitional fines content beyond which the system changes are drastic. Researchers have broadened the concepts of traditional soil mechanics with concepts and theories from other fields of science. One such emergent theory is the percolation theory. Percolation theory mainly finds applications with binary mixes (John and Ernest 2010). It has wide applications from forest fires to banking and even in soil mechanics. It is a probabilistic approach, whose concepts when closely studied; the probability becomes the mix ratio for binary mixes. Thornton (2000) using the concepts of percolation theory and DEM analysis established the theory of force chains. Force chains are nothing but a series of interconnected coarse-grained particles that transmit the load. A soil skeleton is formed when there exist a number of force chains. Such systems will have good structural stability and the shear failure of which was explained as the failure of these force chains in buckling.

2 Materials Used

The materials used in this study are (a) quarry dust and (b) bentonite.

The quarry dust is a poorly graded with a Cu value of 1.93 as indicated by the particle size distribution curve (Fig. 1). The specific gravity of the quarry dust was found to be 2.65. The maximum particle size was less than 2 mm. The bentonite

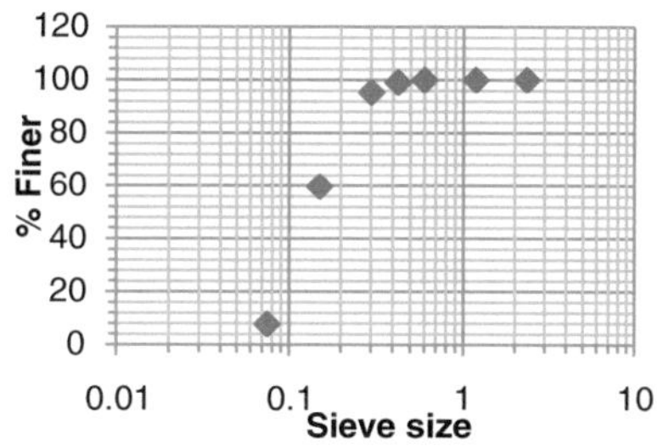

Fig. 1 Particle size distribution curve for quarry dust

Table 1 Index properties of the bentonite used

Property	Bentonite
Specific gravity	2.77
Liquid limit (%)	218
Plastic limit (%)	51
Plasticity index (%)	167
Clay, <2 μm (%)	67
Silt, 75–2 μm (%)	33
USCS soil classification	CH
Cation exchange capacity (meq/100 g)	56.95
Maximum dry density (kN/m^3)	12.6
Optimum moisture content (%)	30

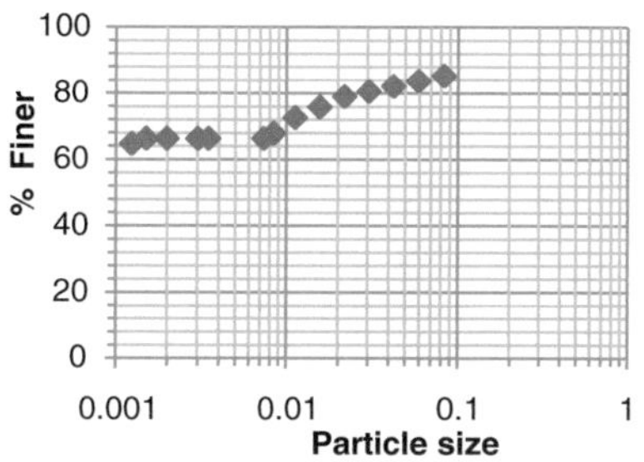

Fig. 2 Hydrometer analysis of bentonite

used in the study had sodium as the major cation this ensured very high swelling nature. The index properties of the bentonite used are shown in Table 1. The result of the hydrometer analysis of bentonite is given in Fig. 2.

3 Experimental Program

The experimental program involved in the establishment of the threshold values in the basic properties such as hydraulic conductivity, compaction as well as in the swelling behaviour of the soils. The experiments were carried out on the samples by varying the bentonite content in proportion to weights. All the samples were mixed at OMC and MDD as determined by the standard mini-compaction apparatus.

3.1 Compaction Behaviour

The compaction behaviour of the quarry dust bentonite mixtures was carried out in a mini-compaction apparatus as shown in Fig. 3. The procedure adopted was as per Sridharan and Sivapullaiah (2005). Although the apparatus was developed for the determination of dry densities of highly fine-grained materials, it could as be applied for materials as per the literature.

3.2 Swell Potential

In the swell potential, all the samples were evaluated using a standard oedometer apparatus. All the samples with varying mix proportions were mixed at OMC, and a period of 24 h was provided for hydration, towards the end of which the samples were compacted to their mean densities. The samples were filled only up to 1.4 cm from the bottom porous stones as per the guidelines mentioned in the IS 2720 Part XLI. The additional 6 mm is to provide for the expected swelling.

Fig. 3 Mini-compaction apparatus

4 Results and Discussion

4.1 Compaction Behaviour

The results of the compaction behaviour are shown in Fig. 4. The results indicate that at lower contents of bentonite, the mean dry densities were increasing this is attributed to the fact that at lower bentonite contents the pores spaces in the quarry dust are being occupied by the bentonites. As a result of this, the packing of the binary mix becomes more compact. At lower percentages of bentonite like 5, 10%, the volume of swelling is not enough to disturb the soil skeleton.

As such, the existing force chains are only being strengthened by the swollen mass of bentonite and also by those stone particles that do not take part in the force chain percolation. This aspect was suggested by Thornton in 2000 based on his DEM studies

At higher bentonite contents, this may not hold valid as the high swelling nature of the bentonite will result in the destruction of this force chains. At higher bentonite replacements, the swollen mass will start to replace the coarser grains from the unit volume (Yang et al. 2006). At this stage, the voids present in the binary mix will primarily be that of the clay content (Simpson and Evans 2015). The soil skeleton slowly loses its integrity and falls apart. This will result in the reduction of the dry density. At 20% bentonite content, this transition can be observed. Beyond 20% bentonite content any compactive effort given to the binary mixture was dissipated by the clay fraction and more or less behaved like stiff clay. This critical fines fraction may be identified as one of the threshold limits of this binary mixture.

4.2 Swell Potential

The swell potential of the binary mixture was examined using a standard oedometer apparatus. The sample mixed at OMC and MDD after 24 h of hydration in a desiccation chamber was filled into the oedometer rings. The sample thickness was kept around 1.4 cm as per IS 2720 Part XLI. The swell pressure tests were conducted for bentonite replacements of 5, 10, 15, 40%, etc. For lower bentonite contents of 5 and 10%, there was no observed swelling. This does not imply that the

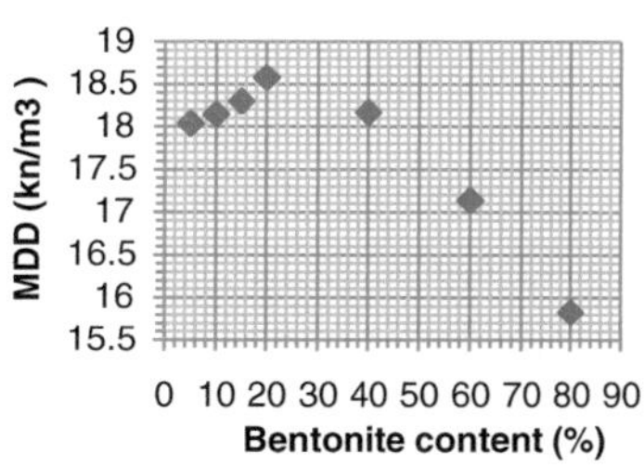

Fig. 4 Dry density versus bentonite content

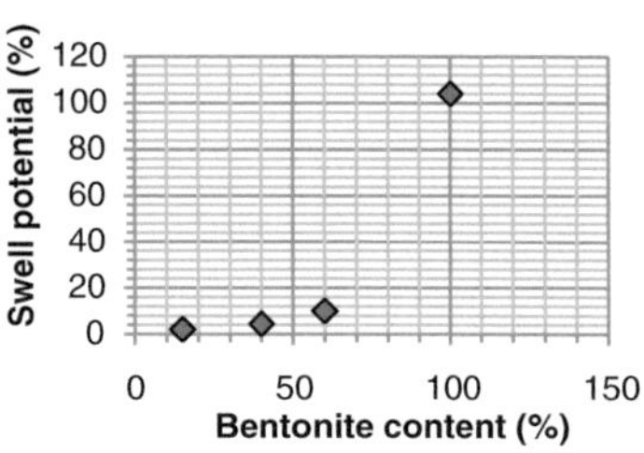

Fig. 5 Maximum swell potential versus bentonite content

bentonite added did not swell, but instead this suggests that the swollen volume is less than the void volume in these mixes. Only at 15% bentonite, any substantial swelling could be identified a critical fines content of 20% meant that the force chains were completely absent as the coarser fractions were displaced from unit volumes

Figure 5 shows the 21-day swell potential data at different bentonite contents. From Fig. 5, it is clearly understood that beyond 50% fines content the swollen volume of bentonite is considerably more as a result of which this can be identified as the transitional fines content for swelling for this binary mixture.

5 Conclusions

This study was mainly intended to analyse and study the behavioural thresholds of binary mixes. A locally available quarry dust and an industrial grade sodium bentonite were selected to form a binary mixture. The bentonite was mixed with quarry dust as 5, 10% replacement by weights, etc., and was subjected to compaction tests as well as swell potential evaluation. From both the tests, critical fines content was able to be identified beyond which the sand-like behaviour changed to clay-like behaviour. In the case of compaction characteristics beyond 20% bentonite content, the mean density started to decrease. As in the case of swell potential, the binary mixture could only behave as a swelling noncohesive soil at 15% bentonite. Thus, these can be identified as the behavioural thresholds were the system behaviour drastically changes. It should be remembered that any variations in the binary mix will result in the variations of the thresholds.

References

IS: 2720 (Part 40). (1977). *Methods of test for soils: Determination of free swell index of soils.* New Delhi, India: Bureau of Indian Standards.

John, F., & Ernest, S. (2010). Percolation threshold of sand-clay binary mixtures. *Journal of Geotechnical and Geoenvironmental Engineering, ASCE, 136*(2). ISSN 1090-0241/2010/2-310-318.

Simpson, D. C., & Evans, T. M. (2015). Behavioural thresholds in mixtures of sand and kaolinite clay. *Journal of Geotechnical and Geoenvironmental Engineering*. ISSN 1090-0241/04015073 (10).

Sridharan, A., & Sivapullaiah, P. V. (2005). Mini compaction test apparatus for fine-grained soils. *Geotechnical Testing Journal*, *28*(3) GTJ12542.

Thornton, C. (2000). Numerical simulations of deviatoric shear deformation of granular media. *Geotechnique, 50*(1), 43–53.

Yang, S., Lacasse, S., & Sandven, R. (2006). Determination of the transitional fines content of mixtures of sand and non-plastic fines. *Geotechnical Testing Journal, 29*(2), 102.

Index and Shear Strength Properties of Clay Soil Contaminated by Tannery Effluent

E. Sibi and S. Karthigeyan

Abstract Most of the industrial effluents are disposed into land containing a variety of combination of chemicals that may bring in considerable changes in the geotechnical properties of soils. The tannery effluent is one such effluent characterized by high BOD and COD, high dissolved solids, high or low pH, presence of heavy metals, calcium salts, chlorides, sulphides, fat, liquor and organic dyes. This paper presents the laboratory results to study the effect of tannery effluent on the index and engineering properties of cohesive soil upon contamination. A series of laboratory tests have been carried out to evaluate the index and engineering properties of tannery contaminated clay soils. The virgin characteristic of clay soils is highly swelling clay of CH classification with differential free swell of 66%. Virgin clay soils have artificially contaminated with varying percentage of tannery effluent collected from Pallavaram in Chennai and tested to evaluate various properties of soils. The results had shown an increase in Atterberg's limits as well as the shear strength with varying percentage of tannery effluent. The nature of the pore fluid has also found to significantly affect the index properties and shear strength.

Keywords Tannery effluent · Clayey soil · Swelling · Shrinkage and shear strength

1 Introduction

India is a leading producer of leather in the world. There are numerous tannery industries located in India. These industries are releasing tons of waste including toxic contents and heavy metals into the adjacent land and water sources.

E. Sibi · S. Karthigeyan (✉)
Department of Civil Engineering, Anna University,
Guindy, Chennai 600025, India
e-mail: mahamaha2001@yahoo.com

E. Sibi
e-mail: sibi737discovery@gmail.com

V. K. Stalin and M. Muttharam (eds.), *Geotechnical Characterisation and Geoenvironmental Engineering*, Lecture Notes in Civil Engineering 16,
https://doi.org/10.1007/978-981-13-0899-4_30

Already, thousands of hectares of land and many valuable water resources have been contaminated with tannery effluent. The soil properties get modified when they mix with these effluents, especially in the case of clay (Rao and Chittaranjan 2012). The modification in soil properties may lead to changes in the engineering behaviour of the soil. The contamination of soil has a potential to affect the stability of the structure resting upon it. Also, the stability of slopes, foundations and piles has a tendency to get affected when the soil surrounding them gets contaminated.

Experimental studies carried out by earlier researchers had shown that the free swell index, liquid limit and plastic limit values of clay soil reduce upon contamination with tannery effluent, whereas the UCC strength reduces upon contamination (Rao et al. 2012a, 2012b; Muthukkumaran 2010). However, the effect of tannery effluent on soil is different and engineering behaviour of soils are to be varied depending upon the chemical composition of the effluent. In this work, tannery effluent obtained from Pallavaram in Chennai was used to contaminate the high swelling clay (CH). The effects of tannery on clay soil and its changes in liquid limit, plastic limit, plasticity index, shrinkage limit, differential free swell and shear strength were studied.

2 Materials Used

Natural clay soil, collected from Velacherry, Chennai, was used for the study. The clay was black in colour and quite stiff. It was then dried, crushed and sieved in IS sieves, and then, its properties were determined as given in Table 1.

The tannery effluent used in the study was collected from Pallavaram. It was black in colour with high viscosity and was generated during the finishing and dyeing process of leather. The properties of the effluent are given in Table 2.

Table 1 Properties of the virgin clay soil

Property	Value
Grain size distribution	
Gravel (%)	3
Sand (%)	10
Silt (%)	17
Clay (%)	70
Atterberg's limit	
Liquid limit (%)	73
Plastic limit (%)	26
Plasticity index (%)	46
Shrinkage limit (%)	14
Differential free swell index (%)	66
Specific gravity	2.68
Unconfined compressive strength (kN/m^2) at 23% water content	196

Table 2 Properties of the tannery effluent

Parameter value	
Colour	Blue
Form	Liquid
pH	3.1
Hexavalent chromium as Cr^{6+}	Below detectable level
Chromium as Cr	985 mg/l
Sulphates as SO_4	12,752 mg/l
Chlorides as Cl	12,484 mg/l
Sodium as Na	16,728 mg/l
Calcium as Ca	561 mg/l
Total dissolved solids	59,810 mg/l

3 Methodology

The various laboratory tests conducted were framed in two series. In the first series of tests, soil was mixed with effluent directly in place of water and mixed soil was not allowed to dry. This was achieved by storing the effluent mixed soil in airtight plastic bags. Then, various laboratory tests were performed on the prepared soil samples immediately after mixing as well as giving some time period of contamination after mixing of 7, 14 and 28 days. It was assumed to simulate the condition that only tannery effluent alone is present within the voids of soils. In the second series of tests, soils were mixed with the effluent and then allowed to dry in the open atmosphere after mixing for various contamination periods of 7, 14 and 28 days as followed in the previous case. At the time of testing, the contaminated soil was crushed, sieved and then added with water for various tests. The various laboratory tests were performed on contaminated clay soils with varying consistency. In this type of soil mixing, the pore fluid was only the presence of water alone.

The liquid limit of the clay soil was determined by incrementally adding tannery effluent to the soil at various percentages. In this case, the pore fluid was effluent as it was tested immediately after mixing. It can be seen from Fig. 1 that the liquid limit values are increased from 73% to 84% with increase in tannery effluent.

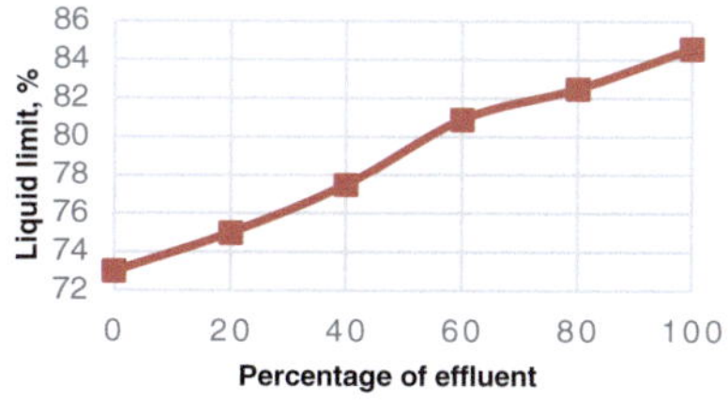

Fig. 1 Liquid limit of soil with varying percentage of tannery effluent

4 Results and Discussion

4.1 *Effect on Liquid Limit*

In order to understand the effect of contamination period on the clay soil, samples were prepared by mixing clay soils with the effluent and stored in an airtight bag for a period of 7, 14 and 28 days. Then, liquid limit tests were conducted at the end of each contamination period. It was observed that the period of contamination is not showing any effect on the liquid limit of clay sample. The liquid limit, at the end of 28 days of contamination period, is the same as that when tested immediately after mixing as shown in Fig. 2.

The liquid limits of the soils were also determined by testing the specimens after drying the contaminated effluent. In this case, soil samples were contaminated, but the pore fluid is water in place of effluent. It observed a slight increase in liquid limit from 73 to 76% with contamination period but not showing any influence with respect to the contamination period. It can be seen from Fig. 3 that the effect of contamination period upon the liquid limit is almost nil.

4.2 *Effect on Plastic and Shrinkage Limit*

Similar trends were observed for plastic limit and shrinkage limit. The magnitude of change in plastic and shrinkage limit of contaminated clay soil was more with an increase in percentage of contamination. But the time period of contamination is not showing any change as compared to the immediate response. However, the plastic

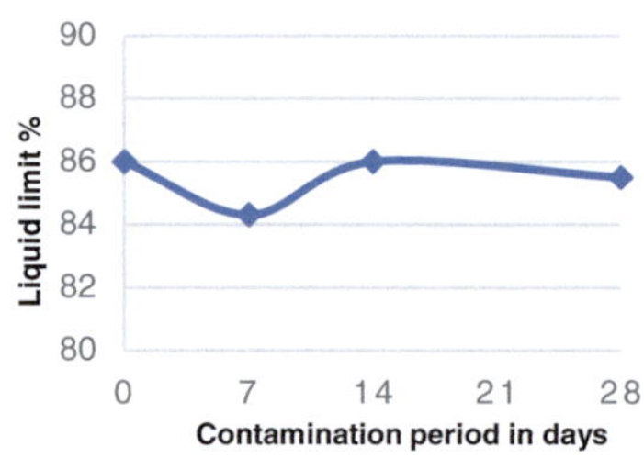

Fig. 2 Liquid limit of soil with contamination period

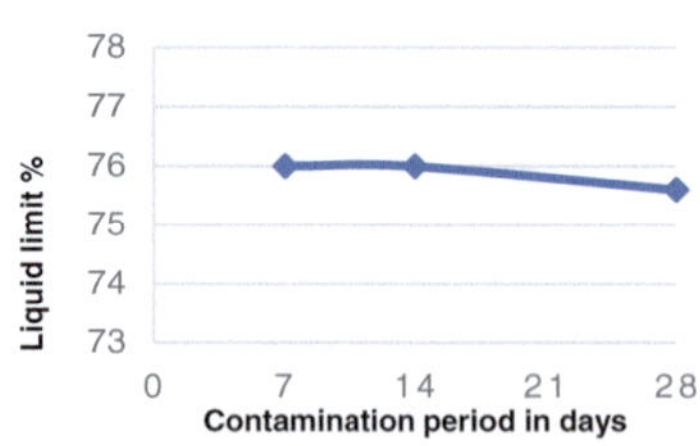

Fig. 3 Liquid limit of dried contaminated clay with water as pore fluid

Fig. 4 Plastic limit of soil with varying percentage of tannery effluent

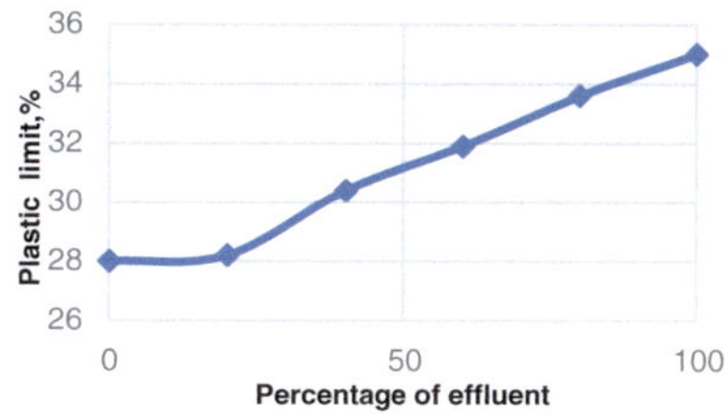

limit increased from 28 to 35% as seen from Fig. 4, whereas the shrinkage limit decreased from 14 to 11.5% as observed from Fig. 5 when the pore fluid was effluent. When the contaminated sample was dried and mixed with water, the plastic limit was 33%, while there was no change in the shrinkage limit as it remained in 14%.

4.3 *Effect on Differential Free Swells*

Differential free swell tests were carried out on the contaminated clay sample after dried. The differential free swell values of contaminated soil were increased from 66 to 86% with contamination. In order to understand the swelling nature of clay, free swell tests were carried out on the contaminated clay with pore fluid inside as an effluent.

It can be seen from Fig. 6 that the free swell of contaminated soil with pore fluid as effluent has shown a higher increase in free swell up to about 200%. It indicates the effect of tannery effluent is very severe on the swelling characteristics of clay.

4.4 *Dry Density of Contaminated Samples*

The maximum dry densities of the contaminated clay samples and the virgin clay were evaluated by conducting the standard Proctor compaction test. The maximum dry density of virgin clay was found to be 15.2 kN/m^3 and an optimum moisture

Fig. 5 Shrinkage limit of soil with varying percentage of tannery effluent

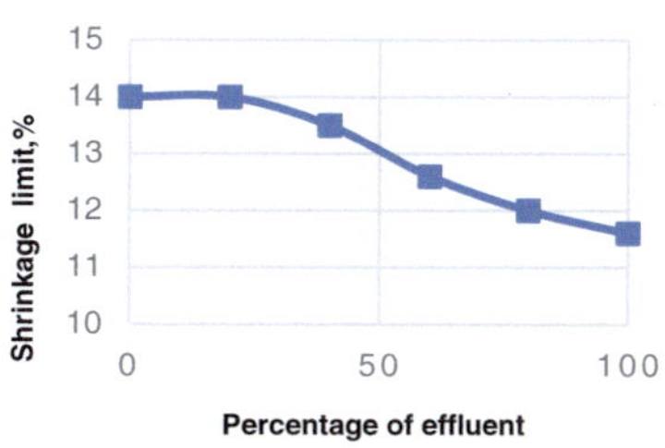

limit of soil with varying percentage of

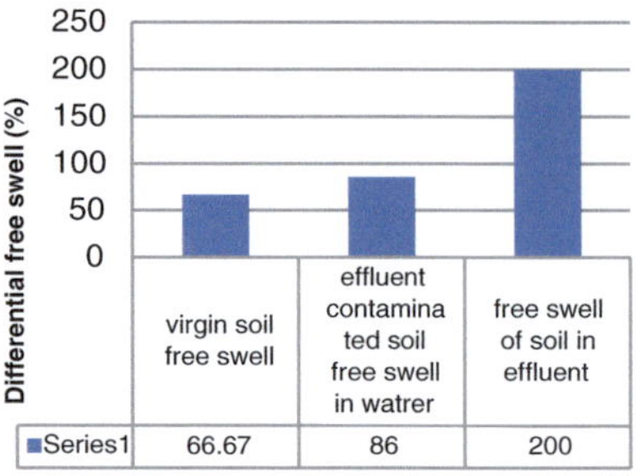

Fig. 6 Differential free swell of contaminated soil in water and in effluent

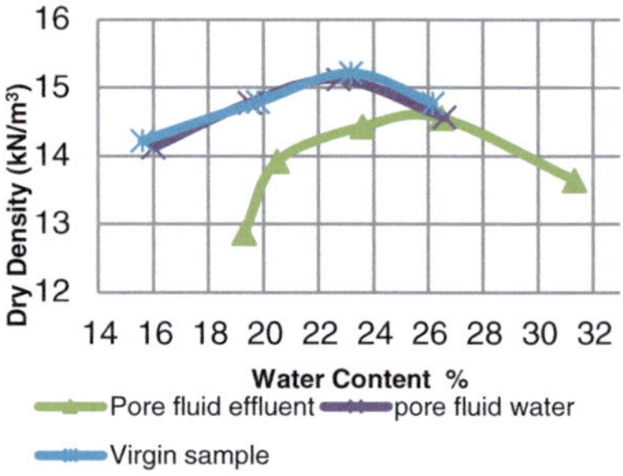

Fig. 7 Dry density–water content relationship curves for the virgin and contaminated soil samples

content of 23%. The contaminated samples with water as pore fluid have shown a maximum dry density of 15.1 kN/m^3 at 22.7% water content. It can be seen from Fig. 7 that the contaminated samples with effluent as the pore fluid have shown a maximum dry density of 14.55 kN/m^3 at 25.5% moisture content.

4.5 *Unconfined Compressive Strength of Contaminated Soil*

The unconfined compressive strength of tannery effluent contaminated clay soil samples was determined by conducting UCC test on samples prepared at an optimum moisture content of 23%. The stress–strain curves for the virgin sample and the contaminated samples with pore fluid effluent as well as water were plotted. It is seen from Fig. 8 that the UCC values increased for both the cases as compared to that of the virgin soil. However, The UCC strength value is more in case of the contaminated clay samples with effluent as the pore fluid than that of the contaminated samples with water as a pore fluid. Also, it is observed that the effect of contamination period has not shown any influence on the UCC value of contaminated clay.

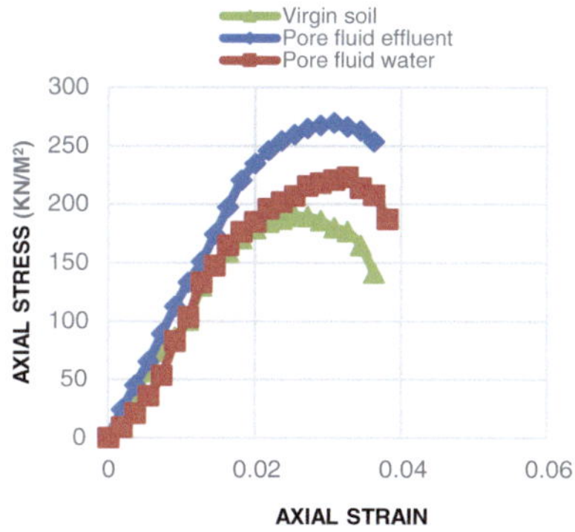

Fig. 8 Stress–strain curves for contaminated samples with effluent and water as pore fluid

5 Conclusions

This paper has presented laboratory test results to study the effect of tannery effluent on clayey soil. The following conclusions can be drawn from the study;

1. Liquid limit and plastic limit of soil increase with increase in contamination, while the shrinkage limit reduces with increase in contamination.
2. The shear strength of the clay increases when the clay is contaminated with the tannery effluent.
3. The presence of tannery effluent in pores slightly reduces the maximum dry density of the clay and increases the optimum moisture content. This may be due to the viscous nature of effluent.
4. The tannery effluent effect is more for contaminated clay with the pore fluid as an effluent than with the pore fluid as water.
5. The results presented in the paper are pertaining to the Pallavaram tannery. However, a change in characteristics of contaminated clay by tannery is mainly depending on the type of tannery effluent. This mainly depends upon the nature and composition of the effluent. Therefore, for general conclusion, it needs to study in detail with different types of tanneries on the change in characteristics of clay.

References

Rao, A. V. N., Naik, K. V. N. L., & Bali Reddy, S. (2012a). A study on the geotechnical properties of tannery effluent on black cotton coil mixes. In *Proceedings of Advances in Civil Engineering and Infrastructure Development*.

Rao, A. V. N., & Chittaranjan, M. (2012). Effect of industrial effluents on the compaction characteristics of expansive soil—A comparative study. *International Journal of Engineering Inventions, 1*(7), 22–28.

Rao, A. V. N., Chittaranjan, M., & Naik, K. V. N. L. (2012b). Undrained shear strength characteristics of an expansive soil treated with certain industrial effluents at different pore fluid content ratios. *International Journal of Innovative Research in Science, Engineering and Technology, 1*(1), 58–65.

Muthukkumaran, K. (2010). Effect of liquid waste on the index and engineering behaviour of cohesive soils. *Indian Geotechnical Journal, 40*(3), 195–200.

Study on the Effect of pH on the Atterberg Limits of Kaolinitic and Montmorillonitic Clay

K. Nivedya

Abstract In the present scenario, due to the urbanization and industrialization, much of the air, water and land have become polluted. Any kind of pollution has direct or indirect effect on the properties of soil. So in the present study, the effect of pH on the properties of two soils was studied. In the present study, effect of pH on the Atterberg limits of two soils was investigated using different pore fluids at different pH. Different pore fluids such as NaCl, KCl, NaOH, and KOH at different pH were selected such that pH varying from 6.5 to 13 could be obtained. From the present study, it could be seen that for the same solution with increase in pH the liquid limit was found to increase with decrease in pH both in the case of kaolinite and bentonite. Plastic limit is of lesser importance with regard to pH, both in the case of kaolinite and bentonite. The effect of pH on shrinkage limit in the case of kaolinite and bentonite showed extremely opposite behavior.

Keywords pH · Kaolinite · Montmorillonite · Plasticity characteristics Shear strength

1 Introduction

In recent years, due to population growth, progressive living standard, and industrial progress, much of the air, water, and land have become polluted. Open dumps, chemical and industrial wastes, and many other sources cause these problems. All types of pollution have direct or indirect effects on ground soil properties.

In the past few decades, a great deal of research has been conducted to understand the influence of pore water chemistry on the strength characteristics of fine-grained soils, while little has been done to understand the effects of pH, especially acidic water, despite the fact that acid contamination has become an increasing concern around the world. Acidic contamination of soil can occur

K. Nivedya (✉)
SOE, Cusat, Kochi 682022, India
e-mail: nivedyabalakrishnan@gmail.com

V. K. Stalin and M. Muttharam (eds.), *Geotechnical Characterisation and Geoenvironmental Engineering*, Lecture Notes in Civil Engineering 16,
https://doi.org/10.1007/978-981-13-0899-4_31

through natural processes such as weathering of mudstone or due to anthropogenic activities such as municipal waste storage, accidental spills, or acidic rains. This problem has been recognized recently, and a few studies have been performed to investigate the effects of pH on soil properties. Yet the effect of pH of pore fluids on the strength of soil remains unclear. This study seeks to shed light on this issue by providing laboratory data concerning the variations in Atterberg limits of two soils subjected to pore fluids at different pH.

Montmorillonite and kaolinite are two most common clay minerals observed in natural clays. The various studies show that the behavior of kaolinite and that of montmorillonite are extremely different when subjected to same environmental conditions. So in this study, kaolinite and bentonite have been chosen to investigate the effect of pH on the geotechnical properties.

2 Materials Used

2.1 Kaolinite

Commercially available kaolinite is used in this study. This was collected from English India Clay Limited, Kochuvely, Trivandrum, Kerala. The kaolinite which procured for the complete investigation was thoroughly mixed for uniformity and then preserved in double layer of polythene bags.

2.2 Bentonite

Bentonite was procured from Associate Chemicals P (Ltd), Kochi, Kerala. The percentage of water present in bentonite depends upon the climatic conditions, since bentonite belongs to the montmorillonite group, which has great affinity for moisture. Hence, the bentonite, which was procured for the complete investigations, was thoroughly mixed for uniformity and then preserved in double layer of polythene bags. These bags were stored in airtight bins.

2.3 Pore Fluids Used

Chlorides are widely distributed in nature usually in the form of sodium, potassium, and calcium salts. Pore fluids such as NaCl, KCl, NaOH, and KOH were selected since they represent many of the industrial effluents. Different solutions of NaCl, KCl, NaOH, and KOH at different normalities were prepared in the laboratory, and their pH values were measured using pH meter. The concentration of the solutions was finally chosen such that the pH values of at least two solutions were close to each other. Concentration of solutions was fixed such that pH ranging from 6 to 13 was obtained.

3 Preparation of Samples

Solutions of NaCl, KCl, NaOH, and KOH at different pH were prepared in the laboratory. Kaolinite and bentonite were soaked in chosen solutions for 3 months. The containers were made airtight in order to prevent the evaporation of the solution. Samples were collected from each container at 15th day, 30th day, and 60th day, and required tests were conducted

4 Effect on Liquid Limit

From the results obtained, it is observed that for the same solution, an increase in pH resulted in the decrease in liquid limit (Fig. 1). The trend is similar in both acidic range and alkaline range. Also with increase in curing period, the liquid limit is found to be decreasing. Liquid limit is increasing with decrease in pH. Also with increase in curing period, liquid limit decreases. At lower pH, flocculation of particles increases; thus, liquid limit increases. In the acidic range, a decreasing trend in the values of liquid limit is observed with increase in the hydrated ion radius. In the alkaline range with decrease in hydrated ion radii, liquid limit is found to increase in both lower pH and higher pH. The trend was found to be similar in the case of bentonite also.

In the case of kaolinite besides the edge to face contacts which contributes to increased resistance when pore fluids of lower dielectric constants are used, the presence of natural cementing agents and cementing compounds formed by stabilization process can influence the liquid limit.

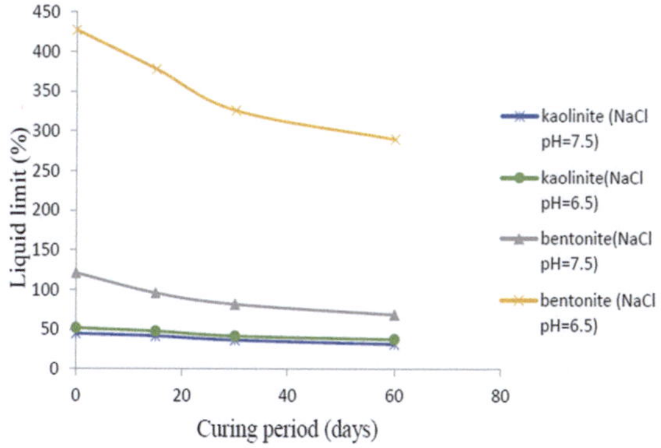

Fig. 1 Variation in liquid limit with curing period

However, different situations can arise from prolonged action of chemicals and particle aggregation may lower the liquid limit. The variations in the liquid limit of kaolinite and bentonite could be explained in the following manner. In the case of bentonite, the surface area is much larger, and for the same weight of clay, the double-layer water is large compared to the free water, and there is very little edge to face contact in the normal case. Attractive forces are not much significant at larger spacing. Hence, it is the variation of the double layer thickness which affects the fluid content at liquid limit. For kaolinite, it has lower surface area, and because of the positive edged areas, a tendency to form edge face contact is more evident, except at highly alkaline pH. The amount of pore fluid at pore space at liquid limit is composed of the double-layer fluid and the free fluid. Since the amount of double-layer fluid per unit weight of soil is small in kaolinite, most of the fluid is free fluid and the weak edge to face bonds allows the flow of the particles at a lower fluid content.

5 Effect on Shrinkage Limit

One of the important physical properties of the clay in which engineers are interested is the volume change in water content on climatic changes. Shrinkage limit is an extremely useful parameter in quantitatively identifying the soil fabric. For a curing period of 60 days, at higher pH, 53% decrease in shrinkage limit was observed in the case of kaolinite, while 150% increase was observed in the case of bentonite when NaCl is used (Fig. 2). Also at lower pH, 50% decrease in shrinkage limit is observed for kaolinite, while 200% increase was noted for bentonite. When KCl is used at higher pH, 56% decrease in shrinkage limit was observed for kaolinite, while 112% increase was observed in the case of bentonite. Similarly at lower pH, 52% decrease in shrinkage limit was observed for kaolinite, while 200% increase was observed for bentonite. In the case of kaolinite interparticle arrangements exist and in the case of montmorillonitic clay particle arrangements are nearly parallel, so there is greater shrinkage with lower shrinkage limit values at higher liquid limit. predominates. The similar behavior was also observed in the case of addition of lime to such soils. The increase in liquid limit with ionic changes is in relation to higher shrinkage limits.

The shrinkage limit of kaolinite generally follows the mechanism of liquid limit. It is observed that for the same solution with increase in pH, there is decrease in both liquid limit and shrinkage limit in the case of kaolinite (Fig. 3).

In contrast to increase in pH of the same solution, liquid limit decreases, while shrinkage limit increases for bentonite. The variations in liquid limit and shrinkage limit are exactly opposite.

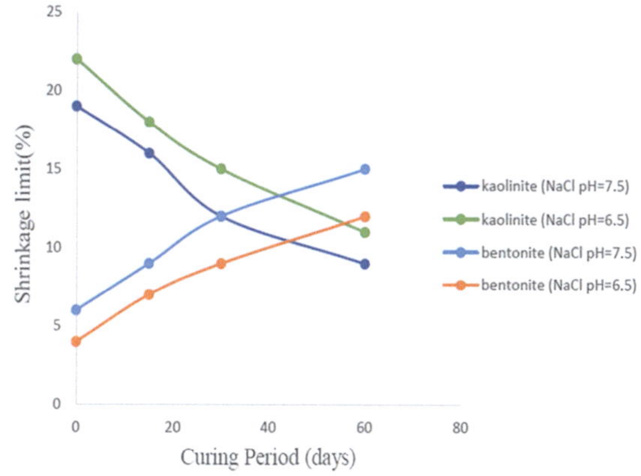

Fig. 2 Variation in shrinkage limit with curing period

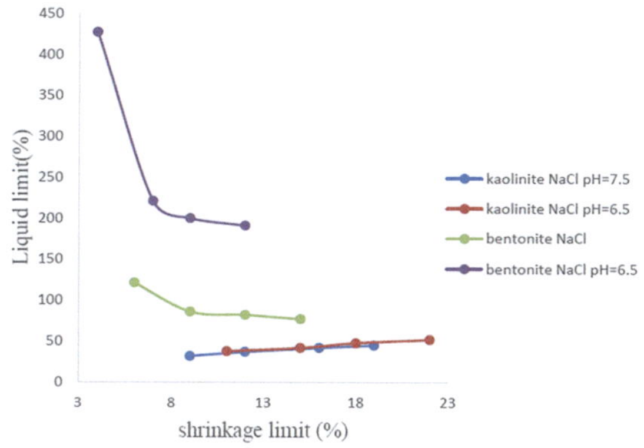

Fig. 3 Variation in liquid limit with shrinkage limit

6 Conclusions

The conclusions drawn from the present study can be summarized as follows:

- In the case of kaolinite, liquid limit was found to be decreased as pH increased when same pore fluids are used. In kaolinite, van der Waals force of attraction between the interparticles plays major role in the variation of liquid limit. Also the amount of double-layer fluid per unit weight of soil is small in kaolinite, so most of the fluid is free fluid and the weak edge to face bonds allows the flow of the particles at a lower fluid content. This could be the reason for the decrease in liquid limit.
- For the same solution, liquid limit was found to be decreased as pH increased in the case of bentonite. This could be due to the reason that as the pH increases, the thickness of the diffuse double layer decreases, and thus, reduction in liquid limit occurs.

- The plastic limit is of lesser importance. There is no distinct trend in changes of plastic limit with pH values were observed. Change in pH does not cause significant variations in the plastic limit of both kaolinite and bentonite. The plasticity index is thus more directly related to liquid limit than plastic limit.
- In the case of bentonite, as the period of curing increased, the liquid limit was found to be decreased. With time particle flocculation and adsorption on the diffuse double layer increases, this may cause the reduction in liquid limit with increase in curing period. Variations are more significant in the case of bentonite.
- For the same solution, it was found that shrinkage limit decreased as pH increased in the case of kaolinite. In kaolinite, shrinkage limit follows the mechanism of liquid limit. Also with increase in curing period, shrinkage limit was found to be decreased.
- On contrary for the same solution, shrinkage limit was found to be increased with increase in pH in the case of bentonite. In bentonite, shrinkage limit does not follow the mechanism of liquid limit. Also with increase in curing period, shrinkage limit was found to be increased for same solution.

Influence of Geomembrane Stiffness on the Integrity of Landfill Covers: Centrifuge Modelling

P. V. Divya, B. V. S. Viswanadham and J. P. Gourc

Abstract The influence of geomembrane in maintaining the sealing efficiency of a landfill cover system at the onset of flexural distress was studied by conducting a series of centrifuge tests performed at 40 gravities. The tests were performed using a 4.5 m radius beam centrifuge having a capacity of 2,500 g-kN available at IIT Bombay on model composite barriers of 0.6 m thickness. The thickness and tensile stiffness of the geomembrane were varied. All the models were instrumented with pore pressure transducers (PPTs), linear variable differential transformers (LVDTs), a digital camera and a charge-coupled device (CCD) video camera to study the performance of the barriers during centrifuge tests. The sealing efficiency of the composite barriers was found to be maintained even after the formation of full-depth cracks within 0.6 m thick clay barriers subjected to a maximum distortion level of 0.125. This observed behaviour is attributed to the downward thrust exerted by the deformed geomembrane at the zone of maximum curvature which hinders the infiltration of water through the cracks. The downward thrust exerted by the geomembrane on the clay barrier was found to be more for barriers provided with stiffer and thicker geomembrane.

Keywords Geomembrane · Composite barriers · Landfills · Digital image analysis · Differential settlements

P. V. Divya · B. V. S. Viswanadham (✉)
Department of Civil Engineering, Indian Institute of Technology Bombay, Mumbai 400076, India
e-mail: viswam@civil.iitb.ac.in

P. V. Divya
e-mail: divya.pv.nair@gmail.com

J. P. Gourc
LTHE, University of Joseph-Fourier, F-3804 Grenoble, France
e-mail: jean-pierre.gourc@ujf-grenoble.fr

V. K. Stalin and M. Muttharam (eds.), *Geotechnical Characterisation and Geoenvironmental Engineering*, Lecture Notes in Civil Engineering 16,
https://doi.org/10.1007/978-981-13-0899-4_32

1 Introduction

Engineered barriers of waste containment systems can be in the form of compacted clay barriers (CCBs), sand bentonite barriers, geosynthetic clay liners (GCLs), composite barriers (CB), polymer amended sand–bentonite barriers, geomembranes or a combination of the above. Compacted clay barriers (CCBs) are economically viable barrier material wherever these soils are abundantly and easily available at sites (Gourc et al. 2010). In modern landfills, geosynthetics are used as lining systems (Bouazza 2002). Geomembranes (GMs) are widely used in landfill construction (Rowe 2005) and are laid above the clay barrier to form composite barriers (CBs). The thickness of GM ranges from 0.75 to 3 mm (Qian et al. 2002).

Differential settlements are inevitable in landfill covers due to the biodegradation of underlying waste in municipal solid waste (MSW) landfills (Sivakumar Babu et al. 2010) and due to the voids among the waste storage containers in low-level radioactive waste (LLRW) landfills (Gourc et al. 2010). An effective barrier should be able to accommodate the differential settlements and maintain its sealing efficiency. Differential settlements of clay barriers have been studied by several investigators by conducting either full-scale model tests (Edelmann et al. 1999; Gourc et al. 2010) or centrifuge model tests (Jessberger and Stone1991; Rajesh and Viswanadham 2011; Divya et al. 2012). However, studies pertaining to the deformation behaviour of composite barriers subjected to differential settlements in a centrifuge are limited. Considering the limitations of reduced scale model tests for simulating stress-dependant phenomenon, centrifuge model tests were used in the present study. Hence, the motivation behind this study is primarily to evaluate the deformation behaviour and sealing efficiency of geomembrane overlying clay-based landfill covers subjected to flexural distress due to differential settlements. As the geomembrane has much higher allowable elongation compared to the clay barrier material, there is a possibility that the tensile strain developed due to differential settlements will be apportioned by the geomembrane in the case of a composite barrier. Differential settlements can be characterized by the distortion level a/l, which is defined as central settlement a over an influence length l where the settlement becomes negligible.

2 Centrifuge Model Tests

In a centrifuge model, prototype stress conditions are simulated by subjecting a model reduced by $1/N$ to N times acceleration due to gravity in a controlled environment (where N = scale factor or g-level). The centrifuge model tests reported in the present study were conducted at 40 gravities (N = 40). A 4.5 m radius large beam centrifuge available at IIT Bombay was used for conducting centrifuge model tests.

A blend of sand and kaolin in the ratio of 1:4 (i.e. 20:80) by dry weight was selected as the model clay barrier material and was selected in such a way that it represents the properties of clay used in landfill covers (Benson et al. 1999). The maximum dry unit weight of the model barrier material was 15.9 kN/m^3 and optimum moisture content (OMC) was 22%. Average value of coefficient of permeability of the model soil barrier at maximum unit weight and OMC was found to be 4.39×10^{-9} m/s. Model GMs were selected from a number of commercially available polyethylene sheets by measuring their thickness (ASTM D6988) and by conducting tensile tests following the sample dimensions and procedure outlined in ASTM D6693. Two model GMs were selected with different thickness and stiffness. The stiffer GM is referred herein as GM1 and weaker geomembrane as GM2. The scaling considerations for modelling GM were explained in Divya et al. (2012). Based on the scaling considerations for GM, linear dimensions of the GM and the tension developed in the GM per unit width are to be reduced by 1/*N* times that of corresponding prototype values. The thickness (t_{mg}) of GM1 was 0.135 mm which corresponds to 5.4 mm in the prototype at 40 gravities and that of GM2 was 0.05 mm (2 mm in the prototype). The value of initial stiffness for GM1 is 480 kN/m and for GM2 is 280 kN/m in prototype dimensions at 40 g. The ultimate tensile stress (kN/m^2) for GM1 was 28,148 kN/m^2, and ultimate tensile strain (%) was 450%. The corresponding values for GM2 were 21,569 kN/m^2 and 52%. The stress–strain behaviour of GM1 and GM2 was comparable with the prototype geomembranes used in landfills (Qian et al. 2002), but thickness of GM1 was slightly higher. This is due to the practical difficulty in getting model geomembrane of very small thickness having similar stress–strain characteristics of prototype material. Thickness of GM2 was comparable with the prototype geomembranes.

A strong box (720 mm in length, 450 mm in breadth and 440 mm in height) with front side made with Perspex sheet and back and rear side made with well-machined stainless steel plates with stiffeners was used to prepare the model barriers. A settlement rate of 1 mm/min (equivalent to 36 mm/day in the field at 40 g) was induced to the model barriers in the present study using a motor-based differential settlement simulator (MDSS). The details of the MDSS system were explained by Rajesh and Viswanadham (2011). In MDSS, the rotational movement of the motor shaft is converted to the translational movement of the central platform through a screw jack and series of gears. The MDSS system consists of a central support system and two-side support system on either sides of central support system with a hinged plate resting on central support system symmetrically. Model clay barrier of 15-mm (0.6 m) thickness was prepared on the top of predrained 30-mm-thick coarse sand layer followed by 30-mm-thick fine sand layer. They are used to avoid stress concentration at the onset of differential settlements and to induce smooth continuous differential settlements to the overlying barrier. After the clay barrier of required thickness was prepared, geomembrane was placed on the top without any wrinkles in the case of composite barriers. Water-tight seal (thick bentonite paste) was applied all along the sides of the clay barrier, and thereafter, bunds were constructed along the sides to retain water during centrifuge tests. It also helps to simulate the lateral extent of the GM (i.e. GM covering large areas)

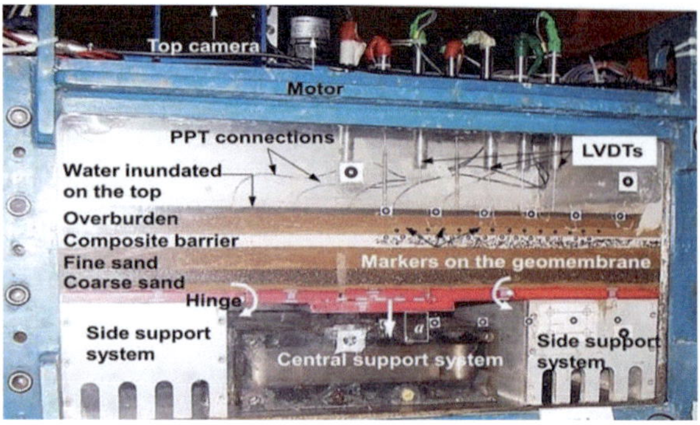

Fig. 1 Centrifuge test package with a CB with GM1

in the field. Discrete markers were embedded 5 mm below the top surface of the model clay barrier at 20 mm centre to centre and also on the GM. These markers were later used for performing digital image analysis (DIA). An overburden of 25 kN/m^2 was induced with the help of a fine sand layer of 27 mm thickness at a dry unit weight of 15 kN/m^3, and a calculated quantity of water was added so that it forms 10-mm free-standing water on the sand surface. Figure 1 shows model test package used in the present study with model composite barrier (CB) with model GM1 before inducing settlements.

Five pore water pressure transducers (PPTs) were installed on the top of the barrier to measure the infiltration of water through the barriers (water breakthrough). Seven linear variable differential transformers (LVDTs) were provided towards the right half section of the model barrier at 50 mm intervals to obtain deformation profiles of the barriers at different settlement stages. One digital camera was mounted along with the model to capture the front elevation of the barrier at different settlement stages, and these images were used for DIA and it also helps to monitor the crack development across the barrier. Also, a CCD video camera was placed in such a way to register the depletion of water retained on the top of the barrier during the centrifuge test. The maximum central settlement $a_{\max}$ induced was 25 mm (1 m at 40 g).

3 Test Results and Discussions

The centrifuge tests results were expressed in terms of average crack width (w_c) and average crack depth (d_c), maximum downward thrust ($q_{g,\max}$) exerted by the GM on the clay barrier and limiting distortion level $a_{\lim}/l$ for all the model barriers, wherever applicable. Limiting distortion level $a_{\lim}/l$ is defined herein as the ratio of central settlement at which a drastic change in water volume above barrier was observed to the influence length l. In the present study, an influence length l of 200 mm in model dimensions was used. The influence of stiffness of the GM on the performance of CB was also discussed. Prototype dimensions are given within the

Table 1 Summary of centrifuge test results

Parameters	Test legend		
	CCB	CB1	CB2
Type of GM	–[a]	GM1	GM2
Cracking pattern	Wide	Narrow	Narrow
w_c (mm)	3.70 (148)[b]	1.20 (48)[b]	1.50 (60)[b]
d_c(mm)	15 (600)[b]	15 (600)[b]	15 (600)[b]
a_{lim}/l	0.056	0.10	0.093
$q_{g,max}$ (kN/m^2)	-[a]	20	12

GM-geomembrane
[a]Not reported/relevant
[b]Prototype values are indicated within the parenthesis; w_c = average crack width; d_c = average crack depth; a_{lim}/l = limiting distortion level

parentheses, wherever relevant, from here onwards. Table 1 summarizes the results of all the centrifuge model tests on clay barriers.

At the onset of differential settlements, the CCB was found to experience wide and full-depth cracking and lose its integrity in terms of sealing efficiency at lower *a*/*l*. The average crack width w_c was 3.7 mm (148 mm) for CCB at the end of centrifuge test. In the case of CB, w_c was reduced to 1.20 mm (48 mm) and 1.50 mm (60 mm) when provided with stiffer GM (GM1) and weaker GM (GM2), respectively. Though full-depth cracks were observed to occur in CB (as in case of CCB), the presence of GM helped in restraining water breakthrough. The sealing efficiency of the barriers can be obtained from the change in height of water on the top of barrier using PPTs. Ratio of initial volume of water v_0 and volume of water at any instance v_a to the initial volume of water v_0 is defined as the infiltration ratio IFR. Figure 2 shows the variation of IFR with *a*/*l* for the barriers.

The a_{lim}/l was determined by back tangent method from the curve showing the variation of IFR with *a*/*l*. The CCB was observed to experience water breakthrough at lower a_{lim}/l of 0.056. In comparison, model CBs of identical thickness were observed to experience negligible changes in infiltration ratio. The value of a_{lim}/l for CB was found to be delayed to 0.1 and 0.093 when provided with GM1 and GM2, respectively. The observed delay in water breakthrough can be attributed to the effect of total downward pressure acting on the CB as explained in the subsequent sections.

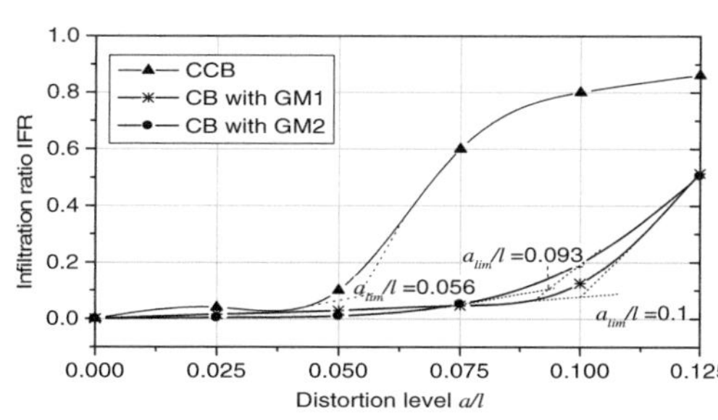

Fig. 2 Variation of IFR with distortion level

Combined bending and elongation method (Tognon et al. 2000) was used to obtain the strain distributions along the GM layer at different settlement stages. Strain in the GM $\varepsilon_g(x) = \varepsilon_{lg}(x) \pm \varepsilon_{\kappa g}(x)$. The elongation strain $\varepsilon_{lg}(x)$ in the GM was obtained from the deformation profile of GM, $w_g(x)$ as $\left[\left(1+\left[w'_g(x)\right]^2\right)^{1/2}-1\right]$.

The deformation profiles are obtained from the co-ordinates of the markers on the GM obtained after performing digital image analysis (DIA). GRAM++ software was used for DIA. The curvature strain in GM $\varepsilon_{\kappa g}(x)$ was obtained as $R_g.\kappa_g(x)t_g$, where R_g = neutral layer coefficient = 1/2, $\kappa_g(x)$ = curvature of GM and t_g = thickness of GM and x = horizontal distance from centre of the composite barrier. The curvature strain in GM was negligible due to small thickness of GM. Further by using the computed strain distributions and the curvature of the deformed GM and stiffness of the GM, the vertical thrust exerted by the GM layer in the cover system was calculated by using membrane theory (Espinoza 1994). The thrust distribution is calculated as $q_g(x) = J_g \cdot \varepsilon_g(x) \cdot \kappa_g(x)$ where, J_g is the stiffness of GM in kN/m, and this was obtained from the tensile load–strain curve of GM1 and GM2 used in the present study. The value of $q_g(x)$ was found to be maximum near the zone of maximum curvature and is denoted herein as $q_{g\max}$. The variation of maximum downward thrust $q_{g\max}$ with distortion level *a*/*l* for the model barriers is shown in Fig. 3.

The $q_{g\max}$ was found to increase with an increase in *a*/*l* and the increase was more beyond *a*/*l* of 0.075 and $a/a_{\max}$ of 0.6. The value $q_{g\max}$ for GM1 is 20 kN/m^2 at a distortion level of 0.125. This implies that a total downward thrust of the order of 45 kN/m^2 (this includes overburden pressure of 25 kN/m^2 in the form of cover) is acting at the zone of maximum curvature of the clay barriers. However, the value of $q_{g\max}$ for model with GM2 is only 12 kN/m^2 at *a*/*l* = 0.125 resulting in a total suppression pressure of 37 kN/m^2. The value of $q_{g\max}$ for model provided with a weaker GM (GM2) was found to be relatively lower than that with GM1. Hence, the stiffness of GM is an important parameter which decides the performance of CB and the stiffer GM can exert more $q_{g\max}$ on the clay barrier surface which hinders the crack widening and infiltration of water through the cracks.

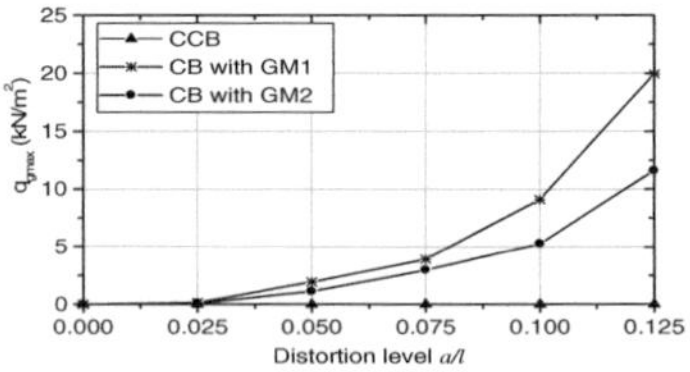

Fig. 3 Variation of $q_{g\max}$ with *a*/*l* and $a/a_{\max}$

4 Conclusions

Performance of CCB and CB at the onset of differential settlements was studied by conducting a series of centrifuge model tests. Tests were also conducted by varying the stiffness of the GM. It was observed that CCB experienced wide (w_c = 148 mm) and full-depth cracking and lose its integrity in terms of sealing efficiency at lower a/l with a_{lim}/l of 0.056. In comparison, w_c was reduced to 48 and 60 mm, and a_{lim}/l was delayed to 0.1 and 0.093, for a CB when provided with stiffer GM (GM1) and weaker GM (GM2), respectively. Though full-depth cracks were observed to occur in also in CB, the presence of GM helped in restraining water breakthrough. The presence of GM imposes a fictitious overburden on the surface of clay barrier due to the mobilization of downward thrust, q_g at the zone of maximum curvature which increase with an increase in a/l. Also, stiffer GM found to exert more q_g which hinders the crack widening and infiltration of water through the cracks at the onset of differential settlements.

References

ASTM D6988. (2008). *Standard Guide for Determination of Thickness of Plastic Film Test Specimens*. Philadelphia, Pa: American Society for Testing and Materials.

ASTM D6693. (2004). *Standard Test Method for Determining Tensile Properties of Nonreinforced Polyethylene and Nonreinforced Flexible Polypropylene Geomembranes*. Philadelphia, Pa: American Society for Testing and Materials.

Benson, C. H., Daniel, D. E., & Boutwell, G. P. (1999). Field performance of compacted clay liners. *Journal of Geotechnical and Geoenvironmental Engineering, ASCE, 125*(5), 390–403.

Bouazza, A. (2002). Geosynthetic clay liners. *Geotextiles and Geomembranes, 20*(1), 1–17.

Divya, P. V., Viswanadham, B. V. S., & Gourc, J. P. (2012). Influence of Geomembrane on the Deformation Behaviour of Clay-based Landfill Covers. *Geotextiles and Geomembranes, 34,* 158–171.

Edelmann, L., Hertweck, M., & Amman, P. (1999). Mechanical behaviour of landfill barrier systems. *Proceedings of the Institution of Civil Engineers-Geotechnical Engineering, 137*(4), 215–223.

Espinoza, R. D. (1994). Soil-geotextile interaction: Evaluation of membrane support. *Geotextiles and Geomembranes, 13*(5), 281–293.

Gourc, J. P., Camp, S., Viswanadham, B. V. S., & Rajesh, S. (2010). Deformation behaviour of clay cap barriers of hazardous waste containment systems: full-scale and centrifuge tests. *Geotextiles and Geomembranes, 28*(3), 281–291.

Jessberger, H. L., & Stone, K. J. L. (1991). Subsidence effects on clay barriers. *Geotechnique, 41*(2), 185–194.

Qian, X., Koerner, R. M., & Gray, D. H. (2002). *Geotechnical aspects of landfill design and construction* (pp. 431–436). New Jersey, USA: Prentice Hall.

Rajesh, S., & Viswanadham, B. V. S. (2011). Hydro-mechanical behavior of geogrid reinforced soil barriers of landfill cover systems. *Geotextiles and Geomembranes, 29*(1), 51–64.

Rowe, R. K. (2005). Long-term performance of contaminant barrier systems. *Géotechnique, 55*(9), 631–678.

Sivakumar Babu, G. L., Reddy, K. R., Chouskey, S. K., & Kulkarni, H. (2010). Prediction of long-term municipal solid waste landfill settlement using constitutive model. *Practice Periodical of Hazardous, Toxic, and Radioactive Waste Management, ASCE, 14*(2), 139–150.

Tognon, A. R., Rowe, R. K., & Moore, I. D. (2000). Geomembrane strain observed in large-scale testing of protection layers. *Journal of Geotechnical and Geoenvironmental Engineering, ASCE, 126*(12), 1194–1208.

Studies on Desiccation Cracking Behavior of Geofiber Reinforced Clay

M. Indu Priya, Uma Chaduvula and B. V. S. Viswanadham

Abstract Expansive clays undergo large volumetric changes with varying moisture content inducing desiccation cracks. Desiccation cracks increase the hydraulic conductivity and reduce the strength of the soil substantially. This can be a setback when expansive clays are used as landfill liners and covers where low hydraulic conductivity is mandatory to ensure a safe barrier. In this study, the effect of discrete and randomly distributed fiber (DRDF) reinforcement on desiccation cracking of soil with varying fiber lengths was investigated. Black cotton soil, a locally available expansive clay, was reinforced with polypropylene fibers of lengths 6, 12, and 18 mm, at 0.5% fiber content (% by dry weight of soil). The slurry samples were allowed to dry at uniform temperature and relative humidity. Various physical and morphological parameters such as cell area, crack width, and moisture content were observed over time. The crack parameters were quantified using digital image analysis (DIA). It was observed that fiber inclusion efficiently reduces desiccation cracking by increasing the tensile strength of the clay.

Keywords Expansive clay · Desiccation cracking · Fiber reinforcement
Geofibers

1 Introduction

Desiccation cracks occur when tensile stresses due to matric suction exceed the tensile strength of the soil (Kodikara et al. 2000). Tensile stresses develop when external or internal restraints are at place in the form of rough layer interfaces or

---

M. Indu Priya
Indian Institute of Technology Guwahati, Guwahati 781039, India
e-mail: m.indupriya1997@gmail.com

U. Chaduvula · B. V. S. Viswanadham (✉)
Indian Institute of Technology Bombay, Mumbai 400076, India
e-mail: viswam@civil.iitb.ac.in

U. Chaduvula
e-mail: uma.chaduvula@iitb.ac.in

V. K. Stalin and M. Muttharam (eds.), *Geotechnical Characterisation and Geoenvironmental Engineering*, Lecture Notes in Civil Engineering 16,
https://doi.org/10.1007/978-981-13-0899-4_33

sections of soil undergoing non-uniform drying. Clayey soils are more susceptible to desiccation due to the possibility of developing high suction from heat. Desiccation cracks affect the hydraulic properties and strength characteristics of the soil significantly. In applications such as clay liners and covers for landfills, low permeability is desired to prevent leachate infiltration (He et al. 2015). Currently, discrete and randomly distributed fiber (DRDF) reinforcement is being adopted to efficiently reduce the detrimental effects of the cracks due to loss of moisture (Viswanadham et al. 2011). Divya et al. (2014) carried out a series of direct tensile strength tests and concluded that fiber inclusion improved the tensile strength-strain characteristics and ductility of soil. Since desiccation cracks primarily arise due to tension, it can be implied that fiber reinforcement significantly reduces the effects due to these cracks. In this study, the effects of fiber reinforcement and fiber lengths on desiccation cracking of soil were investigated. Black cotton soil was reinforced with fibers of three different lengths. Moisture content and crack parameters such as cell area and crack width were observed for each fiber length over the course of desiccation. The crack parameters were measured using digital image analysis.

It is to be noted that throughout the paper, the term 'water content' (w) refers to gravimetric water content and 'fiber content' (f) refers to the percentage of the weight of the fibers by dry weight of the soil.

2 Materials and Methods

2.1 Materials

Black cotton soil used in this study is an expansive clay widely distributed in the south-central part of India. It is classified as high plasticity clay (as per Unified Soil Classification system) with liquid limit of 86.5%. The physical properties of black cotton soil are as presented in Table 1.

The fibers used to reinforce the soil were polypropylene (PP). It had triangular cross section with an effective diameter of 40 μm. Table 2 lists various physical properties of the fibers. Since the fibers are required to also resist alkaline conditions, chloride and alkali testing was also performed. For this experiment, fiber lengths (l) of 6, 12, and 18 mm were selected. Higher fiber lengths were not selected as it had not resulted in uniform distribution during sample preparation.

2.2 Sample Preparation

For sample preparation, crushed and oven dried soil passing through 425 μ IS sieve was taken. Fibers of desired length were weighed and mixed thoroughly with the dry soil. Fiber content of 0.5% was chosen as an optimum value in this study to

Table 1 Properties of black cotton soil

Properties	Values
Specific gravity, G_s	2.58
Atterberg limits	
Liquid limit (%)	86.8
Plasticity index (%)	35
Size fraction	
Gravel fraction (4.75–20 mm)	1
Sand fraction (0.075–2 mm)	5
Silt fraction (0.002–0.075 mm)	34
Clay fraction (<0.002 mm)	60
Classification[a]	CH

[a]As per USCS

Table 2 Properties of polypropylene (PP) fibers

Properties	Values
Specific gravity	0.91
Average diameter	40 μm
Breaking tenacity	4–6 gpd
Modulus of elasticity	>4000 MPa
Fusion point	160–165 °C
Alkali and chloride resistance	Very good

avoid clustering of fibers while mixing. The required quantity of water equivalent to 90% water content (slightly greater than the liquid limit of the soil) was added gradually to the soil to form a slurry. The slurry was then transferred to the acrylic container of internal diameter 190 mm, which was slightly greased only along the inner walls, to 10 mm thickness. Acrylic was preferred over glass because the thermal conductivity of glass is much higher; hence, it tends to heat up faster. This would result in non-uniform heating of the sample. The container was carefully tapped to expel air bubbles and ensure uniform thickness throughout. The sample was then sealed and allowed to rest for 24 h to attain moisture homogenization.

2.3 Test Setup and Procedure

The saturated samples were allowed to desiccate under two 500 W tungsten halogen lamps (Fig. 1), and the test was conducted in a constant temperature of 50 °C and 20% relative humidity as measured with a thermo-hygrometer. The container was placed on a weighing balance (of accuracy 0.01 g) which logged the variation of weight of the sample with time to obtain water content of the sample at an instance. A rectangular sheet with markers at known distances (200 × 250 mm) was placed beneath the containers to be used as a reference scale during image analysis. The digital camera was set to capture images of drying sample at regular intervals.

The sample was allowed to desiccate for approximately 10 h such that it reached a residual water content.

2.4 *Image Analysis*

Analysis of the surface images obtained was done using ImageJ (Abràmoff et al. 2014), an image processing software. The images were set to scale as well as appropriate units using the markers on the sheet below the sample, and the crack parameters were calculated with reference to the scale. The parameters calculated were, namely average crack width (w_{avg}) and average cell area.

3 Results and Discussion

The variation of water content plotted against time (Fig. 2) shows that the trend does not vary either with fiber inclusion or with fiber length. Two distinct stages of drying can be observed: the constant evaporation stage and falling evaporation stage (Tang et al. 2011). The desiccation tests were continued till the water content reached a residual value of 3–5%.

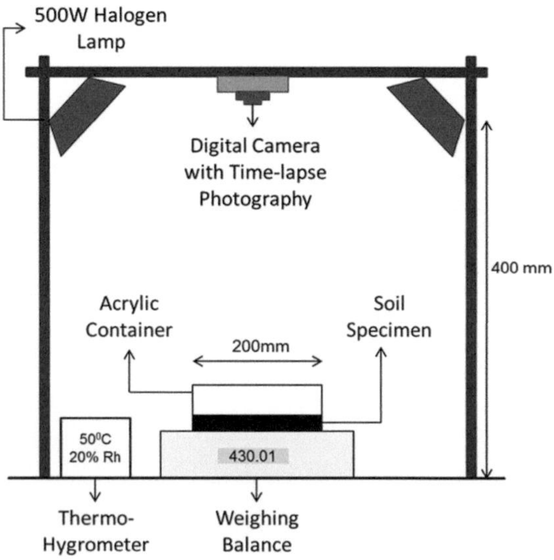

Fig. 1 Test setup for desiccation cracking

Figure 3 shows the average crack widths, w_{avg}, calculated for the samples over the course of their desiccation. In the unreinforced sample, as expected, the crack width increased with time before becoming stable after sufficient moisture is lost. However, in case of the reinforced samples, distinct peaks were observed in the graphs.

The difference in observations was due to the increase in tensile strength with fiber inclusion. In cracking of soil, two opposing forces were in action: tensile strength and friction. While tensile strength of the soil caused shrinkage during desiccation, friction in the soil–container interface resisted it. In the unreinforced sample, the tensile strength of soil was very less. Thus, the friction overcame it easily causing the soil to fail. But the fiber reinforcement added to the soil effectively resisting the friction till a maxima in crack width occurs. Beyond this point, tensile strength of fibers overcame the friction and the sample shrunk resulting in reduced crack width.

It was also observed that the reinforced samples moved as a unit unlike the discrete cells in unreinforced sample which indicated that the former had greater final strength than the latter.

Cell areas (Fig. 4) reduced with time for all the samples. As water evaporated, cracks propagated within the soil. Cell formation took place when the soil failed and formed a distinct boundary. As water loss continued, individual cells shrunk resulting in a reducing trend. At any instant, cell areas of reinforced samples were much lesser than the unreinforced sample. This was because crack propagation was

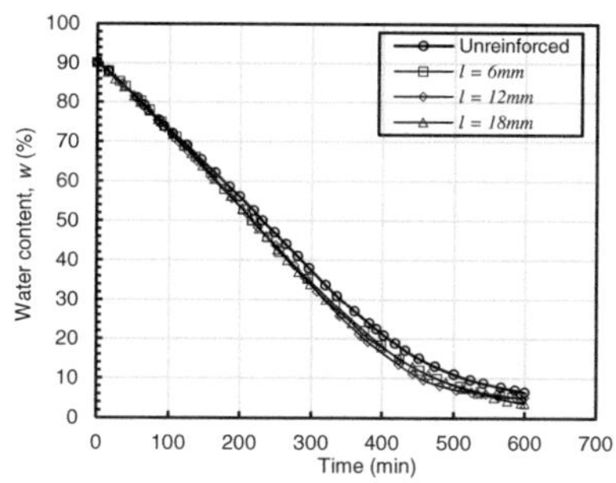

Fig. 2 Moisture content variation with time

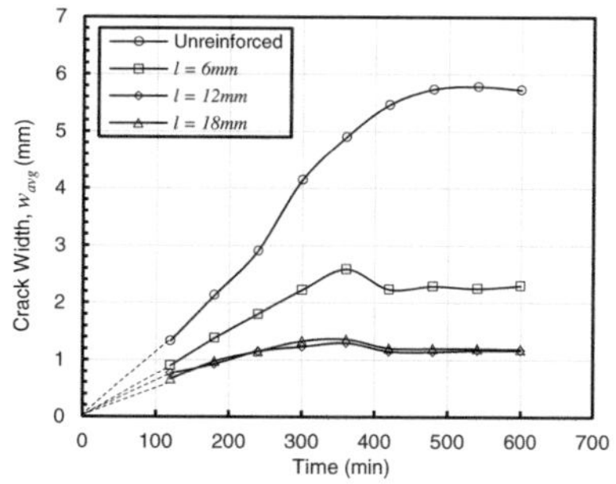

Fig. 3 Variation of average crack width, *w*avg, with time

Fig. 4 Variation of average cell area with time

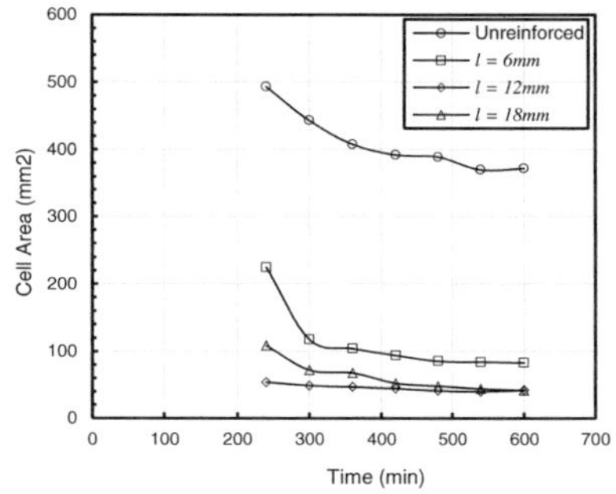

hindered by the randomly distributed fibers leading to smaller cells, while unreinforced sample provided less hindrance.

For field applications, crack parameters are generally desired such that it maintains lower hydraulic conductivity and higher strength. Therefore, crack width and cell area ought to be low as it is indicative of increased performance. On comparing the fiber lengths in Figs. 3 and 4, the average crack parameters were higher for 6 mm fibers compared to 12 and 18 mm fibers. Fibers of 6 mm length got pulled out completely and formed wide cracks indicating that the increase in tensile strength was not that significant in comparison to unreinforced sample. Although the difference between 12 and 18 mm fibers is not significant, 12 mm fibers have lower crack parameters between the two lengths. Moreover, 18 mm fibers formed a cluster while mixing with resulted in non-uniform distribution of fibers. Thus, 12 fibers were more effective as reinforcing the soil.

4 Conclusions

From the laboratory experiments on effects of fiber reinforcement of black cotton soil on desiccation cracking and analysis of the data obtained, the following conclusions can be drawn:

(1) Final average crack parameters were much lower at any instant after fiber inclusion which is representative of their lower hydraulic conductivity in comparison to unreinforced samples.
(2) Tensile strength of fiber reinforced samples was increased as evidenced by the peaks in Fig. 3 which was absent in the unreinforced sample.
(3) Results from effects of variation of fiber length on desiccation cracking showed that 6 mm fibers were ineffective in reinforcing the sample. While the performance of 12 and 18 mm fibers was similar, 12 mm fibers were more uniformly distributed in the sample than 18 mm fibers.

References

Abràmoff, M. D., Magalhães, P. J., & Ram, S. J. (2014). Image processing with image. *Journal Biophotonics International.*

Divya, P. V., Viswanadham, B. V. S., & Gourc, J. P. (2014). Evaluation of tensile strength-strain characteristics of fiber-reinforced soil through laboratory tests. *Journal of Materials in Civil Engineering, 26*(1), 14–23.

He, J., Wang, Y., Li, Y., & Ruan, X. (2015). Effects of leachate infiltration and desiccation cracks on hydraulic conductivity of compacted clay. *Water Science and Engineering, 8,* 151.

Kodikara, J. K., Barbour, S. L., & Fredlund, D. G. (2000). Desiccation cracking of soil layers. In *Proceedings of the Asian Conference in Unsaturated Soils* (pp. 693–698), UNSAT ASIA 2000, Singapore.

Tang, C., Cui, Y., Shi, B., Tang, A., & Liu, C. (2011). Desiccation and cracking behaviour of clay layer from slurry state under wetting–drying cycles. *Geoderma, 166*(1), 111–118.

Viswanadham, B. V. S., Rajesh, S., Divya, P. V., & Gourc, J. P. (2011). Influence of randomly distributed geofibers on the integrity of clay-based landfill covers: A centrifuge study. *Geosynthetics International, 18*(5), 255–271.

Shear Strength Parameters of Granite Rock Mass: A Case Study

G. V. Ramana, Pathak Shashank and Hari Dev

Abstract In situ shear strength parameters of rock mass are important design parameters required for design of structures founded on or constructed in rock mass. Block shear test method is quite popular among practicing engineers for determination of cohesion and friction angle of rock-to-rock and concrete-to-rock interface. Generally, at the stage of preliminary design, a value of desired shear strength parameters can be taken from available literature. However, such values must be used with proper engineering judgement. In case of rock mass, structural features of geology plays a very crucial role in affecting its engineering behavior. Same rock type may have significantly different shear strength parameters depending upon site-specific geological characteristics. In this paper, the variation in rock-to-rock and concrete-to-rock shear strength parameters of rock mass consisting of primarily granites is studied among three different geological variants, namely (a) western part of Himalaya, (b) Garo-Khasi Hills of Meghalaya, and (c) Eastern Ghat Belt of Andhra Pradesh. The variation in shear strength, cohesion, and friction angle is studied in terms of strength ratio, cohesion ratio, and friction angle ratio. These values are also compared with values for granites available in the literature. Site parameters supported with quantitative analysis of in situ data indicate that highest shear strength is expected at Eastern Ghat Belt and least shear strength at Garo-Khasi Hills.

Keywords Shear strength · Granites · Cohesion · Friction angle

G. V. Ramana (✉) · P. Shashank · H. Dev
Central Soil and Materials Research Station, Ministry of Water Resources, RD & GR, Olof Palme Marg, Hauz Khas, New Delhi, India
e-mail: ramana.laxmi@gmail.com

P. Shashank
e-mail: shashankpathaks@gmail.com

H. Dev
e-mail: haridev@nic.in

V. K. Stalin and M. Muttharam (eds.), *Geotechnical Characterisation and Geoenvironmental Engineering*, Lecture Notes in Civil Engineering 16,
https://doi.org/10.1007/978-981-13-0899-4_34

1 Introduction

Quantification of in situ shear strength of concrete over rock and rock over rock interface is important for design of structures founded on or constructed in rock mass. Shear strength of concrete–rock (C/R) and rock–rock (R/R) interfaces is generally represented in terms of cohesion and friction angle on these interfaces. Block shear test method [IS 7746 (1996) and ISRM (2007)] is quite popular for determination of cohesion and friction angle along the predetermined concrete–rock or rock–rock interface.

Geological features such as mineralogical components at the time of rock formation and discontinuities play an important role in deciding the shear behavior of these interfaces. A discontinuity is a plane or surface that marks a change in physical or chemical characteristics in a rock mass. A discontinuity can be in the form of a bedding plane, spacing, schistosity, foliation, joint, fracture, fissure, crack, aperture, filling, or fault plane (Zhang 2005). The properties of discontinuities such as orientation, persistence, roughness, and infilling also play a crucial role in shear strength of rock mass. Friction angle of a rough surface consists of two components: (a) friction of the rock material and (b) interlocking produced by the irregularities of the surface (Barton 2013). The orientation of a major geological discontinuity relative to an underground structure also controls the in situ shear strength parameters of rock mass. In addition to this, shear strength parameters are also influenced by weathering action and alterations of geological discontinuities (Thuro and Scholz 2003). Thus, above discussion indicates that several uncertainties are associated with any geological site (Ramamurthy 2001). These uncertainties cannot be accounted by using any theoretical treatment and therefore, in situ testing, though there are several testing limitations, may provide a better idea of rock-mass behavior.

Presented paper compares the shear strength parameters (cohesion and friction angle) of granite rock mass from three different regions of India, namely (a) Western part of Himalaya (WHI), (b) Garo-Khasi-Jaintia Hills (GKJ) of Meghalaya, and (c) Eastern Ghat Belt (EGB) of Andhra Pradesh. The variation among these parameters is also quantified in terms of ratios of these parameters for the three variants.

2 Geology of Test Sites

Geological information related to the three variants of granite rock mass is given in Table 1 for three variants of granite rock mass.

Based on the above geological parameters, the three variants of granite rock mass can be arranged as EGB>GKJ and WHI>GKJ with respect to RQD, RMR,

Table 1 Geological data for granites of GKJ, WHI, and EGB

Location	Rock type at test location	Available geological parameters	References
GKJ	Slightly weathered granite-gneiss	RQD = 75% RMR = 50–55 Rock class = III	CSMRS (2016)
WHI	Granite-gneiss with thin phyllitic partings	All the joint planes are generally wavy, rough, and unfilled. At few locations joint openings are filled with surface clay. Staining along the joint planes is also noticed at places	CSMRS (2014a, b) and CSMRS—R/R (2014c)
EGB	Garnetiferous-quartzo-feldspathic-biotite-gneiss	RQD = 80–100% RMR = 50 The joints are tight to slightly open, moderately spaced to widely, rough, planar, irregular, continuous to discontinuous, and straight to curvilinear in nature	CSMRS (2014a, b, c) and GSI (2014)

and rock conditions at test locations. However, it is difficult to differentiate clearly between WHI and EGB based on available data. Therefore, highest shear strength is expected at EGB and least shear strength at GKJ.

3 In Situ Shear Test Data

Block shear tests were conducted on predetermined interfaces as described in IS 7746 (1996) and ISRM (2007) on properly constructed rock blocks over rock mass and concrete blocks over rock mass. Generally, the data is collected in the form of shear stress versus shear displacement plots at various normal stress levels. Generally, a clear peak and then a residual strength is observed in all the case. Shear strength parameters are determined by fitting Mohr-Coulomb criterion (Eq. 1) which relates peak shear stress (τ_f) [and residual shear stress (τ_{fr})] to applied normal stress (σ_n) on scatter of peak (and residual) shear stress versus corresponding normal stress.

$$\tau_{\mathrm{f}} = c + \sigma_{\mathrm{n}} \tan \varphi \quad (1)$$

where c is cohesion and φ is friction angle. Based on the in situ testing conducted by authors at three locations mentioned, peak and residual shear strengths are shown in Table 2.

For quantitative comparison of shear strength of interfaces in two different geological variants, ratio of respective shear strength values under same normal stress is defined as shear strength ratio corresponding to that normal stress. Based on the data of Table 2, peak and residual shear strength ratio of EGB to GKJ (EGB/GKJ), WHI to GKJ (WHI/GKJ), and EGB to WHI (EGB/WHI) is plotted against applied normal stress in Fig. 1. It is worth mentioning that necessary linear interpolation is also carried out in order to determine the shear strengths at normal stresses where test data is not available.

A decreasing trend is observed with increasing applied normal stress. The strength ratio tends toward unity on increasing stress. This indicates that at higher normal stresses, behavior of various variants of granite rock mass becomes approximately similar. It may be observed that the strength ratio between EGB and GKJ for R/R interface at applied normal stress of 0.4 MPa is significantly higher and clearly out of the trend of the general scatter. Except this, all other strength ratios lie between 0.9 and 2.5. If strength ratio of three variants are represented as EGB:GKJ:WHI = 1:y:z, then, y varies from 0.40 to 0.83 and z varies from 0.69 to 1.12. Using the data of Table 2, cohesion and friction angles are also determined as shown in Figs. 2, 3, 4, and 5.

Based on analysis carried out in Figs. 2, 3, 4, and 5, Table 3 shows the summary of various shear strength parameters for the three variants of granite rock mass.

Table 2 In situ peak and residual shear strength for three variants of granite rock mass

EGB (CSMRS 2014a, b, c)			GKJ (CSMRS 2016)			WHI (CSMRS 2014a, b, c)		
(σ_n) MPa	(τ_f) MPa	(τ_{fr}) MPa	(σ_n) MPa	(τ_f) MPa	(τ_{fr}) MPa	(σ_n) MPa	(τ_f) MPa	(τ_{fr}) MPa
R/R interface								
0.40	2.09	0.97	0.40	0.54	0.43	–	–	–
0.80	1.55	1.20	0.60	0.73	0.62	–	–	–
0.96	2.40	1.41	0.80	1.12	0.93	–	–	–
1.20	3.09	1.93	1.00	1.24	1.01	–	–	–
C/R interface								
0.34	1.03	0.60	0.40	0.61	0.54	0.58	0.98	0.84
0.42	1.28	0.68	0.47	0.52	0.45	0.91	1.20	1.11
0.56	1.37	0.75	0.79	0.99	0.80	1.02	1.58	1.42
0.66	1.34	0.93	0.98	1.13	1.02	1.43	1.70	1.50
1.09	2.01	1.34	–	–	–	1.48	1.75	1.54

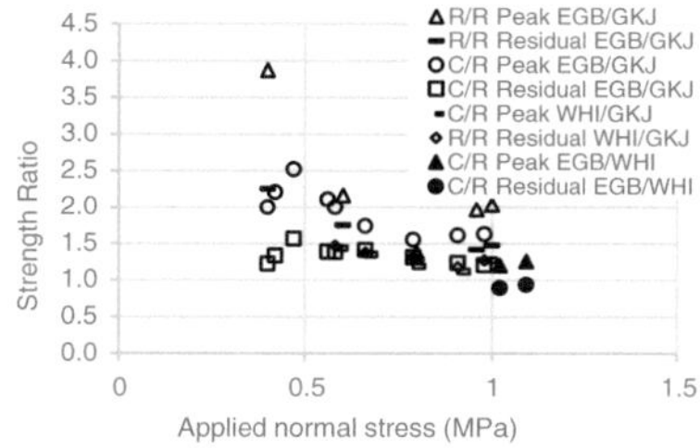

Fig. 1 Shear strength ratio versus applied normal stress

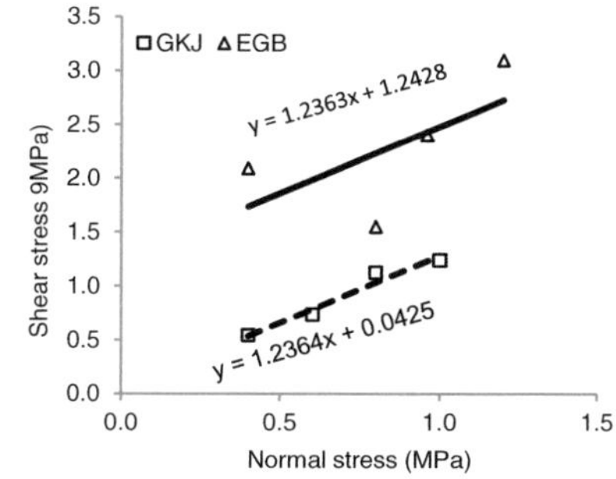

Fig. 2 Peak shear strength parameters at R/R

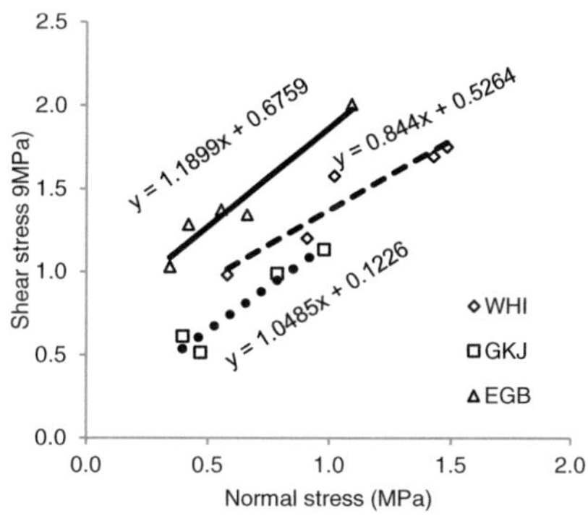

Fig. 3 Peak shear strength parameters at C/R

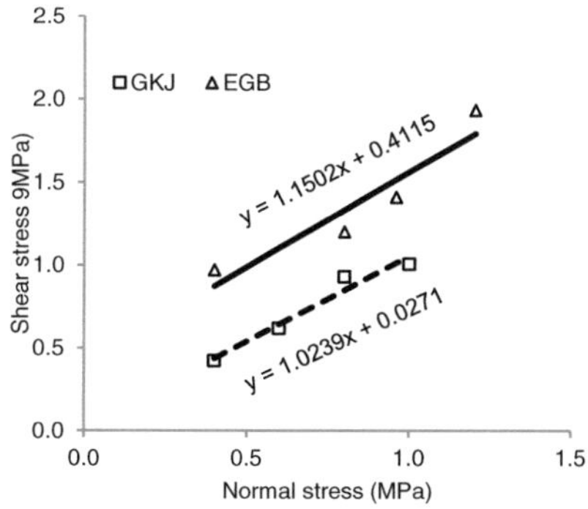

Fig. 4 Residual shear strength parameters at R/R

Fig. 5 Residual shear strength parameters at C/R

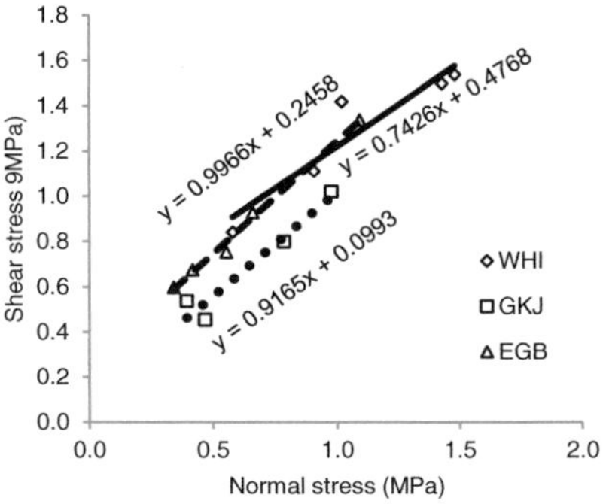

The literature mentions a range of cohesion from 0 to 0.5 MPa (Afrouz 1992) and a range of friction angle from 45° to 50° (Daemen and Schultz 1995) for granite rock mass without any specific mention of C/R or R/R interface. The literature underestimates the cohesion value for peak shear strength of R/R interface at EGB, C/R interface at EGB, and C/R interface at WHI, while 75% values fall within the range as mentioned by Afrouz (1992). In case of friction angle, 33% values fall within the range as mentioned by Daemen et al. (1995). Literature predicts unsafe estimates of friction angle for peak shear strength of C/R interface at WHI and for residual shear strength of C/R interface at EGB, WHI, and GKJ, whereas underestimates the R/R interface friction angle at EGB, WHI, GKJ for peak shear strength and at WHI for residual shear strength. Cohesion values are found to be better covered by literature range, and friction angle may be unsafe, especially in C/R cases. This reflects site-specific variation which is generally not covered in the literature. Therefore, in situ evaluation at carefully selected test location is must.

Ratio of respective cohesion values and ratio of respective friction angles of the three variants of granite rock mass are also evaluated and shown in Table 4 along with their shear strength ratios.

Table 3 In situ shear strength parameters for three variants of granite rock mass

Granite variants	Parameter *c* (MPa), *Φ* (degrees)							
	C, C/R	Φ, C/R	Cr, C/R	Φr, C/R	C, R/R	Φ, R/R	Cr, R/R	Φr, R/R
EGB	0.68	49.93	0.25	44.89	1.24	51.09	0.41	49.11
WHI	0.52	40.16	0.48	36.60	0.49	66.00	0.33	53.81
GKJ	0.12	46.34	0.10	42.49	0.04	51.09	0.03	45.68

Table 4 Cohesion, friction angle, and shear strength ratios for three variants of granite rock mass

Parameter	EGB	WHI	GKJ
Cohesion	1	0.76–1.92	0.03–0.40
Friction angle	1	0.80–1.29	0.93–1.00
Shear strength	1	0.69–1.12	0.40–0.83

As discussed earlier, major reasons behind good shear strength at EGB are probably the better rock conditions. Another factor which may play a crucial role is that testing at EGB was done in an open trench, where loss of normal stress on top reaction pad can be controlled in a much better way as compared to test inside a drift (WHI and GKJ), where deformation/cracking in top concrete reaction pad is difficult to control. Least control was observed at GKJ by authors. Reduction in normal stress is usually expected to reduce the observed peak shear stress. In addition to this, ratios indicate that GKJ rocks are comparatively weakest in cohesion such that overall shear strength is lowest at GKJ. Possibly weathered condition of rock and excessive deformations in top reaction pad may be governing factors behind this. In some cases, rock at WHI has shown a highest value of cohesion. This may be possibly due to clay filling of joints at test locations of WHI. If EGB is taken as a basis then at the other two locations due to all possible uncertainties shear strengths are found to vary from minus 60% to plus 12%. Cohesion varies from minus 97% to plus 92%, and friction angle varies from minus 20% to plus 29%. Major variation is found in cohesion.

4 Conclusions

In this paper, variation in shear strength parameters of rock over rock and concrete over rock interfaces in granite rock mass is studied among three different geological variants, namely (a) Western part of Himalaya, (b) Garo-Khasi Hills of Meghalaya, and (c) Eastern Ghat Belt of Andhra Pradesh. This case study on limited field data has shown more variation in cohesion values as compared to friction angle among different variants of granite rock mass. Remarkable variation in cohesion suggests that behavior of rock mass is significantly different at low normal stresses. Additionally, lesser variation in friction angle indicates that similarity in behavior increases at higher normal stresses. Thus, it is found that cohesion governs the shear strength at low normal stresses. Therefore, more number of tests may be performed at low stresses so that a more reliable estimate of cohesion can be made.

References

Afrouz, Andy. (1992). *Practical Handbook of Rock Mass Classification Systems and Modes of Ground*. Boca Raton, Florida, USA: CRC Press.

Barton, N. (2013). Shear strength response of some geological materials in triaxial compression. *Journal of Rock Mechanics and Geotechnical Engineering, 2013,* 249–261.

CSMRS. (2014a). Report on In-Situ Shear Strength Parameters of Rock Mass at Indira Sagar Polavaram Multipurpose Project, A.P.

CSMRS. (2014b). Report on In-Situ Shear Strength Parameters of Rock Mass in Dam Site Drift of Kirthai Project Stage-II, J&K.

CSMRS (2014c) Report on In-Situ Shear Strength Parameters of Rock to Rock and Concrete to Rock Interfaces in Dam Site Drift of Kirthai Project Stage-II, J&K.

CSMRS. (2016). Report on In-Situ Shear Strength Parameters of Rock Mass at Right Bank Drift at dam Axis of Mawphu–II H.E. Project, Meghalaya.

IS 7746. (1991, 1996). Indian Standard code of practice for in-situ shear test on rock.

ISRM. (2007). The complete ISRM suggested methods for rock characterization, testing and monitoring: 1974–2006. In R. Ulusay & J.A. Hudson (Eds.), *International Society for Rock Mechanics*.

Daemen, J. J. K., Schultz, R.A. (1995) Rock Mechanics: Proceedings of the 35th US Symposium on Rock Mechanics. Edited by University of Nevada/Reno/5-7.

Ramamurthy, T. (2001). Shear strength response of some geological materials in triaxial compression. *International Journal of Rock Mechanics and Mining Sciences, 38*(2001), 683–697.

Thuro, K., Scholz, M. (2003). Deep weathering and alternation in granites-a product of coupled process. In *Geo Proceedings on International Conference at Royal Institute of Technology*, Stockholm, Sweden, 13–15th October 2003.

Zhang, L. (2005). Engineering Properties of Rocks. In *Geo-Engineering Series, Vol. 4*. Elsevier Geo-Engineering series.

GSI. (2014). *Report on Geotechnical investigations for the proposed spillway, Indirasagar Polavaram, multipurpose Project, east and west Godavari districts*. Andhra Pradesh: Engineering geology division, southern region, Hyderabad.

Effect of Gypsum on Strength Behavior of Lime-Modified Pond Ash as an Underground Stowing Material

R. Shanmuga Priya, S. Sreelakshmi and G. Kalyan Kumar

Abstract Backfilling the mine voids with the suitable material can provide additional ground support by restricting lateral deformation of surrounding coal pillars and roof. This type of support mechanism helps to prevent caving, minimizes surface subsidence, and enhances pillar recovery. River sand is widely used as a stowing/backfilling material, as it is easily available and suitable for its geotechnical properties. River sand is in great demand for many construction projects and is in short supply in many areas. So it is highly desirable to find an alternative stowing material. Some alternative materials used for backfilling are waste rock, mill tailings, quarried rock, sand and gravel. Over the last decade, attention has been made on the use of coal combustion products as an alternate filling material for underground works. Usually, pond ash as a backfill material remains loose and serves only as a temporary working platform rather than offering any lateral strength. Adding low percentages of admixtures to backfill mass permits the development of cohesive strength and self-supporting ability during adjacent pillar mining. This self-supporting nature of the backfill permits higher recovery of pillar ore, which in turn improves the utilization of the mining reserve and the economics of the mining operation. In the present study, laboratory mine model was prepared by using pond ash as a backfilling material with different percentages of lime (2, 6, 8, and 10%) and gypsum (0.5, 1, 1.5, and 2%) as additives. To examine the effect of gypsum on the strength characteristics of stowed pond ash, core samples were collected from the model which were cured for different time periods such as 7, 14, and 28 days. To determine the engineering properties of pond ash composites unconfined compression tests as well as triaxial compression tests were performed.

R. Shanmuga Priya (✉) · S. Sreelakshmi · G. Kalyan Kumar
Civil Engineering Department, National Institute of Technology Warangal, Warangal, India
e-mail: priya.gte@gmail.com

S. Sreelakshmi
e-mail: sreelakshmi2192@yahoo.in

G. Kalyan Kumar
e-mail: kalyanu.g@gmail.com

V. K. Stalin and M. Muttharam (eds.), *Geotechnical Characterisation and Geoenvironmental Engineering*, Lecture Notes in Civil Engineering 16,
https://doi.org/10.1007/978-981-13-0899-4_35

The increase in strength of pond ash composites was observed for lime content up to 6% and gypsum of 1.5%.

Keywords Stowed model · Pond ash composites · Compressive strength Pillar recovery

1 Introduction

Coal-based thermal power plants are the major source of power generation in India. The geological coal reserve of India is about 301.56 billion tonnes (Geological Survey of India 2014). About 70% of this is used in power generation. India's energy consumption has been increasing at faster rates due to population growth and industrial development. India being the third biggest hard coal producer after China and the USA faces demand for energy, and hence, coal should be produced in large scale.

In India 81% of the coal is obtained from opencast mining. Every million tonne of coal extracted by surface mining methods damages a surface area of about 4 ha in India. Environmentally, underground mining is much better than opencast because it does not immediately affect the natural habitat (Kundu and Ghose 1998). Also, opencast production may reduce within next 15–20 years. To meet the energy requirement, the Indian mining industry has to adopt mass production technology in underground mining.

Underground mining accounts for about 15–20% of total coal production in India. Mine voids created should be filled or it may cause land subsidence due to collapse of the overlying strata. Most of the subsidence problems are have occurred suddenly and often remain as serious threats to the subsequent development of townships. Thus, to provide ground supports to minimize the land subsidence problems, stowing is the valuable part of mining (Kesimal et al. 2002; Yao et al. 2012). The filling of mine cavities would also release millions of tons of coal blocked in support pillars and thus increases the pillar recovery.

River sand is widely used as a stowing/backfilling material, as it is easily available and suitability of geotechnical properties. The availability of river bed sand as a void filling material is decreasing. So it is important to find an alternative to this which is highly desirable. Some alternative materials used for backfilling are waste rock, mill tailings, quarried rock, sand and gravel. In recent days coal combustion by-products (CCBs) like fly ash or Pond ash are used for filling of mine voids. Bottom ash stowing practices are successfully practiced in the underground coal mines of SCCL. Kumar et al. (2003) have reported the use of pond ash (a mixture of fly ash and bottom ash) of grain size between 75 and +20 mesh with equal percentage of water (by weight) for underground stowing of a coal mine. Palariski (1993) reported the use of fly ash, mill tailings, rock and binding agents to make consolidated backfill material to improve extraction percentage in coal mines. The pond ash water mixture with additives exhibited 100% settlement of solids within 30 min of placement. The settlement time further reduced with an increase in additive concentration. The additives also improved the percolation rate of the mixture with little adverse effect on mine water (Mishra and Das 2014). Fawconnier and Korsten (1982) reported that the use of pulverized fly ash filling had effectively stabilized the coal pillars reducing the risk of pillar failure in areas of low safety.

CCB can be utilized in bulk magnitude for stowing of underground mines in place of sand and filling up abandoned opencast mine voids. But it is often observed that sand or CCBs used as backfill material remains loose and merely serves as a temporary working platform rather than offering any lateral stress on the opening walls to improve the stability situation (Srivastava 1995). Maser et al. (1975) reported successful studies on fly ash–cement mixture for subsidence control.

This paper mainly aims to investigate the effect of addition of various percentages of gypsum to the strength characteristics of laboratory stowed lime-treated pond ash. UU triaxial tests were carried out on the samples to determine the optimum contents of lime and gypsum. Studies were also made to evaluate the effects of various lime contents and different gypsum contents at an optimum lime content of 6% (Sivapullaiyah and Moghal 2011).

2 Material Used

2.1 Pond Ash

Pond ash was collected from the ash pond of NTPC Ramagundam, Warangal district, Telangana State, India. The chemical and geotechnical properties of pond ash samples used in this study are given in Tables 1 and 2, respectively.

2.2 Lime and Gypsum

The addition of lime enhances the pozzolanic reactivity of pond ash. Analytical quality gypsum was chosen to avoid the interference of impurities. Laboratory reagent quick lime [Cao] was supplied by Tarnath chemicals. Gypsum [$CaSO_4 \cdot 2(H_2O)$] supplied by Rajendira Scientifics was used for the experimental purpose.

Table 1 Physical properties of pond ash

Parameters	Pond ash
Specific gravity	2.14
Percentage of fines	56.37
Bulk density (g/cc)	0.89
Coefficient of permeability (10^{-3} mm/s)	7.03
Plasticity	Non plastic
Particle density	1.66
Porosity (%)	53.6

Table 2 Chemical composition of pond ash

Composition	Compound by weight %
Silicon dioxide (SiO_2)	75.06
Aluminum oxide (Al_2O_3)	16.25
Titanium dioxide (TiO_2)	1.30
Iron oxide (Fe_2O_3)	2.12
MgO	0.30
CaO	0.5
Potassium oxide (K_2O)	0.40
Na_2O	0.31

3 Experimental Program and Test Procedure

To study the strength of the stowed pond ash, the experiments were carried out in the following stages: Stage 1: pond ash considered as the stowing material was characterized. Stage 2: A series of unconsolidated undrained triaxial tests were done to determine the optimum gypsum content.

3.1 Mix Preparation

To prepare combinations of pond ash with additives, pond ash is mixed with varying percentage of additives on a dry weight basis. The pond ash was stabilized with quick lime of 2, 4, 6, 8, and 10% of dry weight of pond ash. To accelerate the bonding between particles, laboratory grade gypsum with 0.5, 1, 1.5, and 2% was also used to reduce the stabilized matrix by producing more ettringite.

3.2 Sample Preparation

In the present study, tests on sedimented stowed pond ash with additives were carried out in two stages. In the first stage, the mine model shown in Fig. 1 was stowed with pond ash composite slurry. The slurry consists of 65% pond ash composite and 35% water by its own weight. After drainage of water from the stowed mass, undisturbed core samples of the sediment composites were collected by Shelby tube. Core samples were cured in the desiccators for different time periods, i.e., 7, 14, and 28 days. Figure 2 shows the core samples of stowed pond ash composite collected from the mine model.

Fig. 1 Stowed pond ash composite in the mine model

Fig. 2 Core samples from mine model

4 Results and Discussion

4.1 Grain Size Analysis

The grain size distribution of pond ash was carried out according to IS 2970 (1974), and values are provided in Table 3.

It was observed that more than 50% of pond ash particles are between 0 and 75 microns that indicates a favorable trait for pozzolanic reaction, particles ranging from fine sand to silt sizes. The coefficient of uniformity affects the workability of the pond ash grains. The particle size distribution curve is shown in Fig. 3.

UU triaxial shear tests were carried out at confining pressure of 100 kPa, and a strain rate of 1.25 mm/min was applied on the sample. When 0.5% of gypsum is added to the pond ash lime composite, the deviatoric stress was observed as 1174.3 kPa.

Table 3 Grain size distribution of pond ash

Parameters	Pond ash
Gravel (%)	0
Sand (%)	43
Silt (%)	55
Clay (%)	2
D10 (mm)	0.0055
D30 (mm)	0.032
D60 (mm)	0.083
Cu	15.1
Cc	2.24
IS classification	ML

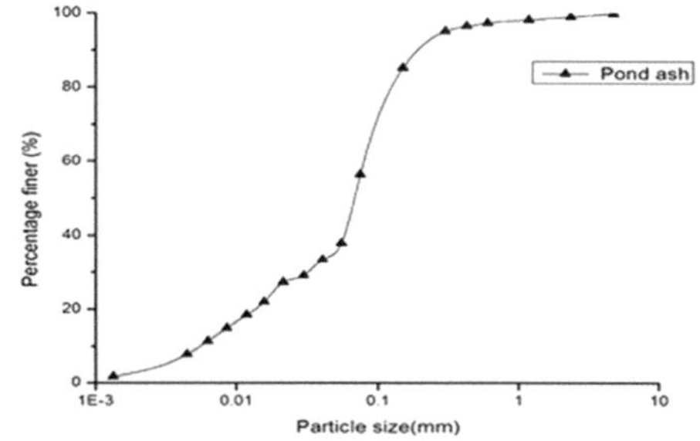

Fig. 3 Particle size distribution of the pond ash

An increase in 1% of lime has shown 35% increase in deviatoric stress. But further increase of 0.5% of gypsum has shown reduction in stress of 8%. It is evident from the results that the deviatoric stress increases up to 1.5% of gypsum content and decreases with further increase in gypsum content. The variation of deviatoric stress and axial strain for the pond ash composites for 7-day and 14-day curing are shown in Figs. 4 and 5. The peak strain for the optimum mix containing 6% lime and 1.5% gypsum for 14-day curing period is found to be 1.68 times the peak strain for 7-day curing period. The failure strain for the composites lies

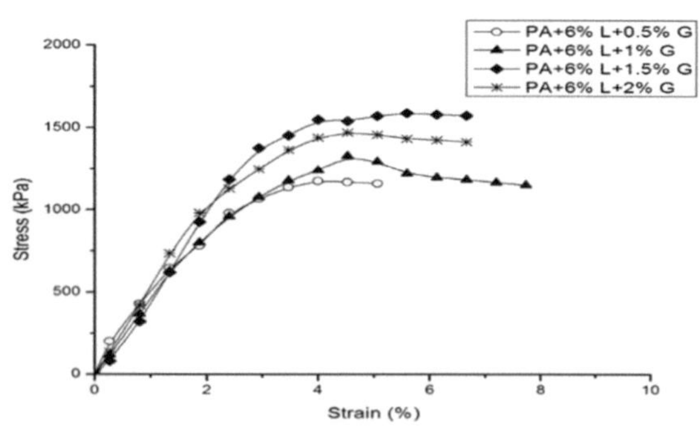

Fig. 4 Stress–strain plot for lime-modified PA with gypsum at 7-day curing period

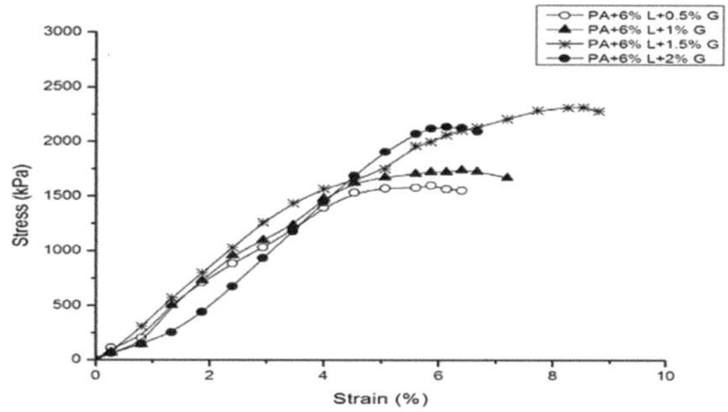

Fig. 5 Stress–strain plot for lime-modified PA with gypsum at 14-day curing period

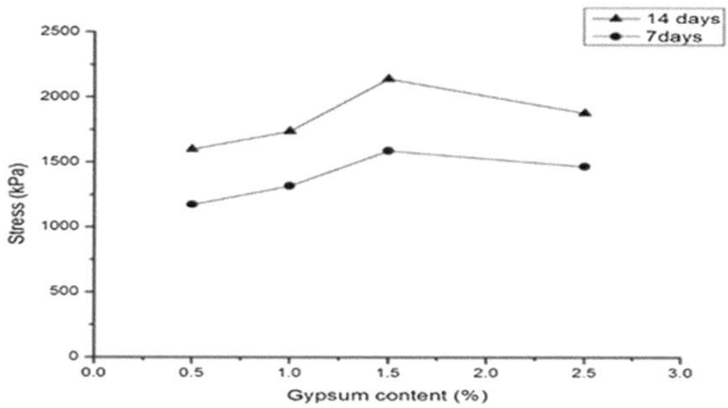

Fig. 6 Deviatoric stress versus gypsum content for lime-modified PA with gypsum

between 4 to 5% and 5.5 to 8.5% for 7 days and 14 days of curing, respectively. Figure 6 shows the variation of deviatoric stress for different % of gypsum in lime-modified PA.

The strength of the composite increases up to the threshold value of 1.5% G. In the mix containing 6% lime and 0.5% gypsum the peak stress for 14-day curing period is found to be 1.35 times the peak stress for 7-day curing period. The peak stress for the optimum mix containing 6% lime and 1.5% gypsum for 14-day curing period is found to be 1.46 times the peak stress for 7-day curing period. Thus, it is evident that the addition of small percentages of gypsum (0.5–2.0%) to pond ash lime mix increases the strength of the stowed specimens.

5 Conclusion

The aim of the present investigation is to study the engineering properties of the developed pond ash composite material which can be used as an alternative to sand in stowing underground mine voids. The improvement of strength of pond ash due to the addition of gypsum is similar to FAL-G technology being used in brick making. The effect of gypsum on the strength properties of pond ash with varying percentage of lime has been investigated, and the following conclusions are drawn:

(1) In the unconsolidated undrained triaxial shear results show that deviatoric stress of Stowed pond ash composites attained peak value at axial strains in the range of 4.5–7.5% for all the specimens.
(2) The major conclusion for the study is that low lime pond ash from NTPC Ramagundam has greater potential to be used as a strong stowing material with the addition of lime and gypsum.

References

Fawconnier, C. J., & Korsten, R. W. O. (1982). Ash fill in pillar design—Increased underground extraction of coal. *The SAIMM Monograph Series, 4,* 277–361.

Kesimal, A., Yilmaz, E., Ercikdi, B., Alp, L., Yumlu, M., & Ozdemir, B. (2002). Laboratory testing of cemented paste backfill. *Journal Chamber Mining Engineering, Turkey, 41*(4), 25–32.

Kumar, V., Ahuja B. P., Dattatryulu, J. V., Bhaskar Rao, B., Ghosh, C. N., & Sharma, A. K. (2003). Hydraulic stowing of pond ash in underground mines of Manuguru. In *India Proceedings 3rd International Conference on Fly ash utilization and Disposal* (pp. VI–1–7). New Delhi, India, February 19–21, 2003.

Kundu, N. K., & Ghose, M. K. (1998). Status of soil quality in the subsided areas caused by underground coal mining. *Indian Journal Environmental Protect, 25,* 110–113.

Maser, K. R., Wallhagen R. E., & Dieckman, J. (1975). Development of fly ash cement mine sealing system. *USBM,* Open File Report 26–76, NTIS-PB-250611.

Mishra, D. P., & Das, S. K. (2014). One-dimensional consolidation of sedimented stowed pond ash and pond ash-lime mixture deposits—A comparative study. *Particulate Science and Technology, 33,* 172–177.

Palariski, J. (1993). The use of fly ash tailings, rock and binding agents as consolidated backfill for coal mines. In *Proceedings of Mine fill 1993* (pp. 403–408). H.W. Gelen, (Ed.), SAIMM.

Sivapullaiyah, P. V., & Moghal, Arif Ali Baig. (2011). Role of gypsum in the strength development of fly ashes with lime. *Journal of Materials in Civil Engineering, 23*(2), 197–206.

Srivastava, S. D. (1995). An investigation into the design of post pillars in cut-and-fill method of mining. *Unpublished Ph.D. Dissertation*, India: I.I.T. Kharagpur.

Yao, Y., Cui, Z., & Wu, R. (2012). Development and challenges on mining technology. *Journal of Materials Science Research, 1*(4), 73.

Stress–Dilatancy Relation of Sea Deposits of Mumbai Coast

T. Aishwarya, N. Siddharth Prabhu and A. Juneja

Abstract It is not unusual for the offshore deposits to carry loads in excess to their ultimate capacity, when these soils bear the loads of heavy oil platforms. These soils contain high calcium carbonate content, which make them prone to particle crushing at these high loads. This paper investigates the effect of particle breakage on the load response of offshore marine deposits. Two sea deposits off Mumbai coast were chosen for the tests. Both the soils had similar mineralogical composition. In this study, dilatancy was studied as a function of particle crushing at crushing stresses up to 4.5 MPa. A series of direct shear tests were performed on samples which were crushed to various loads. The experimental results showed that with increase in extent of crushing, dilatancy decreased up to a point and then increased. It also indicates that crushing takes place to a considerable extent even after settlements have reached a near-constant value. The decrease in dilatancy with increase in particle crushing was more pronounced in angular particles than sub-rounded and rounded particles. This was also confirmed from the SEM images. Stress–dilatancy equations of both the soils were developed using drained shear tests. Particle breakage seems to significantly affect the measured coefficient of friction and hence the dilatancy component.

Keywords Dilatancy · Particle crushing · Shear strength

T. Aishwarya (✉) · A. Juneja
Department of Civil Engineering, Indian Institute of Technology Bombay, Powai 400076, Mumbai, India
e-mail: taishwarya3@gmail.com

A. Juneja
e-mail: ajuneja@iitb.ac.in

N. Siddharth Prabhu
Sri Jayachamarajendra College of Engineering, Mysore 570006, India
e-mail: siddharth0896@gmail.com

V. K. Stalin and M. Muttharam (eds.), *Geotechnical Characterisation and Geoenvironmental Engineering*, Lecture Notes in Civil Engineering 16,
https://doi.org/10.1007/978-981-13-0899-4_36

1 Introduction

Any country's economic growth is closely related to the energy demand which is most often satisfied by the oil and natural gas industry. Offshore platforms have moved to deeper depths, where the mechanical characteristics of the soft soil sediments can altogether be different. Many researchers have worked on the geotechnical investigation and site characterization of offshore construction. Loads of the order 10,000–30,000 tonnes can be expected to be imposed on foundations of these structures. Such heavy loads cause the soil to crush under the foundation because of very high compressive stresses. All this would affect the soil dilatancy. It is an important characteristic of the soil which governs its strength at high confining pressures especially in these cases. Particle crushing is one of the factors which can alter dilatancy of the soil. Crushing influences the peak strength of the soil which in turn is a function of stress–dilatancy and the friction angle at critical state. Hence, study on effect of crushing on dilatancy of soil would be empirical to understand its strength behaviour.

1.1 Background

Reynold (1885) coined the term dilatancy after his observation on volumetric response of dense materials during shearing. For given stress levels, loose sands contract upon shearing and reach a state of constant shear stress and constant volume called as critical state (Schofield and Worth 1968). However, dense sands show a volumetric increment, i.e. dilation as the shearing progresses, and reach critical state at very large strains. Taylor (1948) was the first to study stress–dilatancy relation. Taylor considered strain energy and developed the following equation (Eq. 1) based on conservation of energy during shearing.

$$\frac{\tau}{\sigma} + \frac{d\varepsilon_v}{d\gamma} = \mu \quad (1)$$

where τ = shear stress mobilized on the shear plane, σ = effective normal stress across shear plane, $d\varepsilon_v$ = volumetric strain increment, $d\gamma$ = shear strain increment, and μ = coefficient of friction.

Houlsby (1991) carried out theoretical and numerical analysis on effect of dilatancy on various practical problems such as slope stability, bearing capacity of surface footings and piles. A fivefold increase in bearing capacity of pile was observed when dilation angle was varied from 0° to 20° indicating the significance of dilatancy at high confinement pressure.

Many researchers have worked on the methods to quantify the particle crushing. Marsal (1973) used grain-size distribution before and after crushing and defined a term called breakage factor (B_g) that represented the sum total of increase in the

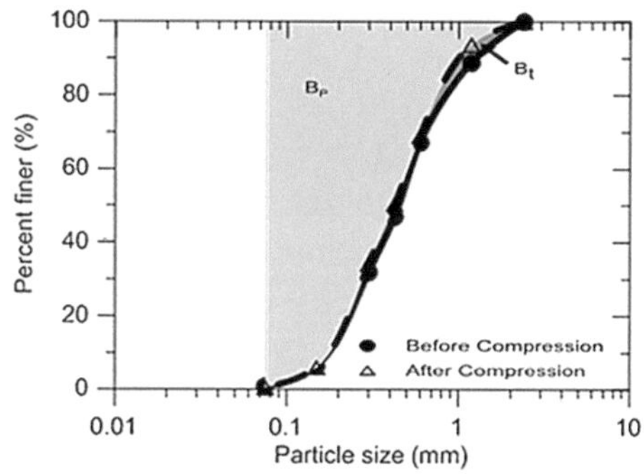

Fig. 1 Grain-size distribution curves of Sand A, before and after crushing

mass retained on each sieve used to perform gradation. Miura and O-hara (1979) defined breakage as change in specific surface area as he assumed soil particles to be spherical in shape and calculated specific surface area using mean diameter of each sieve.

1.2 *Objective*

Objective of the study is to find the effect of particle crushing on dilatancy characteristics of offshore deposits. This was accomplished by inducing particle crushing in the soil by the way of applying high compressive stress under a compression frame and then determining its strength and dilatancy property by direct shear testing. Particle crushing was quantified based on a term called relative breakage (B_r) introduced by Hardin (1985). B_r is defined as

$$B_r = \frac{B_t}{B_p} \tag{2}$$

where B_t is the total breakage defined as area between initial and current grain-size distribution and B_p is the breakage potential defined as the area between initial distribution and cut-off distribution of silt-size particles (75 micron). Dark grey and light grey portions in Fig. 1 represent the terms B_t and B_p, respectively.

2 Experimental Investigation

Two offshore samples obtained from the Arabian Sea off Mumbai coast were used. These samples are labelled as Sand A and B, respectively, in this study. Physical properties of the test materials are summarized in Table 1.

Table 1 Physical properties of soils

Soil	Gs	emax	emin	D10	D50	Cu	Cc
Sand A	2.81	0.802	0.453	0.185	0.495	3.08	1.12
Sand B	2.78	0.834	0.437	0.078	0.19	2.94	0.87

Sand A consisted of angular particles, and Sand B comprised of rounded to sub-rounded particles as observed in Fig. 2. Test samples were prepared by placing the sand in a 50-mm diameter, 24-mm-high cylindrical mould. About 75 g of the soil was taken for the test. The samples were crushed by applying compressive load of 1–4.5 MPa. In each case, the gradation of the soil was determined prior and after the application of the crushing load using a set of sieves. Pressure was applied after the settlement reached a constant value and until 1–3 days. However in some samples, the load was applied up to 5 days. Sieve analysis was conducted on the sample after the test also to determine the change in particle-size distribution due to crushing. Hardin's total and relative breakage (B_r) was then obtained from the particle gradation curves.

Crushed sample was then subjected to direct shear testing for determination of angle of dilatancy. The cylindrical mould is used for crushing had smaller dimensions; therefore, the mass of sample available for direct shear test was lesser than usual test. So, sample thickness in the apparatus had to be reduced. This was achieved by using acrylic blocks of dimensions 60 mm × 60 mm × 13 mm as spacer, both at the top and bottom which reduced the thickness of sample to about 10 mm. The modified apparatus was calibrated to account for the difference in the measured angle of friction. For this purpose, test was carried out with and without the blocks and the difference in the stress–strain response indicated an overestimation of measured friction angle. It is noteworthy that the samples were failed under a normal load of 50–150 kN/m^2 which is the usual confining stress experienced by the soils below and surrounding the heavy offshore foundations. This difference was deducted from stress–strain curve of modified apparatus to obtain true stress–strain behaviour of the sample. Bolton's relation Eq. (3) was used to determine angle of dilatancy.

$$\Phi_{ps} - 0.8\psi = \Phi_{cv} \tag{3}$$

Fig. 2 SEM images of offshore deposits

(a) Sand A **(b)** Sand B

where ϕ_{ps} is peak angle of friction, ψ is angle of dilation, and ϕ_{cv} is critical state angle of friction.

3 Results and Discussions

Two soil samples were subjected to compressive stresses up to 4.5 MPa and tested using direct shear apparatus for determination of its strength and dilatancy behaviour. On comparing the grain-size distribution curves before and after compression, Hardin's total and relative breakage term was obtained. The total breakage was found to be dependent on both the magnitude and duration of stress applied. Figure 1 shows the initial and current grain-size distribution curves of Sand A when subjected to compressive stress of 4.5 MPa for 5 days. It is evident that the coarser particles had undergone crushing as there is a drift in the gradation curve towards left at larger particle size which gradually coincides with the virgin curve on moving downwards. On the other hand in Sand B which comprised of mainly rounded to sub-rounded particles, the reverse was true (Fig. 3).

The variation of angle of dilatancy with total breakage was studied. The evolution of Br with crushing load for both the sample is shown in Table 2. It is obvious that relative breakage increases with increase in crushing load. However, it is noticeable that considerable crushing takes place even after the settlement has reached a near-constant value. Sand A exhibited comparatively higher crushing than Sand B which can be attributed to angularity in its particle shape.

The stress–strain behaviour of soil samples post-crushing was obtained from direct shear testing. The results were calibrated to account for the modification done to the direct shear apparatus. Figure 4 shows the variation of shear stress normalized by vertical stress and the ratio of vertical strain increment to horizontal strain increment for both the sands. $d\varepsilon_v/d\gamma$ component in Taylor's stress–dilatancy relation is equal to $-dy/dx$.

Figure 5 shows the variation of angle of dilatancy with total breakage for both the samples. It can be inferred that for small increment in total breakage B_t, a good

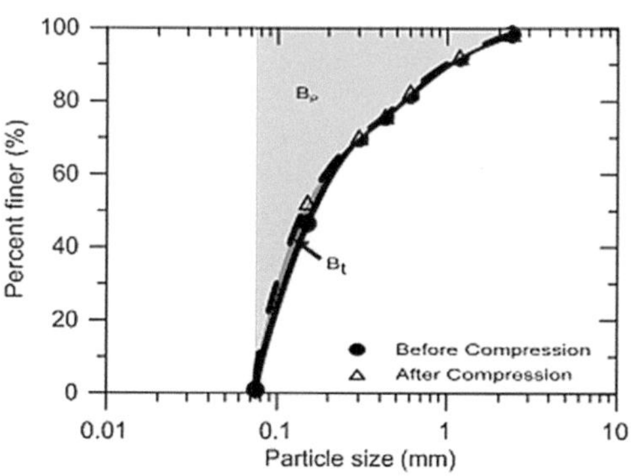

Fig. 3 Grain-size distribution of Sand B, before and after crushing

Table 2 Variation of relative breakage with crushing

Crushing stress (MPa)	Duration of crushing (days)	Total breakage, B_t	Relative breakage, B_r
Sand A			
1	5	0.2927	0.001954
2	1	0.032	0.000213
2	3	0.329	0.002196
4.5	1	0.322	0.002149
4.5	5	2.67	0.017825
Sand B			
1	5	0.001	6.67E-06
2	1	0.0095	6.34E-05
2	3	0.256	0.001709
4.5	1	0.3653	0.002438
4.5	5	2.224	0.014847

Fig. 4 Stress–dilatancy relation for offshore soils

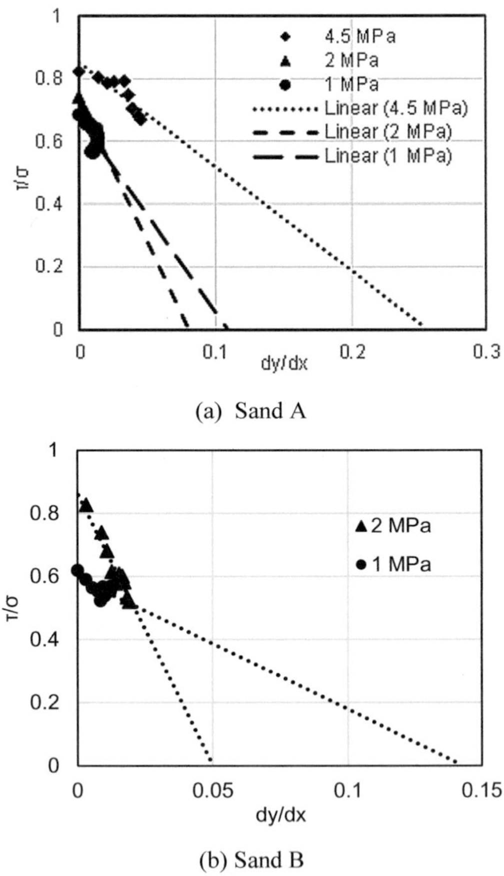

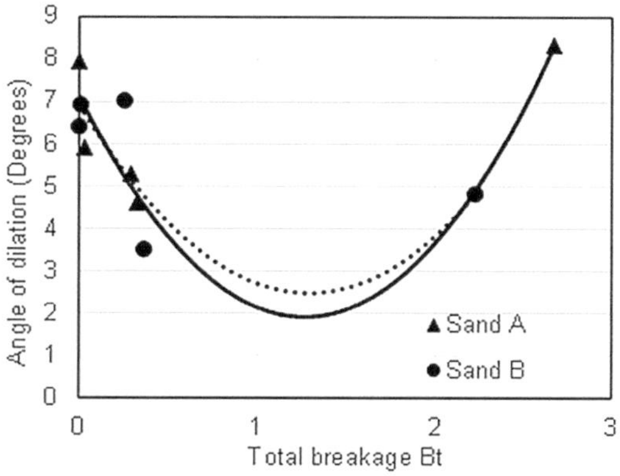

Fig. 5 Variation of dilation angle with total breakage

deal of decrease in angle of dilation was found. This could be due to decrease in the angularity of the particles. Also, the fall in dilation angle was found to be more for angular Sand 'A' than sub-rounded Sand 'B'. This confirms that the initial drop was due to decrease in the angularity of the particles. However, in both cases, it could be seen that the dilation angle goes on decreasing, reaches a minimum value, and there after increases. Although the decrease was expected, yet it is difficult to explain the increase in the dilation after the transient drop, especially since both the samples exhibit similar behaviour. One possible reason for this behaviour is that the particle crushing created additional contact points which offset the otherwise decreasing trend in the dilatancy component. These additional contact points helped increase the shear strength of both the soils which in the figure is reflected by the increase in the dilatancy. This effect perhaps also shadowed the otherwise decreasing dilatancy component when the soil roughness reduced by particle crushing. This needs further investigation.

4 Conclusions

Particle breakage was induced on two offshore deposits of Mumbai at high compressive loads. Since this is preliminary work, no definite relations can be drawn at this stage. Change in the particle size as a consequence of soil crushing was very less due to low range of crushing stress adopted in the testing. Although with the increase in crushing, the dilatancy decreased initially owing to decrease in the angularity of the particle. The decrease was followed by an increase in dilatancy. Further study on soils crushed at higher stresses would help to better understand the relation between particle crushing and dilatancy. The breakage continued even after the settlement reached a near-constant value.

References

Hardin, B. O. (1985). Crushing of soil particles. *Journal of Geotechnical Engineering, 111*(10), 1177–1192.

Houlsby, G. T. (1991). How the dilatancy of soil affect their behavior. In *Proceedings of 10th European Conference in soil mechanics and foundation Engineering* (pp. 1189–1202).

Marsal, R. J. (1973). *'Mechanical Properties of Rockfill', Embankment-dam Engineering*. New York: Wiley.

Miura, N., & O-hara, S. (1979). Particle crushing of decomposed granite soil and shear stresses. *Soils and Foundations, 19*(4), 1–14.

Reynolds, O. (1885). On the dilatancy of media composed of rigid particles in contact, with experimental illustrations. *Philosophical Magazine, 20*(5), 469–481.

Schofield, A. N., & Wroth, C. P. (1968). *Critical State Soil Mechanics*. New York: McGraw-Hill Book Company.

Taylor, D. W. (1948). *Fundamentals of Soil Mechanics*. New York: Wiley.

Evaluation of Soil Parameters by Using Light Cone Penetrometer

S. P. Raghu Prasanth, A. Arun Perumal and K. Ilamparuthi

Abstract Conventional investigation methods such as standard penetration test and cone penetration test fail to identify weak deposits like soft clay, and designing the foundation without identifying the presence of such layer and its strength is risky. Under the conditions stated, a simple field testing equipment is required to operate in such deposits and is also to be well suited for projects like road works, simple buildings. One such equipment is a light cone penetrometer (LCPT). Tests are done using LCPT at a few sites, and resistances are recorded over a depth of 6 m at each location. Resistances are also measured by conducting standard penetration test at location wherever LCPT was done. Correlation between LCPT and SPT resistances is developed. Simultaneously soil samples are collected from each location of LCPT and tested in laboratory for index properties and parameters such as shear strength, compressibility and field CBR. These soil parameters are correlated with the LCPT resistance individually.

Keywords Penetrometer test · Light cone penetrometer · Soil parameters
Correlation

1 Introduction

Field tests play a major role in subsurface investigation in assessing strength and compressibility properties of various deposits at a location of a project. There are different methods of investigation, and among them penetration tests such as standard penetration test (SPT), static cone penetration test (SCPT), cone penetration test

S. P. Raghu Prasanth · A. Arun Perumal · K. Ilamparuthi (✉)
Anna University, Guindy, Chennai 600025, India
e-mail: kanniilam@gmail.com

S. P. Raghu Prasanth
e-mail: prasiprasanth8@gmail.com

A. Arun Perumal
e-mail: arunperumal2000@yahoo.co.in

V. K. Stalin and M. Muttharam (eds.), *Geotechnical Characterisation and Geoenvironmental Engineering*, Lecture Notes in Civil Engineering 16,
https://doi.org/10.1007/978-981-13-0899-4_37

(CPT), dynamic cone penetration test (DCPT) are widely employed in geotechnical investigation in general to assess relative strength of deposits. There are penetrometers with special arrangement for measuring in situ stress, pore water pressure, dynamic properties, etc. Overall it can be said that a subsurface investigation is incomplete without a penetration test. It is because not only the relative strength of deposits is known at in situ but also many empirical correlations between penetration resistance and various properties (shear strength parameters, compressibility parameters, over consolidation pressure, swell properties, subgrade modulus, etc.) are developed over a period based on numerous field data (Meyerhof 1956; Ilamparuthi 1981; Zaki et al. 1990; Aroosa 2000; Moayed and Naeini 2006; Ravisharma and Ilamparuthi 2009; Deepika and Chakravarthy 2012; Nwobasi and Paul 2013). Despite well-established correlation and a lot of improvisation over operation of penetrometers, researchers are working on developing simple device, which can be employed effectively in poor deposits wherein undisturbed sampling is next to impossible, and thus, penetrometers like T-bar penetrometers, ball penetrometer and light cone penetrometer are developed and are put in use. In this study, a light cone penetrometer is developed and its operation is standardised. This equipment is developed in such a way that both static (push the cone under pressure as in SCPT) and dynamic driving as in the case of DCPT. Tests are conducted at four sites by employing LCPT and SPT, and the resistances are correlated so as to use the correlations which are based on SPT '*N*' values. Soil samples both undisturbed and disturbed samples are collected and tested. The results thus obtained are correlated with LCPT resistances. The details of equipment, operation procedure and correlations developed are presented in this paper.

2 Methodology

The light cone penetrometer is adopted in the field along with the standard penetration test equipment. Boreholes were drilled in the field by rotary drilling method, and standard penetration test were conducted at intermittent depths in the borehole. Tests using light cone penetrometer were conducted around the borehole at a radial distance of 1 m. Field tests were also done, to determine the CBR value of the soil at its natural state, at a radial distance of 1 m. The resistances recorded in the light cone penetration test were then correlated individually with the resistance measured in the standard penetration test and the field CBR value obtained from the tests. Simultaneously soil samples were collected from the site to determine their shear strength parameters and compressibility characters. These parameters were then correlated individually with the penetration resistance obtained from the light cone penetration test.

Table 1 Dimensions of various components of LCPT

Component	Dimension
Cone	Dia—25 mm, apex angle—30°
Driving rod	Dia—20 mm, length—500 mm
Guide stand	1000 × 600 × 500
Hammer	Weight—14.5 kg
Hammer guide rod	Dia—50 mm, length—700 mm
Anvil	Dia—100 mm, length—150 mm

3 Light Cone Penetrometer

The light cone penetrometer used in this study was designed and fabricated by the authors. The components of the light cone penetrometer are cone, detachable driving rods, guide stand, hammer, hammer guide rod and anvil. The dimensions of the components of LCPT are as presented in Table 1.

4 Procedure

There is no standard procedure for LCPT test, and a procedure is adopted which is explained as follows.

At a selected test location, the surface is cleaned and leveled to fix the guide frame assembly. The frame is fixed firmly on to the ground with anchor rods driving through the base plate of the frame. The driving rod is made to pass through the hole of the cross bar which is connecting the frames. The cone is fixed at the bottom of the driving rod, and the top of the driving rod is connected to the anvil with the help of adapter. The hammer rod is fixed to the anvil through which the hammer is allowed to fall. The verticality of the driving rod assembly is checked before driving the rod. Figure 1 shows the operation of LCPT at a site.

The cone and drive rod assembly is driven into the soil by dropping the hammer on the anvil at the rate of 15 blows/min. The blows required for every 5 cm penetration of drive rod, and cone assembly is recorded. After every 50 mm penetration of drive rod and cone into the soil, the anvil and the hammer rod are removed and the length of the drive rod is extended by attaching another graduated rod. The driving rod assembly is driven further depths by attaching the anvil and by dropping the hammer on the anvil. The procedure is repeated till the number of blows required for 50 mm penetration exceeds 20. This condition is exercised to avoid damage to the cone and drive rod assembly. The test can also be terminated if the depth of penetration reaches 6 m. After the test is completed, the driving rod assembly is removed by back hammering technique. The hammer rod and the anvil are detached and the hammer is inserted through the driving rod above which the anvil is fixed for back hammering.

Fig. 1 Light cone penetrometer

The resistance thus recorded is plotted with depth to know the relative strength of the deposit. The cumulative number of blows for every 30 cm penetration is calculated and plotted in the graph.

5 Test Locations and Details of Tests

The sites selected for conducting field tests are Muttukadu, Poonamalle, Mogappair West and Perumbakkam. SPT tests were conducted at these sites, and the soil strata were identified. The soil strata present in Muttukadu, Poonamalle and Mogappair were identified as poorly graded sand of loose- to medium-dense state. The soil present in Perumbakkam was high plastic clay of soft to medium stiff consistency. The field tests conducted in the sites were SPT, LCPT and CBR tests. At each site one SPT and three LCPTs at a radial distance of 1 m from the SPT location, three field CBR tests and field density tests were conducted. Simultaneously soil samples were collected from the sites and were tested for index, strength and compressibility properties.

6 Results and Discussion

As stated LCPT was conducted at three locations at each site, the average value was determined for every 50 mm penetration, and the values determined over the layers of penetration of 6 m were compared with SPT values recorded. The SPT values recorded are taken as N_{60} and the values are not corrected for overburden since LCPT and SPT are conducted under identical conditions of the deposits. These two values are correlated, and relation between them for sand and clay deposits is determined independently. Similarly the resistances of LCPT is correlated with angle of shearing resistance in case of test in sandy soil and correlated with unconfined compression strength in case of test in clayey soil. The correlation established is presented in the subsequent sections.

6.1 Correlation Between SPT and LCPT

From the field tests conducted the resistances recorded in the standard penetration tests and the light cone penetration tests in sand strata are plotted in Fig. 2. The correlation between the SPT and the LCPT resistance for sand of loose to medium dense condition is found as,

$$N_{\mathrm{SPT}} = 0.446 * N_{\mathrm{LCPT}} \tag{1}$$

Similarly, resistances of SPTs and LCPTs recorded in clay deposits are plotted in Fig. 3, and the correlation obtained between them is given in the following equation.

$$N_{\mathrm{SPT}} = 0.55 * N_{\mathrm{LCPT}} \tag{2}$$

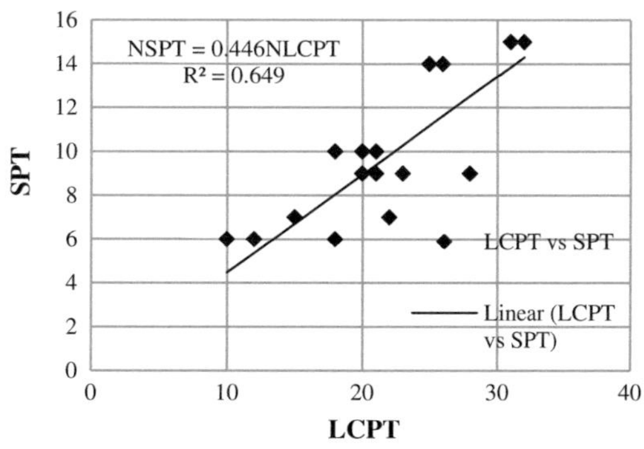

Fig. 2 LCPT resistance versus SPT-N value for sand

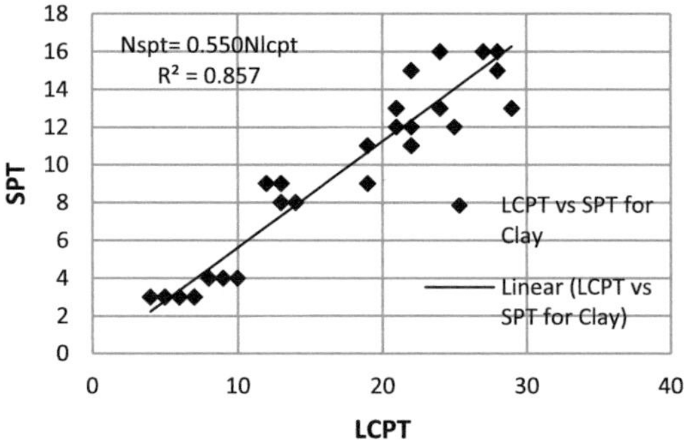

Fig. 3 LCPT resistance versus SPT-N value for clay

6.2 *Correlation Between LCPT and ϕ*

Direct shear tests were conducted on the soil samples of sand collected from Muttukadu, Poonamalle and Mogappair, and angle of shearing resistances of sand of respective locations are presented in Table 2 along with average N_{LCPT} values and relative densities. The angle of shearing resistance obtained are 29° for loose dense condition of sand, 33°–34° for medium dense condition and 36° for dense condition. The respective N_{LCPT} values are 10 blows, 18 blows and 22 blows. Both N_{LCPT} and ϕ values obtained from direct shear test are related, and the relation between them is as given in Eq. 3.

$$\phi\ (deg) = \sqrt{20 * N_{\mathrm{LCPT}}} + 15^{\circ} \tag{3}$$

where ϕ = angle of shearing resistance of sand.

The ϕ-values thus obtained from Eq. 3 compares well with the value obtained from direct shear test for loose and medium dense condition of sand.

6.3 *Correlation Between LCPT and C_u*

The natural moisture content of the samples is varied between 33 and 42%. The undrained cohesion value of the clay of Perumbakkam is found to vary between 29 and 39 kN/m^2. The higher value is for the clay sample with lower moisture content wherein N_{LCPT} equal to 18 blows.

Table 2 Variation of ϕ value with N_{LCPT}

N_{LCPT}	Relative density	ϕ (°)	Equation (3)
10	Loose	29	29.10
17	Medium	33	330
19	Medium	34	34.50
22	Dense	36	35.90

Table 3 Comparison of C_u values

N_{LCPT}	Undrained cohesion (C_u) (kN/m^2)		
	Experiment	Equation (4)	Terzaghi (based on N value)
12	38.8	39.6	41.2
12	36.2	39.6	41.2
10	28.7	33	34.3
10	31.1	33	34.3

The undrained cohesion values thus obtained are related with N_{LCPT} values, and the following empirical relation is obtained.

$$C_u = 0.033 * N_{LCPT} * P_a \tag{4}$$

where N_{LCPT} = number of blows for 30 cm penetration, C_u = undrained cohesion and P_a = atmospheric pressure ($\cong$100 kN/m^2). The value of C_u obtained from Eq. 4 is compared with experimental values as well as Terzaghi's recommendation based on SPT value in Table 3.

The C_u values obtained by the empirical relation (Eq. 4) compares reasonably with the experimental values as well as with the values obtained from the relation suggested by Terzaghi.

6.4 Correlation Between LCPT and Compressibility Character

The compressibility characteristics of clay samples collected from the field were determined by conducting consolidation tests in the laboratory. It is well known that the compressibility of clay soil depends on content of clay fractions and its plasticity, and therefore, a correlation between coefficient of volume change (m_v), plasticity index and N_{LCPT} as in Eq. (5) is formed, which is an identical equation as suggested by Stroud (1974) except that the above Eq. (5) relates the m_v with N_{LCPT} instead of SPT '*N*'. From the equation multiplication factor f is determined using experimental m_v value and field N_{LCPT} value, and the values thus determined are plotted in Fig. 4.

$$f = \frac{1}{(m_v * N_{LCPT} * P_a)} \tag{5}$$

where f = multiplication factor, m_v = coefficient of volume change and P_a = 100 kN/m^2.

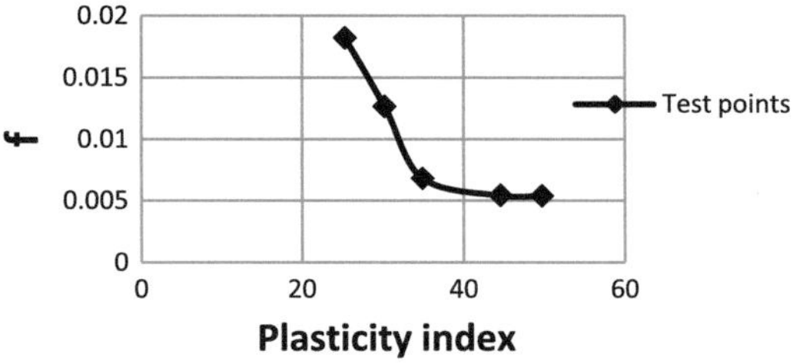

Fig. 4 Correlation between LCPT resistance and coefficient of volume change

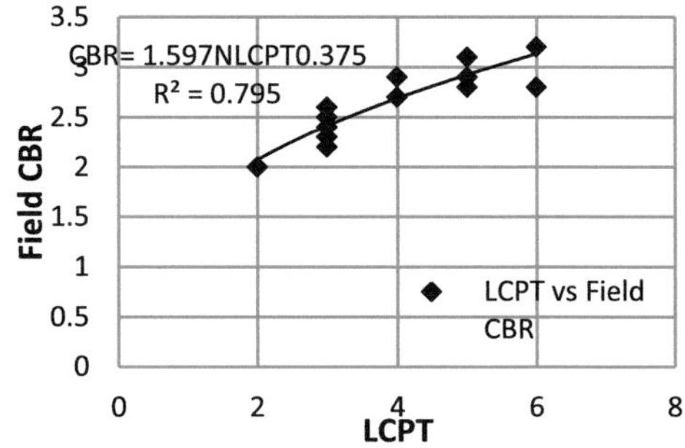

Fig. 5 LCPT versus field CBR

6.5 Correlation Between LCPT Resistance and Field CBR Value

The strength of the subgrade soil was determined by conducting field CBR test as per IS 2720-1970. The surcharge load applied 100 N. It was found that the CBR value varied marginally with the LCPT resistance. The field CBR values are then plotted against the LCPT resistance in Fig. 5. The correlation is obtained by curve fitting technique in Eq. 6.

$$\text{CBR} = 1.59 * (N_{\text{LCPT}})^{0.375} \tag{6}$$

7 Conclusion

LCPT and SPT resistances recorded with depths showed almost similar trend irrespective of nature of deposit; however, the LCPT resistances are higher than SPT resistances recorded irrespective of the type of deposit. The resistances of LCPT are correlated with SPT '*N*' values and strength parameters of sand and clay. The empirical relations obtained are presented. However, these relations need further examination by conducting tests at more locations.

References

Aroosa, H. (2000). Correlation of static cone penetration test results and dynamic probing test results. Research study for the data of South Limburg (Vol. 1), Netherlands.

Deepika, C., & Chakravarthy, V. K. (2012). Evaluation of properties of soil subgrade using dynamic cone penetration index. *International Journal of Engineering Research and Development, 4*, 7–15.

Ilamparuthi, K. (1981). *Interpretation of SCPT data*. M.E. thesis, Anna University, Chennai.

Meyerhof, G. G. (1956). Penetration tests and bearing capacity of cohesionless soils. *Journal of Soil Mechanics and Foundation Engineering Division, ASCE, 82*(1), 866–876.

Moayed, R. Z., & Naeini S. A. (2006). *Evaluation of modulus of subgrade reaction (K_s) in gravely soils based on SPT results*. Imam Khomeini International University.

Nwobasi, & Paul, A. (2013). Estimation of undrained shear strength of soil using cone penetration test. *International Journal of Scientific and Engineering Research, 4*, 409–420.

Ravisharma, M. S., & Ilamparuthi, K. (2009). Interpretation of electropierocone data of Chennai coast. *Journal of Ocean Engineering, 36*(6–7), 511–520.

Stroud, M. A. (1974). The standard penetration test in insensitive clays and soft rocks. In *Proceedings of the 1st European Symposium on Penetration Testing* (Vol. 1, pp. 372–373).

Zaki, B. A., Fouad, G. M., & Ahmed, K. M. (1990). Density prediction by static cone penetrometer. *JKAU Engineering Sciences, 2*, 19–33 (Jeddah).